The Editorial Policy for Proceedings

The series Lecture Notes in Physics reports new developments in physical research and teaching – quickly, informally, and at a high level. The proceedings to be considered for publication in this series should be limited to only a few areas of research, and these should be closely related to each other. The contributions should be of a high standard and should avoid lengthy redraftings of papers already published or about to be published elsewhere. As a whole, the proceedings should aim for a balanced presentation of the theme of the conference including a description of the techniques used and enough motivation for a broad readership. It should not be assumed that the published proceedings must reflect the conference in its entirety. (A listing or abstracts of papers presented at the meeting but not included in the proceedings could be added as an appendix.)

When applying for publication in the series Lecture Notes in Physics the volume's editor(s) should submit sufficient material to enable the series editors and their referees to make a fairly accurate evaluation (e.g. a complete list of speakers and titles of papers to be presented and abstracts). If, based on this information, the proceedings are (tentatively) accepted, the volume's editor(s), whose name(s) will appear on the title pages, should select the papers suitable for publication and have them refereed (as for a journal) when appropriate. As a rule discussions will not be accepted. The series editors and Springer-Verlag will normally not interfere with the detailed editing except in fairly obvious cases or on technical matters.

Final acceptance is expressed by the series editor in charge, in consultation with Springer-Verlag only after receiving the complete manuscript. It might help to send a copy of the authors' manuscripts in advance to the editor in charge to discuss possible revisions with him. As a general rule, the series editor will confirm his tentative acceptance if the final manuscript corresponds to the original concept discussed, if the quality of the contribution meets the requirements of the series, and if the final size of the manuscript does not greatly exceed the number of pages originally agreed upon. The manuscript should be forwarded to Springer-Verlag shortly after the meeting. In cases of extreme delay (more than six months after the conference) the series editors will check once more the timeliness of the papers. Therefore, the volume's editor(s) should establish strict deadlines, or collect the articles during the conference and have them revised on the spot. If a delay is unavoidable, one should encourage the authors to update their contributions if appropriate. The editors of proceedings are strongly advised to inform contributors about these points at an early stage.

The final manuscript should contain a table of contents and an informative introduction accessible also to readers not particularly familiar with the topic of the conference. The contributions should be in English. The volume's editor(s) should check the contributions for the correct use of language. At Springer-Verlag only the prefaces will be checked by a copy-editor for language and style. Grave linguistic or technical shortcomings may lead to the rejection of contributions by the series editors. A conference report should not exceed a total of 500 pages. Keeping the size within this bound should be achieved by a stricter selection of articles and not by imposing an upper limit to the length of the individual papers. Editors receive jointly 30 complimentary copies of their book. They are entitled to purchase further copies of their book at a reduced rate. As a rule no reprints of individual contributions can be supplied. No royalty is paid on Lecture Notes in Physics volumes. Commitment to publish is made by letter of interest rather than by signing a formal contract. Springer-Verlag secures the copyright for each volume.

The Production Process

The books are hardbound, and the publisher will select quality paper appropriate to the needs of the author(s). Publication time is about ten weeks. More than twenty years of experience guarantee authors the best possible service. To reach the goal of rapid publication at a low price the technique of photographic reproduction from a camera-ready manuscript was chosen. This process shifts the main responsibility for the technical quality considerably from the publisher to the authors. We therefore urge all authors and editors of proceedings to observe very carefully the essentials for the preparation of camera-ready manuscripts, which we will supply on request. This applies especially to the quality of figures and halftones submitted for publication. In addition, it might be useful to look at some of the volumes already published. As a special service, we offer free of charge LaTeX and TeX macro packages to format the text according to Springer-Verlag's quality requirements. We strongly recommend that you make use of this offer, since the result will be a book of considerably improved technical quality. To avoid mistakes and time-consuming correspondence during the production period the conference editors should request special instructions from the publisher well before the beginning of the conference. Manuscripts not meeting the technical standard of the series will have to be returned for improvement.

For further information please contact Springer-Verlag, Physics Editorial Department II, Tiergartenstrasse 17, D-69121 Heidelberg, Germany

Series homepage – http://www.springer.de/phys/books/lnpp

Lecture Notes in Physics

BOOK SALE

Springer
Berlin
Heidelberg
New York
Barcelona
Hong Kong
London
Milan
Paris
Singapore
Tokyo

Physics and Astronomy 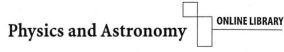 ONLINE LIBRARY

http://www.springer.de/phys/

Bernard Brogliato   (Ed.)

Impacts
in Mechanical Systems

Analysis and Modelling

Springer

Editor

Bernard Brogliato
Laboratoire d'Automatique de Grenoble
UMR, CNRS-INPG 5528, ENSIEG
BP 46, Domaine Universitaire
38402 Saint Martin d'Hères, France

Library of Congress Cataloging-in-Publication Data applied for.

Die Deutsche Bibliothek - CIP-Einheitsaufnahme

Impacts in mechanical systems : analysis and modelling / Bernard
Brogliato (ed.). - Berlin ; Heidelberg ; New York ; Barcelona ; Hong
Kong ; London ; Milan ; Paris ; Singapore ; Tokyo : Springer, 2000
   (Lecture notes in physics ; Vol. 551)
   (Physics and astronomy online library)
   ISBN 3-540-67523-X

ISSN 0075-8450
ISBN 3-540-67523-X Springer-Verlag Berlin Heidelberg New York

Springer-Verlag Berlin Heidelberg New York
a member of BertelsmannSpringer Science+Business Media GmbH

© Springer-Verlag Berlin Heidelberg 2000
Printed in Germany

The use of general descriptive names, registered names, trademarks, etc. in this publication does not imply, even in the absence of a specific statement, that such names are exempt from the relevant protective laws and regulations and therefore free for general use.

Typesetting: Camera-ready by the authors/editors
Cover design: *design & production*, Heidelberg

Printed on acid-free paper
SPIN: 10719287   55/3141/du - 5 4 3 2 1 0

Preface

This volume contains extended versions of the five plenary lectures given at the Euromech Colloquium 397 "Impact in Mechanical Systems", at Grenoble, from June 30 until July 1–2, 1999, France. As the title indicates, it is devoted to the study of rigid multi-body mechanical systems subject to nonsmooth effects, such as impacts, Coulomb friction, constraints addition and deletion. Actually, this represents a large and important class of nonsmooth mechanical systems. Its study can be traced back to the ancient Greeks. The 17th and 18th centuries also witnessed a great deal of (scientific) excitement about shock dynamics. Prestigious names like Descartes, Newton, Poisson, Gauss, Huygens, Bernoulli, to name a few, have long been attached to the study of collisional effects between two rigid bodies. Later, scientists like Darboux, Routh, and Carnot, also contributed significantly to the field. The earliest studies on impact dynamics were essentially motivated by fundamental scientific questions in physics (what is the role of hardness in the rebound phenomenon, is springiness necessary for a rebound to occur, use in light models), as well as more practical goals (calculation of bullet trajectories).

Interest in such a class of dynamical systems today is certainly much more related to engineering, and in particular the development of simulation software for virtual prototyping, a topic of great importance in industry. However, it still possesses strong connections with physics: the study of granular matter, planetary rings, may benefit from using the models described in this book (let us also recall that so-called billiards, which are a particular class of impacting lossless mechanical systems, have motivated intense mathematical studies). In particular, the study of Newton's cradle is closely related to what one needs to properly understand and predict phenomena such as clusterization and fluidization, which are well known in granular matter dynamics. In addition, numerical simulations are quite important in these fields. Furthermore, scientific communities like computer science (virtual reality), robotics, systems and control, applied mathematics and, evidently, applied mechanics find various fields of investigation in multi-body systems. As the reader will see throughout this book, nonsmooth mechanical systems with unilateral (or inequality) constraints represent a very interesting class of dynamical systems. They are not a simple extension of systems with bilateral constraints, or of systems with impulses. To express it in a language that has recently become fashionable in the computer science and systems and control communities, they constitute a class of hybrid dynamical systems; in other words, they merge continuous as well as discrete-event phenomena (roughly, their state space may be seen as the product of $\mathbf{R}^n$ with a finite set of symbols). Contrary to some widely held opinion, their dynamics is quite complex.

Many important problems associated with the dynamics of multi-body mechanical systems with unilateral constraints still remain open: mathematical problems (existence, uniqueness, continuous dependence on initial data, bifurcations, chaos), numerical analysis problems (how to discretize such a complex mixture of differential equations and algebraic conditions), mechanics (some phenomena, such as multiple impacts, with or without friction, still require much study on the modelling side), systems analysis (controllability, stabilizability). The five chapters in this book contain contributions related to mathematics, modelling and numerical simulations.

- **Mathematical Analysis** The first chapter, by M. Kunze (Mathematics Dept., Cologne University) and M.D.P. Monteiro-Marques (Mathematics Dept., Lisbon University), is devoted to presenting the so-called Moreau's sweeping process. This evolution problem, invented by Moreau in the 1960s, applies to quasistatics as well as to dynamics. It was first motivated by applications in fluid mechanics and later on in nonsmooth mechanical systems. The focus of this chapter is on mathematical analysis.
- **Numerical Analysis and Simulation** The second chapter, by M. Abadie (Schneider Electric Research Center, Grenoble), is dedicated to numerical simulation problems. It describes the work done at the company Schneider Electric to improve the virtual prototyping of electrical devices. It also contains an overview of the existing tools for simulation of nonsmooth mechanical systems. As with other analyses, our nonsmooth systems require very specific numerical tools and cannot be accomodated by classical software and algorithms. The algorithms presented in this chapter have been inspired by the discretization of the sweeping process as done by Moreau (see the first contribution), with appropriate modifications to comply with industrial needs (they are to be used directly by Schneider's engineers), whereas Moreau's scheme was primarily devoted to the simulation of granular matter.
- **Stability and Bifurcations** The third chapter by A. Ivanov (University of Moscow), deals with stability and bifurcation phenomena. It is a fact that systems with unilateral constraints possess specific sorts of bifurcations which are not encountered in smooth dynamics. They occur with grazing trajectories and have therefore been called grazing bifurcations. Also, the stability of trajectories requires new analytical tools.
- **Energetical Restitution Coefficient** The contribution of the fourth chapter by W.J. Stronge (Mechanical Engineering Dept., Cambridge University), focuses on collisions between two bodies using elasto-plastic models. It concentrates on the study of an energy coefficient of restitution. It also contains some developments on multiple impacts (the so-called Newton's cradle).
- **Multiple Impacts** The final chapter, by Y, Hurmuzlu and V. Ceanga (Mechanical Engineering Dept., Southern Methodist University), concentrates on

multiple impacts without friction. It proposes a completely new way to attack the multiple impact problem, by using a new set of physical coefficients (including the energetical coefficient presented in the previous chapter) to describe the shock phenomenon. Newton's cradle and the rocking block are used to develop the theoretical analysis. Experiments confirm the analysis.

In summary, this volume is an advanced introduction to the field of analysis, modelling and numerical simulation of rigid body mechanical systems with unilateral constraints. It will be worthwhile reading for anybody interested in this topic, be it at the mathematical, mechanical or numerical level. In fact, all these fields of investigation feed one another and it is almost compulsory to have a general view of the problems in the other fields to be able to propose sound solutions in a particular domain. The book contains some established (although not always very well known outside the nonsmooth dynamics community) results, as well as quite new ideas.

I would like to express warm thanks to my colleagues who kindly accepted to prepare a plenary talk at the Euromech 397, and, most importantly, who made the effort to write these chapters. They are gratefully acknowledged here. I would also like to recall that the Euromech Colloquium 397 was organized within the framework of a European INTAS project coordinated by Bill Stronge. This book rounds off this cooperation nicely.

Saint-Martin d'Hères, April 2000   *Bernard Brogliato*

Contents

An Introduction to Moreau's Sweeping Process
M. Kunze, M.D.P. Monteiro Marquès 1

Dynamic Simulation of Rigid Bodies:
Modelling of Frictional Contact
M. Abadie.. 61

Stability of Periodic Motions with Impacts
A.P. Ivanov.. 145

Contact Problems for Elasto-plastic Impact
in Multi-body Systems
W.J. Stronge 189

Impulse Correlation Ratio
in Solving Multiple Impact Problems
Y. Hurmuzlu, V. Ceanga.. 235

An Introduction to Moreau's Sweeping Process

Markus Kunze[1] and Manuel D.P. Monteiro Marques[2]

[1] Mathematisches Institut, Universität Köln, Weyertal 86, 50931 Köln, Germany
[2] C.M.A.F. and Departamento de Matemática, Faculdade de Ciências,
  Universidade de Lisboa, Av. Prof. Gama Pinto, 2, 1699 Lisboa, Portugal

Abstract. Starting from an elementary level, this rewiew paper summarizes some results and applications concerning first and second order sweeping processes.

1 Introduction

These lecture notes concern the so-called "sweeping process" that plays an important role in elastoplasticity, quasistatics, and dynamics. A sweeping process consists of two main ingredients: one that sweeps and one that is swept. As an example, consider a large ring that contains a smaller ball inside, and the ring will start to move at time $t = 0$. Depending on the motion of the ring, the ball will just stay where it is (in case it is not hit by the ring), or otherwise it is swept towards the interior of the ring. In this latter case the velocity of the ball has to point inwards to the ring in order not to leave; see Fig. 1.

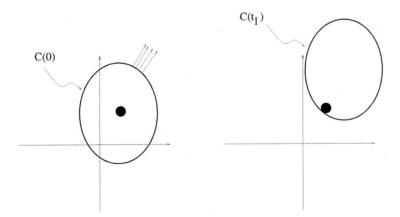

Fig. 1. A sweeping process

In more mathematical terms, this becomes

$$-u'(t) \in N_{C(t)}(u(t)) \quad \text{a.e. in} \quad [0, T], \quad u(0) = u_0 \in C(0), \qquad (1)$$

where $u(t)$ is the position of the ball at time t, whereas $C(t)$ is the (set comprising the) ring at time t. The expression $N_{C(t)}(u(t))$ denotes the outward normal cone to the set $C(t)$ at position $u(t)$, so (1) just says that the velocity $u'(t)$ of the ball has to point inwards to the ring at almost every (a.e.) time $t \in [0, T]$. The restriction to "almost every" is due to the fact that usually we will find no smooth solution $t \mapsto u(t)$ of (1), but only solutions that are differentiable everywhere besides on some subset of $[0, T]$ of measure zero. The initial condition $u(0) \in C(0)$ states that the ball initially is contained in the ring.

Equ. (1) is the simplest instance of the sweeping process, introduced by Moreau in the seventies. In general, the time-dependent moving set $t \mapsto C(t)$ is given, and we intend to prove the existence of a (unique) solution $t \mapsto u(t)$ that will take values in some Hilbert space H (above $H = \mathbb{R}^2$). Note also that $C(t)$ is allowed to change its shape while moving, contrary to the example of the ring which just moved by translation and kept its initial shape.

The paper is organized as follows. In Section 2 we start by introducing some concepts from convex analysis and functional analysis, like normal cones and functions of bounded variation. Then in the first main part of these notes, Section 3, we summarize some important results for the sweeping process (1) concerning existence and uniqueness of solutions. As expected, the better behaved the set will move (e.g. without jumps), the better will be the solution. So we will prove in Section 3.1 quite in some detail the existence of a unique Lipschitz continuous solution to (1) in case the set $t \mapsto C(t)$ moves in a Lipschitz continuous way (in a suitable sense, measured by the Hausdorff distance of sets). Afterwards in Section 3.2 we carefully analyze an example from elastoplasticity and show how the general theory from Section 3.1 applies. Then in Section 3.3 we discuss a variant of the sweeping process which arises naturally in the modeling of more complex problems, like quasi-variational inequalities. Contrary to (1) the moving set here may depend also on the current state $u(t)$, i.e. $C = C(t, u(t))$ instead of $C = C(t)$, and we will see in particular that not all such state-dependent sweeping processes will admit a solution. Finally in Section 3.4 we add some supplementary remarks.

In the second main part (Sections 4 and 5), we turn to problems which are of second order in time. In Section 4 we deal with

$$-q''(t) \in N_{V(q(t))}(q'(t)). \tag{2}$$

Note that this formulation does make sense only if the velocity $u(t) = q'(t)$ were an absolutely continuous function (thus admitting a derivative a.e.). However, under the usual assumptions that the multifunction $q \mapsto V(q)$ is continuous in the sense of Hausdorff distance and has closed convex values with nonempty interior, it turns out that u in general is only a continuous function of bounded variation; therefore the differential inclusion (2) has to be rewritten accordingly. We will prove the existence of a solution under somewhat stronger technical assumptions on V, since this allows to highlight some additional technical difficulties that arise when using a discretization procedure (similar to the one in [38]) which is of great importance for the subsequent applications.

In Section 5, we consider such an application to the dynamics of unilateral contact. In Section 5.1 we present the corresponding mathematical formulation, introduced by Moreau, to model the dynamics of a particle or a system under frictionless constraints. The appropriate mathematical description is given by a differential inclusion which is a second order sweeping process of the above type, with an extra term to account for forces acting on the system. Then we outline the recent proof by Mabrouk [38] of the existence of a solution which is a motion q with bounded variation velocity u. It should be stressed that purely inelastic as well as purely or partially elastic collisions can be handled with this technique. Finally, in Section 5.2, we give a very brief account on how the previous ideas apply to unilateral contact with friction and on some promising recent advances in the field.

2   Some preliminaries from convex analysis and functional analysis

In these notes we will be concerned with a Hilbert space H which may have finite or infinite dimension. The inner product and norm, respectively, are denoted by $\langle \cdot, \cdot \rangle$ and $|\cdot|$, and $\overline{B}_r(x)$ denotes the closed ball of radius r centered at $x \in H$.

For a closed convex $C \subset H$ and $x \in C$ the set

$$N_C(x) = \{\xi \in H : \langle \xi, c - x \rangle \leq 0 \ \forall c \in C\}$$

denotes the outward normal cone to C at x. See Fig. 2 for some examples.

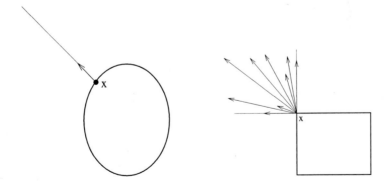

Fig. 2. Examples of normal cones

Observe that in fact we have drawn the sets $x + N_C(x)$ rather than $N_C(x)$.

Remark 1. We always have that $0 \in N_C(x)$, $N_{\{x\}}(x) = H$, and $N_C(x) = \{0\}$ for $x \in \text{int} C$, the interior of C, if $\text{int} C \neq \emptyset$. The latter relation shows the outward normal cone is only interesting for $x \in \partial C$, the boundary of C. Note that $\text{int} C = \emptyset$ might well be the case, think e.g. of $C = \{(x,0) : 0 \leq x \leq 1\} \subset \mathbb{R}^2$. In this example we obtain $N_C((0,0)) = \{(\xi_1, \xi_2) \in \mathbb{R}^2 : \xi_1 \leq 0, \xi_2 \in \mathbb{R}\}$, $N_C((1,0)) = \{(\xi_1, \xi_2) \in \mathbb{R}^2 : \xi_1 \geq 0, \xi_2 \in \mathbb{R}\}$, and $N_C((x,0)) = \{(0, \xi_2) \in \mathbb{R}^2 : \xi_2 \in \mathbb{R}\}$ for $x \in]0,1[$ as normal cones. ◇

If $x \in H$ and $C \subset H$ is closed convex, then there exists a unique $y \in C$ that minimizes the distance of x to C. This y is called the projection of x onto C and written as $y = \text{proj}(x, C)$. Hence the characterizing properties are

$$y = \text{proj}(x, C) \quad \text{if and only if} \quad y \in C, \quad |x - y| = \text{dist}(x, C) \qquad (3)$$

with $\text{dist}(x, C) = \inf\{|x - c| : c \in C\}$. Equivalently,

$$y = \text{proj}(x, C) \quad \text{if and only if} \quad y \in C, \quad \langle y - x, y - c \rangle \leq 0 \ \forall c \in C, \qquad (4)$$

and hence

$$y = \text{proj}(x, C) \quad \text{if and only if} \quad y \in C, \quad x - y \in N_C(y). \qquad (5)$$

Moreover,

$$d_H(C_1, C_2) = \max\left\{ \sup_{x \in C_2} \text{dist}(x, C_1), \ \sup_{x \in C_1} \text{dist}(x, C_2) \right\} \qquad (6)$$

is the Hausdorff distance between the sets C_1 and C_2.

It will also be convenient to introduce the support function of a convex $C \subset H$ defined as

$$\delta^*(x, C) = \sup\{\langle x, c \rangle : c \in C\} \quad \text{for} \quad x \in H. \qquad (7)$$

Remark 2. Note that $x \mapsto \delta^*(x, C)$ is a convex map on C, that is, $\delta^*(\lambda x + (1 - \lambda)\bar{x}, C) \leq \lambda \delta^*(x, C) + (1 - \lambda)\delta^*(\bar{x}, C)$ for $\lambda \in [0, 1]$ and $x, \bar{x} \in H$. If C is unbounded, it might be the case that $\delta^*(x, C) = +\infty$ for some $x \in H$. ◇

We will later need some further elementary geometrical facts.

Lemma 1. *The following assertions hold:*

(a) *If $U \subset H$ is a closed subspace and $V = U^\perp$ is the orthogonal complement of U in H, then $N_V(x) = U$ for $x \in V$.*

(b) *If $C, D \subset H$ are nonempty, closed, and convex with $\text{int} C \cap D \neq \emptyset$, then*

$$N_{C \cap D}(x) = N_C(x) + N_D(x) \quad \text{for} \quad x \in C \cap D.$$

(c) If $C \subset H$ is nonempty, closed, and convex and $x \in H$ is such that dist $(x, H \setminus C) \geq \rho > 0$, then $\rho |y| + \langle y, x \rangle \leq \delta^*(y, C)$ for all $y \in H$.

(d) If $C \subset H$ is nonempty, closed, and convex and $\rho > 0$, then the set $C_\rho = \{x \in H : \text{dist}(x, H \setminus C) \geq \rho\}$ is convex.

Proof: (a) If $u \in U$, then $\langle u, v - x \rangle = 0$ for all $v \in V$, since also $x \in V$. Conversely, if $u \in N_V(x)$, then

$$\langle u, v \rangle \leq \langle u, x \rangle \quad \text{for all} \quad v \in V. \tag{8}$$

Assume $\langle u, v_0 \rangle \neq 0$ for some $v_0 \in V$. Then we may suppose $\langle u, v_0 \rangle > 0$, since otherwise we could replace v_0 by $(-v_0) \in V$. But then we can use $v = \lambda v_0 \in V$ in (8) and let $\lambda \to \infty$ to obtain a contradiction. Thus $\langle u, v \rangle = 0$ for all $v \in V$, i.e., $u \in V^\perp = \overline{U} = U$. (b) Firstly, for $\xi \in N_C(x)$, $\eta \in N_D(x)$, and $c \in C \cap D$ we find $\langle \xi + \eta, c - x \rangle = \langle \xi, c - x \rangle + \langle \eta, c - x \rangle \leq 0$, whence $\xi + \eta \in N_{C \cap D}(x)$. On the other hand, let $\xi \in N_{C \cap D}(x)$ for $x \in C \cap D$. Since generally $N_C(x) = N_{C-x}(0)$ for $x \in C$ we can assume without loss of generality that $x = 0 \in C \cap D$. Define

$$\mathcal{D} = D \times [0, \infty[\quad \text{and} \quad \mathcal{C} = \{(c, t) \in C \times \mathbb{R} : t \leq \langle \xi, c \rangle\}.$$

We claim that $\mathcal{D} \cap \text{int}\, \mathcal{C} = \emptyset$. Indeed, otherwise we would find $c_0 \in C \cap D$, $t_0 \in [0, \infty[$ and $\varepsilon > 0$ such that $|c - c_0| \leq \varepsilon$ and $|t - t_0| \leq \varepsilon$ for $c \in H$ and $t \in \mathbb{R}$ implies $t \leq \langle \xi, c \rangle$. As $\xi \in N_{C \cap D}(0)$, we have $\langle \xi, c_0 \rangle \leq 0$, and hence $t_0 \in [0, \infty[$ and also $t \leq \langle \xi, c_0 \rangle \leq 0$ for all t with $|t - t_0| \leq \varepsilon$, a contradiction. On the other hand, $\text{int}\, \mathcal{C} \neq \emptyset$, since e.g. $(\hat{c}, \langle \xi, \hat{c} \rangle - 1) \in \text{int}\, \mathcal{C}$ for $\hat{c} \in \text{int}\, C$. Since both $\mathcal{D}$ and $\mathcal{C}$ are convex subsets of $H \times \mathbb{R}$ that contain $(0, 0) \in H \times \mathbb{R}$, by Mazur's separation theorem [26, Thm. 7.12 (b)] we find $\eta \in H$ and $r \in \mathbb{R}$ such that

$$\mathcal{D} \subset \{(u, t) \in H \times \mathbb{R} : \langle \eta, u \rangle + rt \geq 0\},$$
$$\text{int}\, \mathcal{C} \subset \{(u, t) \in H \times \mathbb{R} : \langle \eta, u \rangle + rt < 0\}. \tag{9}$$

Since by assumption $\hat{c} \in \text{int}\, C \cap D$ for some $\hat{c} \in H$, we can use $(\hat{c}, 0) \in \mathcal{D}$ and $(\hat{c}, \langle \xi, \hat{c} \rangle - 1) \in \text{int}\, \mathcal{C}$ to conclude from relations (9) that $r \neq 0$. As $0 \in D$ we find $(0, t) \in \mathcal{D}$ for all $t \in [0, \infty[$, and thus $rt \geq 0$ for all $t \in [0, \infty[$. This shows that in fact $r > 0$. Define $\xi_1 = -r^{-1}\eta$ and $\xi_2 = \xi + r^{-1}\eta$. Then $\xi = \xi_1 + \xi_2$. Moreover, if $d \in D$ then $(d, 0) \in \mathcal{D}$, and hence $\langle \xi_1, d \rangle = -r^{-1}\langle \eta, d \rangle \leq 0$ by (9), i.e., $\xi_1 \in N_D(0)$. Next, for $c \in C$ we take $(c, \langle \xi, c \rangle) \in \mathcal{C}$ to obtain $\langle \xi_2, c \rangle = r^{-1}(r\langle \xi, c \rangle + \langle \eta, c \rangle) \leq 0$, and therefore $\xi_2 \in N_C(0)$, i.e., $\xi \in N_D(0) + N_C(0)$.
(c) Fix $y \neq 0$ and $\varepsilon > 0$ small. Then we obtain $x + (1 - \varepsilon)\rho \frac{y}{|y|} \in C$ by assumption, and hence $\delta^*(y, C) \geq \langle y, x + (1 - \varepsilon)\rho \frac{y}{|y|} \rangle = \langle y, x \rangle + (1 - \varepsilon)\rho|y|$. Thus $\varepsilon \to 0$ yields the claimed estimate. (d) Assume on the contrary that $|\lambda x + (1 - \lambda)\bar{x} - z| \leq \rho - \varepsilon$ for some $\lambda \in]0, 1[$, $x, \bar{x} \in C_\rho$, $z \in H \setminus C$, and $\varepsilon > 0$. Let $x_1 = z + (1 - \lambda)(x - \bar{x})$ and $\bar{x}_1 = z + \lambda(\bar{x} - x)$. Then $|x - x_1| \leq \rho - \varepsilon$ and $|\bar{x} - \bar{x}_1| \leq \rho - \varepsilon$. As $x, \bar{x} \in C_\rho$ it follows that $x_1, \bar{x}_1 \in C$. But since C is convex, then also $z = \lambda x_1 + (1 - \lambda)\bar{x}_1 \in C$, a contradiction. $\qquad \square$

The variation of a function $u : [0, T] \to H$ is defined as

$$\text{var}\,(u) = \text{var}\,(u, [0, T])$$

$$= \sup \left\{ \sum_{i=0}^{N-1} |u(t_{i+1}) - u(t_i)| : \; 0 = t_0 < t_1 < \ldots < t_{N-1} < t_N = T \right.$$

$$\left. \text{is a partition of } [0, T] \right\}, \qquad (10)$$

and u is called of bounded variation, if $\text{var}\,(u) < \infty$.

Example 1. (a) If $u : [0, T] \to H$ is Lipschitz continuous, i.e., for some constant $L > 0$

$$|u(t) - u(s)| \le L|t - s|, \quad t, s \in [0, T],$$

then u is of bounded variation, and $\text{var}\,(u) \le LT$. To see this, consider any partition of $[0, T]$ as in the definition (10). Then

$$\sum_{i=0}^{N-1} |u(t_{i+1}) - u(t_i)| \le L \sum_{i=0}^{N-1} (t_{i+1} - t_i) = L(t_N - t_0) = LT.$$

(b) If $u : [0, T] \to \mathbb{R}$ is increasing, then u is of bounded variation, with $\text{var}\,(u) = u(T) - u(0)$. This follows from

$$\sum_{i=0}^{N-1} |u(t_{i+1}) - u(t_i)| = \sum_{i=0}^{N-1} (u(t_{i+1}) - u(t_i)) = u(T) - u(0).$$

Observe that u is not necessarily continuous here.

(c) Let $u : [0, T] \to H$ be piecewise affine, i.e., there exist $0 = t_0^* < t_1^* < \ldots < t_{N-1}^* < t_N^* = T$ such that

$$u(t) = u_i + \left(\frac{t - t_i^*}{t_{i+1}^* - t_i^*} \right) (u_{i+1} - u_i), \quad t \in [t_i^*, t_{i+1}^*],$$

with $u_i = u(t_i^*)$. Then

$$\text{var}\,(u) = \sum_{i=0}^{N-1} |u_{i+1} - u_i|.$$

Indeed, if we are given any partition of $[0, T]$, then we may add the points $0 = t_0^* < t_1^* < \ldots < t_{N-1}^* < t_N^* = T$ to this partition as this only increases the sum. However, if $t_i^* = s_0 < s_1 < \ldots < s_{M-1} < s_M = t_{i+1}^*$, then

$$\sum_{j=0}^{M-1} |u(s_{j+1}) - u(s_j)| = \sum_{j=0}^{M-1} \left(\frac{s_{j+1} - s_j}{t_{i+1}^* - t_i^*} \right) |u_{i+1} - u_i| = |u_{i+1} - u_i|,$$

and hence no additional contributions appear.

(d) To give an example of a function which is continuous but not of bounded variation, let $u(t) = t\cos(\pi/2t)$ for $t \in {]}0, 1]$ and $u(0) = 0$; see [1, Ch. 8]. Then u is continuous. For the partition

$$t_0 = 0, \quad t_1 = \frac{1}{2n}, \quad t_2 = \frac{1}{2n-1}, \quad \dots, \quad t_{2n} = 1$$

one calculates

$$\sum_{i=0}^{2n-1} |u(t_{i+1}) - u(t_i)| = t_1 + t_1 + t_3 + t_3 + \dots + t_{2n-1} + t_{2n-1} = \sum_{k=1}^{n} \frac{1}{k},$$

and hence the supremum over all possible partitions equals $\sum_{k=1}^{\infty} \frac{1}{k} = \infty$.

(e) A well-known further class of examples of functions that are continuous but of unbounded variation even on every subinterval are (almost every) paths of a Brownian motion; see [5, Cor. 12.27]. $\diamond$

In order to prove the existence of solutions to certain variants of the sweeping process, we mostly will construct a sequence of approximating solutions whose variations are uniformly bounded. The following compactness theorem then allows to select a subsequence which converges towards some limit function, and the task is afterwards to show the limit function is indeed the desired solution.

Theorem 1. *Let H be a Hilbert space and $(u_n)_{n \in \mathbb{N}}$ a sequence of functions $u_n : [0, T] \to H$ that is bounded uniformly in norm and variation, i.e.,*

$$|u_n(t)| \leq M_1, \quad n \in \mathbb{N}, \ t \in [0, T], \quad \text{and} \quad \text{var}(u_n) \leq M_2, \quad n \in \mathbb{N}, \quad (11)$$

for some constants $M_1, M_2 > 0$ independently of $n \in \mathbb{N}$ and $t \in [0, T]$. Then there exists a subsequence $(u_{n_k})_{k \in \mathbb{N}}$ and a function $u : [0, T] \to H$ such that $\text{var}(u) \leq M_2$ and $u_{n_k}(t) \to u(t)$ weakly in H for all $t \in [0, T]$, i.e.,

$$\langle u_{n_k}(t), z \rangle \to \langle u(t), z \rangle \quad \text{for all} \quad z \in H \quad \text{as} \quad k \to \infty.$$

Proof: See [42, p. 10]. $\square$

The next lemma summarizes some facts related to weak convergence.

Lemma 2. *Let $u_n \to u$ weakly in H.*

(a) $|u| \leq \liminf_{n \to \infty} |u_n|$ holds.
(b) If $u_n \in C + \overline{B}_{\varepsilon_n}(0)$ for some closed convex $C \subset H$ and some sequence $\varepsilon_n \to 0$, then $u \in C$.

Proof: (a) See [71, Thm. V.1.1 ii)]. (b) If $u \notin C$, then by the separating hyperplane theorem, cf. [64, Thm. 3.4], there exists $z \in H$ and $\gamma \in \mathbb{R}$ such that $\langle z, u \rangle < \gamma < \langle z, c \rangle$ for all $c \in C$. For every $n \in \mathbb{N}$ choose $r_n \in H$ and $c_n \in C$ with $|r_n| \leq \varepsilon_n$ and $u_n = c_n + r_n$. Then according to the weak convergence and since $|\langle z, r_n \rangle| \leq |z||r_n| \to 0$, $\langle z, c_n \rangle = \langle z, u_n \rangle - \langle z, r_n \rangle \to \langle z, u \rangle$. Thus from $\gamma < \langle z, c_n \rangle$ we obtain the contradiction $\langle z, u \rangle < \gamma \leq \langle z, u \rangle$. $\square$

Later we will also need a result on the behaviour of certain integral functionals w.r. to weak convergence of functions taking values in H.

Lemma 3. *Let* $(v_n)_{n \in \mathbb{N}}$ *be a sequence of functions* $v_n : [0, T] \to H$ *such that* $v_n \to v_*$ *in the weak-star topology of* $L^\infty([0, T]; H)$, *i.e.,*

$$\int_0^T \langle v_n(t), \varphi(t) \rangle \, dt \to \int_0^T \langle v_*(t), \varphi(t) \rangle \, dt \quad as \quad n \to \infty$$

for all $\varphi \in L^1([0, T]; H)$. *[The latter means that* $\varphi : [0, T] \to H$ *is (strongly) measurable and* $\int_0^T |\varphi(t)| dt < \infty$.*] Assume that for each* $t \in [0, T]$ *the set* $C(t) \subset H$ *is nonempty, closed, and convex such that* $d_H(C(t), C(s)) \leq L|t - s|$ *for* $t, s \in [0, T]$ *with some constant* $L > 0$. *Let*

$$\Phi(v) = \int_0^T \delta^*(v(t), C(t)) \, dt \quad for \quad v \in L^\infty([0, T]; H).$$

Then Φ *is lower semicontinuous, i.e.,*

$$\Phi(v_*) \leq \liminf_{n \to \infty} \Phi(v_n).$$

Proof: See [63, Corollary, p. 227]. The proof rests on the fact that $f(v, t) = \delta^*(v, C(t))$ is measurable and $f(\cdot, t)$ is convex for all $t \in [0, T]$. One may also consult [23] for similar results in more general settings. $\square$

Lemma 4. *Let* $u : [0, T] \to H$ *be an absolutely continuous function. Then*

$$\int_0^T \langle u'(t), u(t) \rangle dt = \frac{1}{2} |u(T)|^2 - \frac{1}{2} |u(0)|^2.$$

Proof: Under the assumption made it can be shown by consideration of the difference quotient that

$$\frac{1}{2} \frac{d}{dt} \left(|u(t)|^2 \right) = \langle u'(t), u(t) \rangle \tag{12}$$

at every $t \in]0, T[$ where u has a derivative. Integration yields the claim. $\square$

We remark that absolute continuity of a function is a somewhat technical mathematical notion which for our purpose can be stated as "u is (up to a constant) the indefinite integral of its derivative".

3 First-order problems

3.1 The Lipschitz continuous sweeping process

In this section we assume that $t \mapsto C(t)$ is Lipschitz continuous, i.e.,

$$d_H(C(t), C(s)) \leq L|t - s|, \quad t, s \in [0, T], \tag{13}$$

for some constant $L > 0$. Let us first consider an example where the set $C(t)$ is driven by a point $\xi(t)$.

Example 2. Let $C \subset H$ be a fixed nonempty, closed, and convex set and $\xi : [0, T] \to H$ a function which is Lipschitz continuous with constant $L > 0$. Let $C(t) = C + \xi(t)$ for $t \in [0, T]$. Then $t \mapsto C(t)$ satisfies (13). To see this, fix $x \in C(s)$. Then $y = [x - \xi(s)] + \xi(t) \in C + \xi(t) = C(t)$, and thus dist $(x, C(t)) \leq |x - y| = |\xi(t) - \xi(s)| \leq L|t - s|$. Hence also $\sup_{x \in C(s)}$ dist $(x, C(t)) \leq L|t - s|$, and similarly $\sup_{y \in C(t)}$ dist $(y, C(s)) \leq L|t - s|$. Therefore by the definition of d_H in (6) we obtain the claim. ◇

Our aim is to show that for a Lipschitz continuous moving set there exists a unique solution to (1). Let us first fix what will be called a solution.

Definition 1. A function $u : [0, T] \to H$ is a solution of (1), if

(a) $u(0) = u_0$;
(b) $u(t) \in C(t)$ for all $t \in [0, T]$;
(c) u is differentiable at almost every point $t \in]0, T[$;
(d) $-u'(t) \in N_{C(t)}(u(t))$ for almost every $t \in]0, T[$.

The following theorem is a very basic existence result.

Theorem 2 (Existence). *Assume the map $t \mapsto C(t)$ satisfies (13) and $C(t) \subset H$ is nonempty, closed, and convex for every $t \in [0, T]$. Let $u_0 \in C(0)$. Then there exists a solution $u : [0, T] \to H$ of (1) which is Lipschitz continuous of constant L. In particular, $|u'(t)| \leq L$ for almost every $t \in]0, T[$.*

Proof: The following discretization scheme lies at the heart of many proofs for sweeping processes. We fix $n \in \mathbb{N}$ and choose a time discretization

$$0 = t_0^n < t_1^n < \ldots < t_{N-1}^n < t_N^n = T, \text{ with } (t_{i+1}^n - t_i^n) \leq \frac{1}{n}, \ 0 \leq i \leq N - 1. \tag{14}$$

E.g. we can set $t_i^n = i/n$, but we need not fix the discretization explicitly. The value of $N \in \mathbb{N}$ will depend on n, and $N \to \infty$ for $n \to \infty$. We define the step approximations $u^n : [0, T] \to H$ as follows. Let

$$u_0^n = u_0, \quad u_{i+1}^n = \text{proj}\,(u_i^n, C(t_{i+1}^n)) \in C(t_{i+1}^n), \quad 0 \leq i \leq N - 1. \tag{15}$$

10

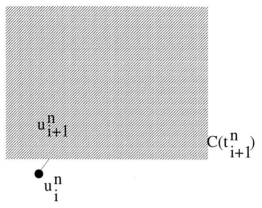

Fig. 3. The catching up algorithm

This is the "catching up" algorithm, since the approximation u_{i+1}^n is made to catch up with the set $C(t_{i+1}^n)$ through projection; see Fig. 3. Recall that we have to achieve $u(t) \in C(t)$ for the solution.

The u_n are defined via linear interpolation,

$$u_n(t) = u_i^n + \left(\frac{t - t_i^n}{t_{i+1}^n - t_i^n}\right)(u_{i+1}^n - u_i^n), \quad t \in [t_i^n, t_{i+1}^n]; \qquad (16)$$

see Fig. 4 below.

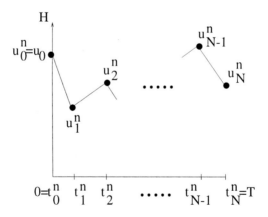

Fig. 4. The approximate solution u_n

We intend to find a subsequence of $(u_n)_{n \in \mathbb{N}}$ that converges to a solution of (1). To this end we wish to apply Theorem 1, and we have to derive the uniform bounds in norm and variation in (11). First note that by (15) and since $u_i^n \in C(t_i^n)$,

$$|u_{i+1}^n - u_i^n| = |\text{proj}\,(u_i^n, C(t_{i+1}^n)) - u_i^n| = \text{dist}\,(u_i^n, C(t_{i+1}^n))$$
$$\leq d_{\mathrm{H}}(C(t_i^n), C(t_{i+1}^n)) \leq L|t_i^n - t_{i+1}^n|, \tag{17}$$

where we also have used (3) and (13). Together with Example 1(c) this leads to

$$\text{var}\,(u_n) = \sum_{i=0}^{N-1} |u_n(t_{i+1}^n) - u_n(t_i^n)| = \sum_{i=0}^{N-1} |u_{i+1}^n - u_i^n| \leq L \sum_{i=0}^{N-1} (t_{i+1}^n - t_i^n)$$
$$= LT =: M_2. \tag{18}$$

From (17) we get $|u_{i+1}^n| \leq |u_i^n| + L(t_{i+1}^n - t_i^n)$, and thus by induction $|u_i^n| \leq |u_0| + Lt_i^n$. According to (16) and (17) this yields

$$|u_n(t)| \leq |u_i^n| + L|u_{i+1}^n - u_i^n| \leq |u_0| + Lt_i^n + L(t_{i+1}^n - t_i^n) = |u_0| + Lt_{i+1}^n$$
$$\leq |u_0| + LT =: M_1 \tag{19}$$

for $t \in [t_i^n, t_{i+1}^n]$, whence for all $n \in \mathbb{N}$ and $t \in [0, T]$, since the bound is independent of n and t. Summarizing (18) and (19), we have shown that (11) in Theorem 1 holds. Thus we may extract a subsequence of $(u_n)_{n \in \mathbb{N}}$ (still indexed by $n \in \mathbb{N}$ for simplicity) and we find u with var $(u) \leq LT = M_2$ such that $u_n(t) \to u(t)$ as $n \to \infty$ weakly in H for all $t \in [0, T]$.

To show u is a solution of (1), note first that $u_n(0) = u_0$ implies $u(0) = \mathrm{w} - \lim u_n(0) = u_0$. Thus (a) of Definition 1 is satisfied; $\mathrm{w} - \lim$ denotes the weak limit. To exploit (17) further, fix $t, s \in [0, T]$. Then $t \in [t_j^n, t_{j+1}^n]$ and $s \in [t_i^n, t_{i+1}^n]$ for some $0 \leq j, i \leq N - 1$. Without loss of generality we can assume $i \leq j$, i.e., $t_i^n \leq t_j^n$. From (16) and (17) we obtain

$$|u_n(t) - u_n(s)|$$
$$\leq |u_n(t) - u_n(t_j^n)| + \sum_{k=i+1}^{j-1} |u_n(t_{k+1}^n) - u_n(t_k^n)| + |u_n(t_{i+1}^n) - u_n(s)|$$
$$\leq \left(\frac{t - t_j^n}{t_{j+1}^n - t_j^n} \right) |u_{j+1}^n - u_j^n| + \sum_{k=i+1}^{j-1} |u_{k+1}^n - u_k^n| + \left(\frac{t_{i+1}^n - s}{t_{i+1}^n - t_i^n} \right) |u_{i+1}^n - u_i^n|$$
$$\leq \sum_{k=i}^{j} |u_{k+1}^n - u_k^n| \leq L \sum_{k=i}^{j} (t_{k+1}^n - t_k^n) = L(t_{j+1}^n - t_i^n)$$
$$= L([t - s] + [s - t_i^n] + [t_{j+1}^n - t]) \leq L\left(|t - s| + \frac{2}{n} \right), \tag{20}$$

the latter by (14). Since $\mathrm{w} - \lim_{n \to \infty} (u_n(t) - u_n(s)) = u(t) - u(s)$, Lemma 2(a) and (20) yield

$$|u(t) - u(s)| \leq \liminf_{n \to \infty} |u_n(t) - u_n(s)| \leq L|t - s|, \quad t, s \in [0, T].$$

12

Consequently, u is Lipschitz continuous with constant L, and hence in particular differentiable almost everywhere in $]0, T[$; cf. [6, Appendix, Sect. A.2]. Thus we have verified (c) of Definition 1. To see (b), we infer from (17) and (14) that

$$\left(\frac{t - t_i^n}{t_{i+1}^n - t_i^n}\right) |u_{i+1}^n - u_i^n| \le L(t - t_i^n) \le L(t_{i+1}^n - t_i^n) \le \frac{L}{n}$$

for $t \in [t_i^n, t_{i+1}^n]$. Therefore by (16), (15), and (13) for $t \in [t_i^n, t_{i+1}^n]$

$$u_n(t) \in C(t_i^n) + \overline{B}_{L/n}(0) \subset C(t) + \overline{B}_{L(t-t_i^n)}(0) + \overline{B}_{L/n}(0)$$
$$\subset C(t) + \overline{B}_{2L/n}(0), \tag{21}$$

where in the last step we have used (14). Hence (21) in fact holds for all $n \in \mathbb{N}$ and $t \in [0, T]$, and thus Lemma 2(b) yields $u(t) \in C(t)$ for all $t \in [0, T]$. Finally we need to verify (d) of Definition 1, the most difficult step of the proof. First note that

$$\langle u_{i+1}^n - u_i^n, u_{i+1}^n - \hat{c}\rangle \le 0, \quad \hat{c} \in C(t_{i+1}^n), \tag{22}$$

by (15) and (4). From (16), (17), and (14) we obtain

$$|u_n(t) - u_{i+1}^n| = |u_n(t) - u_n(t_{i+1}^n)| \le L(t_{i+1}^n - t) \le \frac{L}{n}, \quad t \in [t_i^n, t_{i+1}^n].$$

Since by (13), $C(t) \subset C(t_{i+1}^n) + \overline{B}_{L(t_{i+1}^n - t)}(0) \subset C(t_{i+1}^n) + \overline{B}_{L/n}(0)$ for $t \in [t_i^n, t_{i+1}^n]$, we find from (22) and (17) that

$$\langle u_{i+1}^n - u_i^n, u_n(t) - c\rangle$$
$$= \langle u_{i+1}^n - u_i^n, u_{i+1}^n - \hat{c}\rangle + \langle u_{i+1}^n - u_i^n, [u_n(t) - u_{i+1}^n] + [\hat{c} - c]\rangle$$
$$\le |u_{i+1}^n - u_i^n|\left(\frac{L}{n} + \frac{L}{n}\right) \le \frac{2L}{n}(t_{i+1}^n - t_i^n) \tag{23}$$

for $t \in [t_i^n, t_{i+1}^n]$ and $c \in C(t)$. In the interior $]t_i^n, t_{i+1}^n[$, u_n is differentiable with derivative $u_n'(t) = (t_{i+1}^n - t_i^n)^{-1}(u_{i+1}^n - u_i^n)$, and hence by (23),

$$\langle u_n'(t), u_n(t) - c\rangle \le \frac{2L}{n}, \quad t \in]t_i^n, t_{i+1}^n[, \quad c \in C(t). \tag{24}$$

The estimate (17) also shows

$$|u_n'(t)| \le L, \quad t \neq t_i^n, \quad \text{hence} \quad |u_n'|_{L^\infty([0,T];H)} \le L, \quad n \in \mathbb{N}.$$

Since $L^\infty([0, T]; H)$ is the dual space of $L^1([0, T]; H)$, it is a consequence of the Banach-Alaoglu theorem [64, Thm. 3.15] that we may extract a further subsequence (again indexed by $n \in \mathbb{N}$) such that $u_n' \to v_*$ for some

$v_* \in L^\infty([0,T]; H)$, the convergence being in the weak-star topology on $L^\infty([0,T]; H)$. This means that for all $\varphi \in L^1([0,T]; H)$

$$\int_0^T \langle u_n'(t), \varphi(t) \rangle \, dt \to \int_0^T \langle v_*(t), \varphi(t) \rangle \, dt \quad \text{as} \quad n \to \infty.$$

Since u_n is absolutely continuous,

$$u_n(t) = u_0 + \int_0^t u_n'(s) \, ds, \quad t \in [0,T]. \tag{25}$$

For fixed $t_0 \in [0,T]$ and $z \in H$ define

$$\varphi(t) = \begin{cases} z & : \quad 0 \le t \le t_0 \\ 0 & : \quad t_0 < t \le 1 \end{cases}.$$

Then $\varphi \in L^1([0,T]; H)$ since $\int_0^T |\varphi(t)| dt = t_0 |z|$. Taking the inner product of (25) with z and using the weak convergence $u_n(t) \to u(t)$ it follows as $n \to \infty$ that

$$\langle u(t_0), z \rangle \longleftarrow \langle u_n(t_0), z \rangle = \langle u_0, z \rangle + \left\langle \int_0^{t_0} u_n'(t) \, dt, z \right\rangle$$

$$= \langle u_0, z \rangle + \int_0^{t_0} \langle u_n'(t), z \rangle \, dt$$

$$= \langle u_0, z \rangle + \int_0^T \langle u_n'(t), \varphi(t) \rangle \, dt$$

$$\longrightarrow \langle u_0, z \rangle + \int_0^T \langle v_*(t), \varphi(t) \rangle \, dt = \left\langle u_0 + \int_0^{t_0} v_*(t) \, dt, z \right\rangle.$$

As this holds for arbitrary $z \in H$ we can conclude that

$$u(t) = u_0 + \int_0^t v_*(s) \, ds, \quad t \in [0,T].$$

This again shows that $u : [0,T] \to H$ is differentiable at almost every point $t \in]0,T[$, and moreover $u'(t) = v_*(t)$ for almost every $t \in]0,T[$. In particular, $(-u_n') \to (-u')$ in the weak-star topology on $L^\infty([0,T]; H)$. According to Lemma 3 this yields

$$\int_0^T \delta^*(-u'(t), C(t)) \, dt \le \liminf_{n \to \infty} \int_0^T \delta^*(-u_n'(t), C(t)) \, dt; \tag{26}$$

recall the definition of δ^* from (7). Note also that by Lemma 4 and Lemma 2, and as a consequence of the weak convergence $u_n(T) \to u(T)$,

$$\int_0^T \langle u'(t), u(t) \rangle \, dt = \frac{1}{2} \left(|u(T)|^2 - |u_0|^2 \right) \le \liminf_{n \to \infty} \frac{1}{2} \left(|u_n(T)|^2 - |u_0|^2 \right)$$

$$= \liminf_{n \to \infty} \frac{1}{2} \Big(|u_n(T)|^2 - |u_n(0)|^2 \Big)$$

$$= \liminf_{n \to \infty} \int_0^T \langle u_n'(t), u_n(t) \rangle \, dt. \tag{27}$$

Next, taking the supremum w.r. to $c \in C(t)$ in (24) and integrating over $[0, T]$ we find that

$$\int_0^T \Big[\delta^*(-u_n'(t), C(t)) + \langle u_n'(t), u_n(t) \rangle \Big] \, dt \leq \frac{2LT}{n} \tag{28}$$

for $n \in \mathbb{N}$. Since $\big(\liminf_{n \to \infty} a_n \big) + \big(\liminf_{n \to \infty} b_n \big) \leq \liminf_{n \to \infty} (a_n + b_n)$ for general sequences $(a_n)_{n \in \mathbb{N}}$ and $(b_n)_{n \in \mathbb{N}}$, (28) together with (26) and (27) leads to

$$\int_0^T \Big[\delta^*(-u'(t), C(t)) + \langle u'(t), u(t) \rangle \Big] \, dt \leq 0 \tag{29}$$

by taking $\liminf_{n \to \infty}$ on both sides of (28). We have already shown that $u(t) \in C(t)$ for $t \in [0, T]$. Therefore by definition of δ^*, $\delta^*(-u'(t), C(t)) \geq \langle -u'(t), u(t) \rangle$, or stated differently,

$$\delta^*(-u'(t), C(t)) + \langle u'(t), u(t) \rangle \geq 0.$$

Whence (29) shows

$$\delta^*(-u'(t), C(t)) + \langle u'(t), u(t) \rangle = 0 \quad \text{for almost every} \quad t \in]0, T[.$$

Thus for any $c \in C(t)$, $\langle -u'(t), u(t) \rangle = \delta^*(-u'(t), C(t)) \geq \langle -u'(t), c \rangle$, i.e., $\langle -u'(t), c - u(t) \rangle \leq 0$. By definition of the normal cone we infer $-u'(t) \in N_{C(t)}(u(t))$ for almost every $t \in]0, T[$. This completes the proof of the theorem. □

Remark 3. (a) In the context of Theorem 2, the solution to (1) cannot be expected to be differentiable at really every point $t \in]0, T[$. Consider $H = \mathbb{R}$,

$$C(t) = \begin{cases} [t, 1] & : \quad 0 \leq t \leq \frac{1}{2} \\ [1 - t, 1] & : \quad \frac{1}{2} \leq t \leq 1 \end{cases},$$

and $u_0 = 0 \in C(0) = [0, 1]$. Then the solution is

$$u(t) = \begin{cases} t & : \quad 0 \leq t \leq \frac{1}{2} \\ \frac{1}{2} & : \quad \frac{1}{2} \leq t \leq 1 \end{cases}, \quad \text{with} \quad -u'(t) = \begin{cases} -1 & : \quad 0 \leq t < \frac{1}{2} \\ 0 & : \quad \frac{1}{2} < t \leq 1 \end{cases},$$

and hence belongs to $N_{C(t)}(u(t))$, as $N_{[t,1]}(t) =]-\infty, 0]$ for $t \in [0, \frac{1}{2}]$ and $N_{[1-t,1]}(\frac{1}{2}) = \{0\}$ for $t \in]\frac{1}{2}, 1]$, recall Remark 1. We obtain $d_H(C(t), C(s)) = |t - s|$ for $t, s \in [0, \frac{1}{2}]$ and $t, s \in [\frac{1}{2}, 1]$. Moreover,

$$d_H(C(t), C(s)) = |t - (1 - s)| \leq |t - s|, \quad t \in [0, \frac{1}{2}], \quad s \in [\frac{1}{2}, 1].$$

Hence (13) is satisfied with $L = 1$. On the other hand, u is not differentiable at $t = \frac{1}{2}$, but Lipschitz with constant 1. In the sweeping process interpretation, u_0 is swept by $C(t)$ to $u = \frac{1}{2}$ and then remains resting at $\frac{1}{2}$, since the boundary of $C(t)$ starts to move in the opposite direction at time $t = \frac{1}{2}$.

(b) Note that the proof of Theorem 2 is constructive. The catching-up algorithm can be easily implemented numerically. See [52] for more on numerical methods for the sweeping process.

(c) In the proof of Theorem 2 it could be even shown that the sequence of approximations converges to a solution uniformly. However, this would not have led to great simplification, and the proof actually given explains a general technique useful also for other purposes. ◊

Theorem 3 (Uniqueness). *The solution of (13) is unique in the class of absolutely continuous functions.*

Proof: Let u be a solution of $-u'(t) \in N_{C(t)}(u(t))$, $u(0) = u_0$, and let v be a solution of $-v'(t) \in N_{C(t)}(v(t))$, $v(0) = v_0$. Since $u(t) \in C(t)$ and $v(t) \in C(t)$ for $t \in [0, T]$, it follows that $\langle -u'(t), v(t) - u(t) \rangle \leq 0$ and also $\langle -v'(t), u(t) - v(t) \rangle \leq 0$ for almost every $t \in\,]0, T[$. Due to Equ. (12) in Lemma 4 we find

$$\frac{1}{2}\frac{d}{dt}\left(|u(t) - v(t)|^2\right) = \langle u'(t) - v'(t), u(t) - v(t) \rangle \leq 0$$

almost everywhere in $]0, T[$. Integration yields

$$|u(t) - v(t)|^2 \leq |u(0) - v(0)|^2 = |u_0 - v_0|^2, \quad t \in [0, T]. \qquad (30)$$

In particular, if $u_0 = v_0$, we get the claimed uniqueness. □

From (30) we also get

Corollary 1 (Dependence on data). *Under the above assumptions, if u and v are two solutions with $u(0) = u_0$ and $v(0) = v_0$, then*

$$|u(t) - v(t)| \leq |u_0 - v_0|, \quad t \in [0, T].$$

Theorem 4 (Dependence on the moving set). *Let $t \mapsto C(t)$ and $t \mapsto D(t)$ be two moving sets which satisfy (13) with Lipschitz constants L_C and L_D, respectively. Assume that both $C(t)$ and $D(t)$ are nonempty, closed, and convex for every $t \in [0, T]$. Then, if u denotes the solution to the sweeping process with $t \mapsto C(t)$ and initial value $u(0) = u_0$, and if v denotes the solution to the sweeping process with $t \mapsto D(t)$ and initial value $v(0) = v_0$, the estimate*

$$|u(t) - v(t)|^2 \leq |u_0 - v_0|^2 + 2(L_C + L_D)\int_0^t \Delta(s)\,ds, \quad t \in [0, T], \qquad (31)$$

holds, where

$$\Delta(t) = d_{\mathrm{H}}(C(t), D(t)), \quad t \in [0, T].$$

Proof: For fixed $t \in [0, T]$ we have $u(t) \in C(t) \subset D(t) + \overline{B}_{\Delta(t)}(0)$. Hence there exist vectors $d(t) \in D(t)$ and $r(t) \in H$ such that $u(t) = d(t) + r(t)$ and $|r(t)| \leq \Delta(t)$. It can be shown that it is possible to choose the maps $t \mapsto d(t)$ and $t \mapsto r(t)$ as being measurable. Similarly, we find $v(t) = c(t) + s(t)$ with $c(t) \in C(t)$ and $|s(t)| \leq \Delta(t)$. Arguing as in the proof of Theorem 3 we obtain

$$\frac{1}{2} \frac{d}{dt} \left(|u(t) - v(t)|^2 \right)$$
$$= \langle u'(t) - v'(t), u(t) - v(t) \rangle$$
$$= \langle u'(t), u(t) - c(t) \rangle + \langle v'(t), v(t) - d(t) \rangle - \langle u'(t), s(t) \rangle - \langle v'(t), r(t) \rangle$$
$$\leq -\langle u'(t), s(t) \rangle - \langle v'(t), r(t) \rangle \leq (|u'(t)| + |v'(t)|) \Delta(t).$$

According to Theorem 2, $|u'(t)| \leq L_C$ and $|v'(t)| \leq L_D$ almost everywhere, thus integration yields (31). $\qquad\square$

3.2 An application: Evolution of a quasi-static mechanical system

This section concerns a problem from elastoplasticity, for which we want to show how the sweeping process arises and how the theory developed in Section 3.1 can be successfully applied to yield a solution of the underlying evolution problem. Most of the material is taken from [43, Sect. 6].

The system we will deal with consists of two main ingredients: (displacements of) states and forces. The force f acting on the velocity u' of a state u gives rise to $\langle u', f \rangle$, the power of the force f. Then

$$\int_{t_0}^{t_1} \langle u'(t), f \rangle \, dt = \int_{t_1}^{t_2} \frac{d}{dt} \langle u(t), f \rangle \, dt = \langle u(t_2), f \rangle - \langle u(t_1), f \rangle \cong \langle \delta u, f \rangle$$

is the work of the force f corresponding to the displacement δu. The classical approach to plasticity requires introducing, besides the "visible" elements describing the configuration, additional internal variables (also called hidden parameters). Thus a state u in fact consists of $u = (x, p)$, denoting the "visible" component by x and the hidden (or "plastic") component by p. Usually x is the part that experiences external forces or that may be submitted to constraints. Let us consider an example, see Fig. 5.

The system is comprised by two particles x and p which move in the plane $H = \mathbb{R}^2$. Let $U = \mathbb{R} \times \{0\}$ be the horizontal axis through the origin. This axis is driven by some function $g(t)$ to move parallel to U, and x is constrained to remain in this set, i.e., we require

$$x \in U + g. \tag{32}$$

The point p influences the motion of x, since it is assumed to be coupled to x by means of a spring. In addition, an external force $c(t)$ is applied to x,

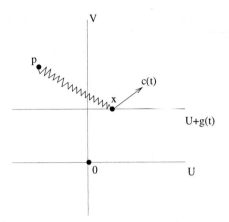

Fig. 5. An example system

and moreover, x moves without experiencing dry friction. We suppose that at each instant t the various forces equilibrate each other. In particular, the force applied to p by the spring has to counter-balance the dry friction resistance of the plane to p. Let us now consider in more detail the forces acting on p and x.

(p–1) As just mentioned, there is a force $f \in H$ describing "plastic resistance" against the velocity p' of p. Usually one imposes the law

$$p' \in N_C(-f), \tag{33}$$

with $C \subset H$ being a fixed nonempty closed convex subset of H consisting of the negatives $(-\hat{f})$ of forces $\hat{f}$; the minus sign is a little confusing but conventionally used here. The law (33) is motivated as follows: By definition of the normal cone this means $\langle p', c + f \rangle \leq 0$ for all $c \in C$. Taking $c = -\hat{f}$ with another force $\hat{f}$, we then ensure by (33) that

$$\langle p', \hat{f} \rangle \geq \langle p', f \rangle \quad \text{for all forces} \quad \hat{f}.$$

Thus f is to be taken as one of those forces acting on p' that minimize the power $\langle p', \hat{f} \rangle$ over all forces $\hat{f}$. Since the quantity $-\langle p', \hat{f} \rangle$ has the interpretation of the dissipated power of $\hat{f}$ acting on p', (33) often is called the "principle of maximal dissipation".

(p–2) The second force that p experiences is the force s of "elastic restoring towards x", described by a law

$$s = A(e), \quad e = x - p. \tag{34}$$

The difference e is called the deviation. Here $A : H \to H$ is chosen an invertible linear matrix, generalizations being possible.

(x–1) The first force that acts on x is the "reaction" r of the constraint $U + g(t)$: In order for x not to leave this set, cf. (32), we require that

$$-r \in N_{U+g(t)}(x). \tag{35}$$

(x–2) There is also a "load" c on x.

(x–3) Finally x experiences the force $-s$ of "elastic restoring towards p" according to (p–2).

Since we analyzed the forces acting at some fixed instant of time t, the given functions g and c will depend on time. We intend to understand the evolution of the state variables $x(t)$ and $p(t)$ under the basic assumption that the motion is sufficiently slow for no inertia to being accounted for, i.e., the acting forces just equilibrate each other. For that reason the system is called quasi-static. Due to (p–1)–(p–2) and (x–1)-(x–3) we therefore assume

$$s + f = 0, \quad r + c - s = 0. \tag{36}$$

Now we are going to show how the whole problem can be conveniently recast and solved as a sweeping process. For simplicity we let $A = I$, the identity matrix, in (34). Thus $s = e = x - p$. The more general case $A \neq I$ could be treated by taking as inner product in H the expression $\langle u, v \rangle = (Au, v)$, with $(\cdot, \cdot)$ the usual Euclidean inner product; the quantity $\frac{1}{2}(Au, u)$ is the elastic energy of u. Then (32)-(36) can be summarized to

$$\left. \begin{array}{ll} x(t) \in U + g(t), & p'(t) \in N_C(s(t)), \\ x(t) = p(t) + s(t), & c(t) - s(t) \in N_{U+g(t)}(x(t)) \end{array} \right\} \tag{37}$$

for (almost every) $t \in [0, T]$, with some fixed $T > 0$. Introducing $V = U^\perp$, the orthogonal complement of U in the Hilbert space H, we find the equivalent form

$$x(t) \in U + g(t), \quad s(t) \in V + c(t), \quad x(t) = p(t) + s(t), \quad p'(t) \in N_C(s(t)) \tag{38}$$

for (almost every) $t \in [0, T]$. Indeed, omitting the variable t for the moment, assume first that (37) holds. Then $x - g = u_0 \in U$ and $0 \geq \langle c - s, u + g - x \rangle = \langle c - s, u - u_0 \rangle$ for all $u \in U$, whence

$$\langle u_0, c - s \rangle \geq \langle u, c - s \rangle, \quad u \in U. \tag{39}$$

Suppose there exists $u_1 \in U$ such that $\langle u_1, c - s \rangle \neq 0$. Without loss of generality we may assume $\langle u_1, c - s \rangle > 0$, since otherwise we may replace u_1 by $-u_1 \in U$; recall U is a subspace of H. But then also $u = \lambda u_1 \in U$ for $\lambda > 0$, and this gives a contradiction in (39) as $\lambda \to \infty$. Thus $\langle u, s - c \rangle = 0$ for all $u \in U$, i.e., $s - c \in V$. Conversely, if $s \in V + c$ and $u \in U$, then $\langle c - s, u \rangle = 0$, in particular $\langle c - s, x - g \rangle = 0$. Therefore if $u \in U$, then

$\langle c - s, u + g - x \rangle = 0$, and consequently $c - s \in N_{U+g}(x)$. Hence we have shown that in fact (38) is equivalent to (37).

To solve (38) for the unknown functions $x(t)$ and $p(t)$, we make the following general hypotheses.

(H1) H is a separable Hilbert space, $U \subset H$ a subspace, and $V = U^{\perp}$;

(H2) $C \subset H$ is nonempty, closed, convex, and bounded;

(H3) both $g : [0, T] \to V$ and $c : [0, T] \to U$ are given Lipschitz continuous functions;

(H4) $\text{int } C \cap (V + c(t)) \neq \emptyset$ for $t \in [0, T]$.

The assumption in $(H3)$ that $g(t) \in V$ and $c(t) \in U$ is no real restriction, since e.g. for a general $g(t)$ we may decompose $g(t) = g_U(t) + g_V(t)$ with $g_U(t) \in U$ and $g_V(t) \in V$, and then deal with $x(t) \in [U + g_U(t)] + g_V(t)$. This amounts to allowing $U(t) := U + g_U(t)$ to move with time in a Lipschitz continuous manner, but this would introduce no serious additional difficulties. Concerning $(H4)$, note that a necessary condition for (38) to have a solution is $C \cap (V + c(t)) \neq \emptyset$ for $t \in [0, T]$, because $s(t) \in C \cap (V + c(t))$. Thus $(H4)$ may be interpreted as a "safe load assumption", since the condition has to hold with a certain "safety margin".

The next step in rewriting (38) consists in a kind of Lyapunov-Schmidt reduction by projection of the equations on U and $U^{\perp} = V$. For this, we introduce new variables z and y by

$$z = \text{proj}_U(p) \quad \text{and} \quad y = -\text{proj}_V(p). \tag{40}$$

Then (38) is equivalent to

$$z'(t) - y'(t) \in N_C(y(t) + c(t) + g(t)), \quad z(t) \in U, \quad y(t) \in V \tag{41}$$

for (almost every) $t \in [0, T]$. To see this, first suppose (38) is satisfied. Then $(x - g) \in U$, $(s - c) \in V$, and $p + c - g = (x - g) + (c - s) \in U \oplus V$. Therefore $x - g = \text{proj}_U(p + c - g)$ and also $c - s = \text{proj}_V(p + c - g)$. Since the projection is linear on subspaces, and due to $g \in V$ and $c \in U$, we find $x - g = \text{proj}_U(p + c - g) = \text{proj}_U(p) + c$, whence $z = \text{proj}_U(p) = x - c - g$, and similarly $y = -\text{proj}_V(p) = s - c - g$. We get $z' - y' = p' \in N_C(s) = N_C(y + c + g)$. On the other hand, let (41) hold and define

$$p = z - y, \quad x = z + c + g, \quad \text{and} \quad s = y + c + g. \tag{42}$$

Then $x - g = z + c \in U$, $s - c = y + g \in V$, $x = p + s$, and in addition $p' = z' - y' \in N_C(y + c + g) = N_C(s)$. Accordingly, we have shown (41) and (38) are equivalent. Through the transformations (40) and (42) we established a one-to-one correspondence between (41) and (38) resp. (37).

Now the connection of (41) to the sweeping process will become obvious.

Lemma 5. *If* $z : [0, T] \to U$ *and* $y : [0, T] \to V$ *are a solution to (41), then* y *solves the sweeping process* $-y'(t) \in N_{C(t)}(y(t))$, *with*

$$C(t) = (C - c(t) - g(t)) \cap V. \tag{43}$$

Proof: Firstly, by (41) we have $y(t) \in V$ and $y(t) + c(t) + g(t) \in C$, i.e., $y(t) \in C(t)$. Next, fix $\hat{c} \in C$ such that $\hat{c} - c(t) - g(t) \in V$. Since the image of z is contained in U and U is a vector space, also $z'(t) \in U$ for almost every $t \in [0, T]$. Hence $\langle z'(t), \hat{c} - c(t) - g(t) - y(t) \rangle = 0$ as also $y(t) \in V$, and therefore by (41)

$$\langle -y'(t), [\hat{c} - c(t) - g(t)] - y(t) \rangle = \langle z'(t) - y'(t), \hat{c} - c(t) - g(t) - y(t) \rangle \leq 0.$$

Consequently, we have shown $-y'(t) \in N_{C(t)}(y(t))$ at times t where $y'(t)$ does exist. $\square$

According to the orthogonal decomposition above, it turns out that by means of the sweeping process for y in Lemma 5 we already can construct a solution (z, y) to the full problem (41). Once this has been done, we have also proved that the original problem (37) has a solution (x, p). In particular this applies to the example model from Fig. 5 above.

Theorem 5. *Assume that the hypotheses (H1)–(H4) are satisfied and that the compatibility conditions*

$$z_0 \in U, \quad y_0 \in V, \quad y_0 + c(0) + g(0) \in C \qquad (44)$$

hold for the initial values z_0 and y_0. Then there exists a Lipschitz continuous solution $z : [0, T] \to U$ and $y : [0, T] \to V$ of (41) with $z(0) = z_0$ and $y(0) = y_0$.

Proof: First we intend to solve the sweeping process

$$-y'(t) \in N_{C(t)}(y(t)) \quad \text{a.e. in} \quad [0, T], \quad y(0) = y_0 \qquad (45)$$

with $C(t)$ from (43), by application of Theorem 2. According to (44) we have $y_0 \in (C - c(0) - g(0)) \cap V = C(0)$. Moreover, since V is a vector space and $g(t) \in V$, (H4) implies $C(t) \neq \emptyset$ for $t \in [0, T]$, and by (H2) each $C(t)$ is closed and convex. By Lemma 6 below, $t \mapsto C(t)$ is Lipschitz continuous with some constant $L > 0$. Thus by Theorem 2, (45) has a Lipschitz continuous solution $y : [0, T] \to H$, and $y(t) \in V$, since $y(t) \in C(t)$ for $t \in [0, T]$. Let

$$F(t) = \{w \in H : \delta^*(w, C) = \langle w, y(t) + c(t) + g(t) \rangle\}, \quad t \in [0, T],$$

with δ^* from (7). We are going to show that

$$\mathcal{F}(t) = F(t) \cap (U - y'(t)) \neq \emptyset \quad \text{a.e. in} \quad [0, T]. \qquad (46)$$

Indeed, let t be a point where (45) holds. Observe that

$$\text{int}(C - c(t) - g(t)) \cap V = [\text{int}(C) - c(t) - g(t)] \cap V \neq \emptyset,$$

as $g(t) \in V$ and $\mathrm{int}(C) \cap (V + c(t)) \neq \emptyset$ due to $(H4)$. Moreover, $y(t) \in C(t) \subset V$. Hence it follows from Lemma 1(b) and (a) that

$$N_{C(t)}(y(t)) = N_{[C-c(t)-g(t)]\cap V}(y(t)) = N_{C-c(t)-g(t)}(y(t)) + N_V(y(t))$$
$$= N_{C-c(t)-g(t)}(y(t)) + U.$$

Accordingly we may decompose $-y'(t) = w - u$. Then $w = u - y'(t) \in U - y'(t)$, and additionally $w \in F(t)$, since for $\hat{c} \in C$ we obtain $\langle w, c(t) + g(t) + y(t) \rangle \geq \langle w, \hat{c} \rangle$, i.e., $\langle w, c(t) + g(t) + y(t) \rangle \geq \delta^*(w, C)$. As $y(t) \in C(t) \subset C - c(t) - g(t)$ or $c(t) + g(t) + y(t) \in C$, we conclude $w \in F(t)$. This ends the proof of (46). By modification of $\mathcal{F}(t)$ on a set of measure zero in $[0, T]$, we may assume that in fact $\mathcal{F}(t) \neq \emptyset$ for all $t \in [0, T]$. Since $F(t)$ is closed for $t \in [0, T]$, we see that in addition $\mathcal{F}(t)$ is closed. Moreover, with some technical efforts it can be shown that $\mathcal{F} : [0, T] \to 2^H$, the power set of H, is measurable, i.e., for any open $B \subset H$ the set $\{t \in [0, T] : \mathcal{F}(t) \cap B \neq \emptyset\}$ is measurable. As H is assumed to be separable, a result from the theory of set-valued maps, cf. [27, Prop. 3.2], shows that there exists a measurable selection $\zeta : [0, T] \to H$ of $\mathcal{F}$. This means that $\zeta(t) \in \mathcal{F}(t)$ for $t \in [0, T]$, or according to the definition of $\mathcal{F}(t)$,

$$\delta^*(\zeta(t), C) = \langle \zeta(t), y(t) + c(t) + g(t) \rangle \quad \text{and} \quad \zeta(t) + y'(t) \in U \qquad (47)$$

for almost all $t \in [0, T]$. We intend to set

$$z(t) = z_0 + \int_0^t [\zeta(s) + y'(s)] \, ds, \quad t \in [0, T], \qquad (48)$$

if the integral exists. In that case, as $z_0 \in U$ and $\zeta(s) + y'(s) \in U$ for almost every s, we would have $z(t) \in U$ for $t \in [0, T]$, since U is a subspace. Moreover, $z'(t) = \zeta(t) + y'(t)$ almost everywhere in conjunction with the first relation in (47) would prove $z'(t) - y'(t) = \zeta(t) \in N_C(y(t) + c(t) + g(t))$ almost everywhere in $[0, T]$, i.e., (z, y) is the desired solution of (41).

Thus it suffices to derive a bound on $\zeta(s)$, since this also will show that the integral in (48) does exist. For $\rho > 0$ let

$$C_\rho = \{w \in H : \mathrm{dist}\,(w, H \setminus C) \geq \rho\} \subset C.$$

We claim that there exists $\rho > 0$ such that

$$C_\rho \cap (V + c(t)) \neq \emptyset \quad \text{for all} \quad t \in [0, T]. \qquad (49)$$

In fact, otherwise we could choose a sequence $\rho_j \to 0$ and find times $t_j \in [0, T]$ such that $C_{\rho_j} \cap (V + c(t_j)) = \emptyset$ for all j. By definition of C_{ρ_j} hence $\mathrm{dist}\,(v + c(t_j), H \setminus C) \leq \rho_j$ for $v \in V$. If we assume without loss of generality that $t_j \to t_0 \in [0, T]$ as $j \to \infty$, we infer from $x \mapsto \mathrm{dist}\,(x, A)$ being even Lipschitz continuous for a general $A \subset H$ that $\mathrm{dist}\,(v + c(t_0), H \setminus C) = 0$ for all $v \in V$. Therefore $v + c(t_0) \in \overline{H \setminus C} = H \setminus \mathrm{int}C$, i.e., $V + c(t_0) \subset H \setminus \mathrm{int}C$, in contradiction to $(H4)$. Consequently, (49) is satisfied for some small $\rho > 0$. Clearly C_ρ is closed, and it is also convex by Lemma 1(d). Moreover, the argument of Lemma 6 below implies that the moving set $t \mapsto H(t) = C_\rho \cap (V + c(t))$ is Lipschitz continuous as $\operatorname{int} C_\rho \neq \emptyset$ for ρ small. By solving the corresponding sweeping process with an arbitrary initial value chosen in $H(0)$, it follows that there exists a (Lipschitz) continuous function $h : [0, T] \to H$ such that

$$h(t) \in C_\rho \cap (V + c(t)) \quad \text{for} \quad t \in [0, T]. \tag{50}$$

This function h will finally help us to estimate ζ from above. We first note that

$$\rho|w| + \langle w, h(t) \rangle \leq \delta^*(w, C), \quad t \in [0, T], \quad w \in H, \tag{51}$$

according to the definition of C_ρ and by Lemma 1(c). Taking $w = \zeta(t)$ in (51) we obtain from (47) that for almost every $t \in [0, T]$

$$\begin{aligned}
\rho|\zeta(t)| &\leq \delta^*(\zeta(t), C) - \langle \zeta(t), h(t) \rangle = \langle \zeta(t), y(t) + c(t) + g(t) \rangle - \langle \zeta(t), h(t) \rangle \\
&= \langle \zeta(t) + y'(t), y(t) + c(t) + g(t) - h(t) \rangle \\
&\quad - \langle y'(t), y(t) + c(t) + g(t) - h(t) \rangle.
\end{aligned} \tag{52}$$

Recall from (47) that $\zeta(t) + y'(t) \in U$. In addition, $c(t) - h(t) \in V$ by (50), and we have $g(t) \in V$ by $(H3)$ and also $y(t) \in C(t) \subset V$. Since U and V are orthogonal the first term in (52) therefore vanishes and we get

$$\begin{aligned}
\rho|\zeta(t)| &\leq -\langle y'(t), y(t) + c(t) + g(t) - h(t) \rangle \\
&\leq |y'(t)| \Big(|y(t)| + |c(t)| + |g(t)| + |h(t)| \Big) \\
&\leq L \Big(|y|_{L^\infty([0,T];H)} + |c|_{L^\infty([0,T];H)} + |g|_{L^\infty([0,T];H)} + |h|_{L^\infty([0,T];H)} \Big).
\end{aligned}$$

The latter is a finite constant, whence it follows that even $\zeta \in L^\infty([0, T]; H)$. In particular, the integral in (48) is well-defined, and z is Lipschitz continuous with constant $|\zeta|_{L^\infty([0,T];H)} + L$. This completes the proof of Theorem 5. $\square$

The following lemma was needed in the proof of Theorem 5.

Lemma 6. *Under the assumptions of Theorem 5, the moving set $t \mapsto C(t)$ from (43) is Lipschitz continuous.*

Proof: Since both g and c are Lipschitz continuous, it is sufficient to prove that $\tilde{C}(t) = C(t) + c(t) + g(t) = C \cap (V + c(t))$ is Lipschitz continuous; recall that $g(t) \in V$. Fix $w \in \tilde{C}(s)$ and choose a (depending on t) such that $a \in V + c(t)$ and $\overline{B}_\rho(a) \subset C$; this is possible by (49). According to a geometrical inequality of Moreau, see [42, p. 24], we have $\operatorname{dist}(w, \tilde{C}(t)) \leq (1 + \rho^{-1}|w - a|)(\operatorname{dist}(w, C) + \operatorname{dist}(w, V + c(t)))$. As $w \in C$, the first distance is zero. Moreover, $[w - c(s)] + c(t) \in V + c(t)$, so we get $\operatorname{dist}(w, \tilde{C}(t)) \leq (1 + \rho^{-1}|w - a|)\operatorname{lip}(c)|t - s|$, with $\operatorname{lip}(c)$ the Lipschitz constant of c. As $w, a \in C$

and C is bounded by $(H2)$, it follows that $\sup_{w \in \tilde{C}(s)} \mathrm{dist}\,(w, \tilde{C}(t)) \leq K|t - s|$ for some constant $K > 0$. Since the role of s and t can be interchanged, $t \mapsto \tilde{C}(t)$ is Lipschitz continuous with Lipschitz constant K. $\qquad\square$

Remark 4. (a) Since the component y of a solution (z, y) to (41) is a solution to the sweeping process (45) according to Lemma 5, and since this solution to the sweeping process is unique as a consequence of Theorem 3, it follows that the y-component of a solution to (41) is uniquely determined. Utilizing the transformation (42), we conclude that $s = x - p$ in (37) is uniquely determined, too.

(b) Quite often models from elastoplasticity can be formulated as a sweeping process which then is coupled to one or more other equations. This e.g. was the case for the variable y above, see Lemma 5, where a full solution additionally required the determination of z in (41). However, this turned out to be not too difficult if we first managed to solve the corresponding sweeping process for y. One may consult [44] for a similar approach to a model for an elastoplastic rectilinear rod with small longitudinal displacements, and cf. also [9,11–13,17,31]. $\qquad\Diamond$

3.3 The state-dependent sweeping process

The classical sweeping process (1) allows the moving set $C(t)$ to depend only on the time $t \in [0, T]$, and as we have seen in Section 3.1, a well-posed problem arises in that each such Lipschitz continuous sweeping process has a unique solution. Here, following [35,36], we study a generalization of the classical sweeping process in that we allow $C(t, u)$ to depend also on the current state $u = u(t)$. Thus we consider

$$-u'(t) \in N_{C(t,u(t))}(u(t)) \quad \text{a.e. in} \quad [0, T], \quad u(0) = u_0 \in C(0, u_0), \qquad (53)$$

and instead of (13) we suppose

$$d_{\mathrm{H}}(C(t, u), C(s, v)) \leq L_1\,|t - s| + L_2\,|u - v|, \quad t, s \in [0, T], \; u, v \in H. \qquad (54)$$

Note that a solution of (53) in particular has to satisfy $u(t) \in C(t, u(t))$ for $t \in [0, T]$.

There is a special case of (53) which also deserves separate attention, namely parabolic quasi-variational inequalities of the form

$$\text{find} \quad v(t) \in \Gamma(v(t)) : \quad \langle v'(t) + f(t), w - v(t) \rangle \geq 0 \quad \forall w \in \Gamma(v(t)), \qquad (55)$$

with some initial value $v(0) = v_0 \in \Gamma(v_0)$, where $v = v(t) : [0, T] \to H$, and $f : [0, T] \to H$ is some inhomogeneity, and $\Gamma(v) \subset H$ is a set of constraints. Written somewhat differently, (55) means that

$$-v'(t) \in N_{\Gamma(v(t))}(v(t)) + f(t) \quad \text{a.e. in} \quad [0, T], \quad v(0) = v_0 \in \Gamma(v_0). \qquad (56)$$

Then, if v is a solution of (56), and if we define

$$u(t) = v(t) + \int_0^t f(s)\, ds \quad \text{and} \quad C(t, u) = \Gamma\left(u - \int_0^t f(s)\, ds\right) + \int_0^t f(s)\, ds,$$
(57)

it follows that u is a solution of (53), with initial value $u_0 = v_0 \in C(0, u_0)$. Thus indeed the quasi-variational inequalities (56) are particular cases of (53).

When dealing with (56), we will always suppose that $v \mapsto \Gamma(v)$ is Lipschitz with constant $L \geq 0$, i.e.,

$$d_H(\Gamma(v), \Gamma(w)) \leq L\,|v - w|, \quad v, w \in H.$$
(58)

In case that $f \in L^\infty([0, T]; H)$ in (56) (which we will assume for simplicity, but $f \in L^1([0, T]; H)$ is sufficient), (58) implies that (54) holds for C defined by (57), with $L_1 = (L + 1)|f|_{L^\infty([0,T];H)}$ and $L_2 = L$. This might be a bad estimate for L_1, but it will turn out that it is only the size of L_2 in (54) which determines the existence of a solution to (53).

Example 3. Choose $H = L^2(0, 1)$, the usual space of square-integrable real-valued functions on $(0, 1)$ with norm defined by $|\phi|^2_{L^2(0,1)} = \int |\phi(x)|^2 dx$, and set

$$\Gamma(v) = \left\{\phi \in H_0^1(0, 1) : |\phi'(x)| \leq \Phi(x, v(x)) \text{ for a.e. } x \in (0, 1)\right\},$$
(59)

for $v \in L^2(0, 1)$, with $H_0^1(0, 1)$ being the Sobolev space of functions $\phi \in L^2(0, 1)$ that possess a (distributional) derivative $\phi' \in L^2(0, 1)$. In addition, belonging to $H_0^1(0, 1)$ somehow requires ϕ to "vanish at the boundary of $(0, 1)$", and $|\phi|^2_{H_0^1(0,1)} = \int |\phi(x)|^2 dx + \int |\phi'(x)|^2 dx$ defines the norm in $H_0^1(0, 1)$. The function $\Phi : (0, 1) \times \mathbb{R} \to [0, \infty[$ in the definition of $\Gamma(v)$ is prescribed.

The study of problems of type (55), (59), is motivated through quasi-variational inequalities arising e.g. in the evolution of sandpiles; see [62]. In the sandpile example (where one has $x \in \Omega \subset \mathbb{R}^2$ rather than $x \in (0, 1)$ as we are going to treat here), without simplifying assumptions the function Φ will be discontinuous, posing serious technical problems to showing the existence of solutions. We will return to this example again later. $\diamond$

There are several other instances where state-dependent sweeping processes as (53) or parabolic quasi-variational inequalities of type (56) yield the correct mathematical description of the underlying practical problem. State-dependent sweeping processes as (53) occur, for example, in the treatment of 2-D or 3-D quasistatic evolution problems with friction, as treated in [25, Ch. II, III] (see also the account given in [42, p. 155-161]); the models investigated there are generalizations of the one introduced in Section 3.2. In a different context, the state-dependent sweeping process is used in microme-chanical damage models (the so-called Gurson-models) for iron materials with memory to describe the evolution of the plastic strain in presence of small damages; cf. [70,7]. Examples of evolutionary quasi-variational inequalities may also be found in [2] and the references therein, cf. in particular p. 242 f.

Returning now to the general problem of proving existence of solutions to (56) and (53) respectively, we start with a negative result.

Example 4. If $L > 1$ in (58), then in general (56) will not have a solution. To see this, let $H = \mathbb{R}$, $\Gamma(v) = [Lv, 1]$ for $v \leq 1/L$, and $\Gamma(v) = [1, Lv]$ for $v > 1/L$, $v \in \mathbb{R}$. Then $v \mapsto \Gamma(v)$ is Lipschitz continuous with constant L. In addition, $v_0 := 0 \in \Gamma(0) = [0, 1]$, but (56) with $f(t) = -1$, $t \in [0, 1]$, has no solution. Indeed, assume $v : [0, 1] \to H$ to be a solution. By continuity, $Lv(t) \leq 1$ for $t \in [0, \delta]$ with a suitable $\delta > 0$, hence $v(t) \in \Gamma(v(t)) = [Lv(t), 1]$ and $L > 1$ imply $v(t) \leq 0$ in $[0, \delta]$. On the other hand, for $x \leq 0$,

$$N_{\Gamma(x)}(x) = \begin{cases}]-\infty, 0] & : \quad x = 0 \\ \{0\} & : \quad x < 0 \end{cases} \subset \,]-\infty, 0],$$

so that (56) yields $v'(t) \geq -f(t) = 1$ in $[0, \delta]$. But this gives the contradiction $v(t) \geq t > 0 \geq v(t)$ in $]0, \delta]$. ◇

As (56) is a special case of (53), this shows that in general we cannot expect to have a solution of either of the two systems for $L_2 > 1$ in (54) or $L > 1$ in (58), respectively. Moreover, the next more advanced example shows that usually we will not have solutions even for $L_2 = 1$ or $L = 1$.

Example 5. Let $H = \mathbb{R}^2$ and $\Gamma(v) = \{|v|\} \times [-1, 1]$ for $v \in \mathbb{R}^2$, and $f(t) = (0, -1)$. Take $v_0 = (0, 0) \in C(0)$, and notice that Γ is Lipschitz continuous with constant $L = 1$. Suppose that (56) has a solution u on some interval $[0, T]$. Then $v(t) = (v_1(t), v_2(t)) \in \Gamma(v(t))$ implies $v_1(t) = |v(t)|$, and there-fore $v_2(t) = 0$ in $[0, T]$. For every $x = (x_1, x_2) \in \Gamma(v(t))$ we have

$$v_1'(t)(x_1 - v_1(t)) + (v_2'(t) + 1)(x_2 - v_2(t)) \geq 0,$$

and since necessarily $x_1 = v_1(t)$, this reduces to $x_2 \geq 0$ for all $x_2 \in [-1, 1]$, a contradiction. ◇

This already shows that (53) can be much more complex than the classical sweeping process (1), as the latter always has a solution. However, in case that $L_2 < 1$ the existence of a solution to (53) can be shown, and by the above transformation (57) this carries over to (56) for $L < 1$.

Theorem 6. *Let (54) hold for C with some $0 \leq L_2 < 1$, and assume that $C(t, u) \subset H$ is nonempty, closed, and convex for $t \in [0, T]$, and $u \in H$. Suppose that*

$$C(t, A) \cap \overline{B}_R(0) = \left(\bigcup_{u \in A} C(t, u) \right) \cap \overline{B}_R(0) \subset H$$

is relatively compact for all bounded $A \subset H$ and all $R > 0$. If $u_0 \in C(0, u_0)$, then (53) has a solution on $[0, T]$.

The compactness assumption is for technical reasons and always satisfied in the finite dimensional case $H = \mathbb{R}^d$. We will not give the complete proof of the theorem, but rather explain what is the difference to the proof of Theorem 2. To avoid problems with compactness (recall that in an infinite dimensional space closed bounded sets need not be compact), we will restrict ourselves to $H = \mathbb{R}^d$. We start again by fixing a time discretization as in (14). Defining $u_0^n = u_0$, we thus have $u_0^n \in C(t_0^n, u_0^n)$. To mimic the catching-up algorithm, we then should define $u_1^n = \text{proj}\,(u_0^n, C(t_1^n, *))$, and the problem is what to replace the "$*$" with. If we let $u_1^n = \text{proj}\,(u_0^n, C(t_1^n, u_0^n))$, then we have an explicit discretization scheme, but it will turn out that there is no easy way to obtain the necessary bound for convergence of the approximations to a solution. Thus we should rather use the implicit scheme $u_1^n = \text{proj}\,(u_0^n, C(t_1^n, u_1^n))$, but then the first problem is to solve this equation for u_1^n. The next lemma contains the key observation in this respect.

Lemma 7. *If $t \in [0, T]$ and $u \in C(s, u)$ for some $s \in [0, T]$, then there exists $v \in H$ such that $v = \text{proj}\,(u, C(t, v))$ and $|v - u| \leq L_1 |t - s| / (1 - L_2)$.*

Proof: Let $r = L_1 |s - t| / (1 - L_2)$, $D = \overline{B}_r(u)$, and $Fv = \text{proj}\,(u, C(t, v))$ for $v \in D$. Then by (54) for $v \in D$

$$|Fv - u| = |\text{proj}\,(u, C(t, v)) - u| = \text{dist}\,(u, C(t, v)) \leq d_{\mathrm{H}}(C(s, u), C(t, v))$$
$$\leq L_1 |s - t| + L_2 |u - v| \leq L_1 |s - t| + L_2\, r = r,$$

and hence $F(D) \subset D$. Moreover, F is continuous as may be seen from (54) and a geometrical inequality of Moreau for projections, cf. [42, Prop. 4.7, p. 26]. Therefore F is a self-map of the ball $D \subset \mathbb{R}^d$, and consequently has a fixed point $v \in D$ by Browder's fixed point theorem; see [26, Thm. 3.2]. $\quad\square$

Hence we may continue above and choose $u_1^n \in H$ such that $u_1^n = \text{proj}\,(u_0^n, C(t_1^n, u_1^n))$, and moreover $|u_1^n - u_0^n| \leq L_1 t_1^n / (1 - L_2)$. Thus in particular, $u_1^n \in C(t_1^n, u_1^n)$, and hence Lemma 7 applies again to yield u_2^n such that $u_2^n = \text{proj}\,(u_1^n, C(t_2^n, u_2^n))$ and also $|u_2^n - u_1^n| \leq L_1 (t_2^n - t_1^n) / (1 - L_2)$. Iterating this procedure, for $i = 1, \ldots, N$ we find u_i^n with

$$u_i^n = \text{proj}\,(u_{i-1}^n, C(t_i^n, u_i^n)) \quad \text{and} \quad |u_i^n - u_{i-1}^n| \leq L_1 (t_i^n - t_{i-1}^n) / (1 - L_2). \tag{60}$$

It turns out that in fact (60) is all that we need to derive the necessary bounds in norm and variation for the sequence $(u_n)_{n \in \mathbb{N}}$; this is shown as in the proof of Theorem 2, where (17) was one of the main ingredients. The passage to the limit $n \to \infty$ in order to prove that the limit of the approximations indeed yields a solution then is a little more technical than before, and we omit the details.

Finally we are going to indicate an application of Theorem 6 in the setting of Example 3; recall that we have also proved the existence of a solution to (55) in case that $L < 1$ in (58), at least under an appropriate compactness condition, see the remarks below. Note that such condition now will be needed, as $H = L^2(0,1)$ is an infinite dimensional space. For the function Φ in (59) we assume that

$$0 < m \leq \Phi(x,v) \leq M, \quad |\Phi(x,v) - \Phi(x,w)| \leq K|v - w|, \qquad (61)$$

for a.e. $x \in (0,1)$ and $v,w \in \mathbb{R}$. Under these hypotheses, we will obtain the existence of a solution to (55) for K small. In addition, the solution will be quite regular.

Theorem 7. *Let (61) be satisfied for Φ and assume that $K < \frac{\sqrt{2}m}{2M+m}$. Then for every $v_0 \in L^2(0,1)$ such that $v_0 \in \Gamma(v_0)$, there exists a Lipschitz continuous solution $v : [0,T] \to L^2(0,1)$ of (55) with initial value $v(0) = v_0$.*

Again we will not give the complete proof, but indicate some steps. Firstly, it is known that the embedding of $H_0^1(0,1)$ into $L^2(0,1)$ is relatively compact. Since Φ is bounded above by M, all sets $\Gamma(v)$ are contained in some fixed ball in $H_0^1(0,1)$, which, by the preceding remark, will be relatively compact in $L^2(0,1)$. Without going into further particularities, this is enough to guarantee that the necessary compactness condition is satisfied. Therefore the main task is to verify (58). For this, first observe that $|\phi|_{L^2(0,1)} \leq \frac{1}{\sqrt{2}}|\phi'|_{L^2(0,1)}$ for $\phi \in H_0^1(0,1)$ due to Hölder's inequality; note $|\phi(x)| = |\int_0^x \phi'(y)\,dy| \leq \sqrt{x}\,|\phi'|_{L^2(0,1)}$ for $x \in (0,1)$. Let

$$\Gamma'(v) = \left\{ \varphi \in L^2(0,1) : \int_0^1 \varphi\,dx = 0, \ |\varphi(x)| \leq \Phi(x,v(x)) \ \text{a.e.} \right\},$$

for $v \in L^2(0,1)$. Then $\{\phi' : \phi \in \Gamma(v)\} = \Gamma'(v)$ holds, since a function $\phi \in H_0^1(0,1)$ has $\phi(0) = \phi(1) = 0$. Hence the aforementioned observation yields

$$d_H(\Gamma(v), \Gamma(w)) \leq \frac{1}{\sqrt{2}} d_H(\Gamma'(v), \Gamma'(w)), \quad v,w \in L^2(0,1). \qquad (62)$$

Suppose for the moment that we can show the following key lemma.

Lemma 8. *Let $f, g : (0,1) \to \mathbb{R}$ be measurable such that*

$$0 < m \leq f(x), g(x) \leq M \ \text{a.e.} \ x \in (0,1), \qquad (63)$$

and define

$$\Gamma'(f) = \left\{ \varphi \in L^2(0,1) : \int_0^1 \varphi\,dx = 0, \ |\varphi(x)| \leq f(x) \ \text{a.e.} \right\}. \qquad (64)$$

Let $v \in \Gamma'(f)$ and $w = \mathrm{proj}\,(v, \Gamma'(g))$. Then

$$|v - w|_{L^2(0,1)} \leq \left(\frac{2M + m}{m}\right) |f - g|_{L^2(0,1)}. \tag{65}$$

Then (62) and (61) give the claim of Theorem 7, since for $v, w \in L^2(0,1)$

$$d_H(\Gamma(v), \Gamma(w)) \leq \frac{1}{\sqrt{2}} d_H(\Gamma'(v), \Gamma'(w))$$

$$\leq \left(\frac{2M + m}{\sqrt{2}m}\right) |\Phi(\cdot, v(\cdot)) - \Phi(\cdot, w(\cdot))|_{L^2(0,1)}$$

$$\leq \left(\frac{2M + m}{\sqrt{2}m}\right) K |u - v|_{L^2(0,1)},$$

so that we can choose $L = \left(\frac{2M + m}{\sqrt{2}m}\right) K < 1$ to find (58) verified.

Hence one has to see why Lemma 8 holds. Since

$$|v - w|_{L^2(0,1)} \leq |v - \varphi|_{L^2(0,1)}$$

for all $\varphi \in \Gamma'(g)$, it is necessary to construct a particular function $\varphi_* \in \Gamma'(g)$ such that

$$|v - \varphi_*|_{L^2(0,1)} \leq \left(\frac{2M + m}{m}\right) |f - g|_{L^2(0,1)}.$$

With some efforts such φ_* can be found through an appropriate modification of the given $v \in \Gamma'(f)$; see [36] for the details. Those are somewhat tedious and hence omitted here.

Remark 5. (a) Note that Theorem 6 does not make an assertion about the uniqueness of solutions to (53) and (56), respectively. First consider the following simple

Example 6. Let $H = \mathbb{R}$, $\Gamma(v) = [v, 1]$ for $v \leq 1$ resp. $\Gamma(v) = [1, v]$ for $v > 1$, $v \in \mathbb{R}$. Define $v_0 = 0$ and $f(t) = 0$, $t \in [0,1]$. Since $N_{[x,1]}(x) =]-\infty, 0]$ for $x < 1$ and $N_{\{x\}}(x) = \mathbb{R}$, all sufficiently regular functions $v : [0,1] \to \mathbb{R}$ with $v(0) = 0$, $v(t) \leq 1$, and $v'(t) \geq 0$ for $t \in [0,1]$ are solutions to (56). ◇

However, this is a counter-example to uniqueness only for $L = 1$ in (58) corresponding to $L_2 = 1$ in (54), and we already know from Example 5 that there might be no solutions at all in that case. Recently examples have been found showing that in general solutions to (53) are non-unique in case that $L_2 < 1$, see [4,10], and these counter-examples even allow for an arbitrary small L_2.

(b) We have seen that (56) may have no solutions for $L = 1$ in (58) for dimensions $d \geq 2$; cf. Example 5. However, it can be shown that (56) with

$L = 1$ has a solution if $H = \mathbb{R}$. This is mainly due to the fact that "there are only two directions in $\mathbb{R}$"; see [35, Thm. 3.8]. This also enlightens a noteworthy difference between (53) and (56): The former in general will not have a solution if $L_2 = 1$ in (54), even for $H = \mathbb{R}$, as may be seen from the example $C(t, u) = [t + u, +\infty[$. Then $u_0 = 0 \in C(0, u_0)$, and (54) holds with $L_1 = L_2 = 1$. Nevertheless, the existence of a solution to (53) would imply $u(t) \in C(t, u(t))$ for $t > 0$, i.e., $u(t) \geq t + u(t)$, and this is impossible. $\diamond$

3.4   Additional remarks

1.) In Section 3.1 we proved the existence of a solution to the sweeping process under the basic hypothesis that the dependence $t \mapsto C(t)$ be Lipschitz continuous in the sense of (13). There are also a lot of results if this assumption is weakened as we are going to summarize in this remark.

The first variant is the rcbv sweeping process, where rcbv means "right-continuous and of bounded variation." A moving set $t \mapsto C(t)$ will be called rcbv, if

$$d_{\mathrm{H}}(C(t), C(s)) \leq r(t) - r(s), \quad 0 \leq s \leq t \leq T, \tag{66}$$

for some right-continuous non-decreasing function $r : [0, T] \to \mathbb{R}$. Since we take the difference on the right-hand side of (66), we may assume without loss of generality that $r(0) = 0$. As for any partition $0 = t_0 < t_1 < \ldots < t_{N-1} < t_N = T$ of $[0, T]$ the estimate (66) yields

$$\sum_{i=0}^{N-1} d_{\mathrm{H}}(C(t_{i+1}), C(t_i)) \leq \sum_{i=0}^{N-1} [r(t_{i+1}) - r(t_i)] = r(T),$$

condition (66) can be interpreted as requiring that $t \mapsto C(t)$ be of bounded variation; recall (10) and also Example 1. Because the function $r(t) = Lt$ is rcbv, we see that in particular a Lipschitz continuous moving set $t \mapsto C(t)$ is rcbv. The converse is false, as is shown by the following example from [42, p. 29].

Example 7. In $H = \mathbb{R}$ define $C(t)$ for $t \in [0, 2]$ as $C(t) = [0, 2]$ for $t \in [0, 1[$ and $C(t) = [1, 2]$ for $t \in [1, 2]$. Then

$$d_{\mathrm{H}}(C(t), C(s)) = \begin{cases} 0 & : \quad s, t \in [0, 1[\text{ or } s, t \in [1, 2] \\ 1 & : \quad s \in [0, 1[, t \in [1, 2] \end{cases},$$

and hence (13) is violated. Nevertheless, (66) is satisfied with $r(t) = 0$ for $t \in [0, 1[$ and $r(t) = 1$ for $t \in [1, 2]$, and r is a rcbv function. $\diamond$

In vague analogy to the results for the Lipschitz continuous sweeping process, one might expect that (1) with a rcbv moving set has a unique rcbv solution. However, this raises a problem, since continuous bv functions

u in general are differentiable almost everywhere in $]0,T[$, but $\int_{t_0}^{t_1} u'(t)dt = u(t_1) - u(t_0)$ usually is verified only at a few points; see [54, Ch. 8] for an example of a Cantor-like function. Hence the formulation (1) of the sweeping process needs to be replaced by a weaker formulation. To do so, we first recall that a function $u : [0,T] \to H$ of bounded variation defines a H-valued measure du on $[0,T]$ (called the differential measure or Stieltjes measure) as follows. Let $\phi : [0,T] \to H$ be a continuous function with compact support in $[0,T]$.

Lemma 9. *There is a unique $I \in H$, written $I = \int_0^T \phi\, du$, such that for every $\varepsilon > 0$ there exists a partition $0 = t_0 < t_1 < \ldots < t_{N-1} < t_N = T$ of $[0,T]$ with the following property: If $0 = s_0 < s_1 < \ldots < s_{M-1} < s_M = T$ is any partition finer than the given one (i.e., with additional partition points) and if θ_k is chosen arbitrarily in $[s_k, s_{k+1}]$, $k = 0,\ldots,M-1$, then*

$$\left| \sum_{k=0}^{M-1} \phi(\theta_k)[u(s_{k+1}) - u(s_k)] - I \right| \leq \varepsilon.$$

Proof: See [51, Prop. 6.1]. $\qquad\qquad\qquad\qquad\qquad\qquad\qquad\qquad\qquad\square$

So we have well-defined integrals w.r. to du of continuous functions with compact support. By a standard procedure in measure theory this can be used to define an H-valued measure du on $[0,T]$. It turns out that for du the following relations hold, with $[a,b] \subset [0,T]$.

$$\int_{[a,b]} du = u^+(b) - u^-(a), \qquad \int_{]a,b]} du = u^+(b) - u^+(a),$$

$$\int_{[a,b[} du = u^-(b) - u^-(a), \qquad \int_{]a,b[} du = u^-(b) - u^+(a). \qquad (67)$$

Here

$$u^+(a) = \lim_{t \searrow a} u(t) \quad \text{and} \quad u^-(a) = \lim_{t \nearrow a} u(t)$$

are the right resp. left limits of u which can be shown for all $a \in [0,T[$ resp. $a \in]0,T]$ to exist for functions of bounded variation u; see [51,42]. In particular, $u^+(a) - u^-(a) = \int_{\{a\}} du$ is the "jump" in H at a point of discontinuity a. For a continuous function u we have of course that $u^+ = u^- = u$.

Now the weak formulation of the sweeping process (1) reads as follows.

Definition 2. A function $u : [0,T] \to H$ of bounded variation is a solution to

$$-du \in N_{C(t)}(u(t)), \quad u_0 \in C(0), \qquad (68)$$

if

(a) $u(0) = u_0$;

(b) $u(t) \in C(t)$ for all $t \in [0, T]$;

(c) there exists a positive measure $d\mu$ on $[0, T]$ relative to which du has a density $u' \in L^1([0, T]; H, d\mu)$ w.r. to $d\mu$ [i.e., $\int \phi \, du = \int \phi(t) u'(t) \, d\mu(t)$ for all reasonable ϕ] such that

$$-u'(t) \in N_{C(t)}(u(t)) \quad \text{for} \quad d\mu - \text{almost every} \quad t \in [0, T].$$

Although this definition appears to be quite technical, it turns out to be natural since solutions to sweeping processes with merely rcbv moving sets $t \mapsto C(t)$ can behave quite irregularly and have "many jumps". The following theorem is a typical result about rbcv moving sets.

Theorem 8. *Let $t \mapsto C(t)$ be rcbv such that every $C(t) \subset H$ is nonempty, closed, and convex, and let $u_0 \in C(0)$. Then (68) has a unique rcbv solution.*

Proof: By the discretization approach (the catching-up algorithm), as in [45, Prop. 3b]. □

Under the additional assumption that

$$C(t) \cap \overline{B}_R(0) \subset H$$

is compact for every $t \in [0, T]$ and $R > 0$, an alternative proof using the so-called Yosida-Moreau approximations can be found in [40] and [42, p. 29].

We now have generalized the Lipschitz continuous sweeping process to the rcbv sweeping process, but there is a further generalization of Lipschitz continuous sweeping processes that deserves attention. A natural question is what can be said if $t \mapsto C(t)$ is merely continuous, i.e.,

$$d_H(C(t), C(s)) \to 0, \quad t \to s. \tag{69}$$

On first sight, the situation is quite bad, as we can take e.g. $C(t) = \{v(t)\}$ with some continuous but nowhere differentiable function $v : [0, T] \to H$ which is not of bounded variation. Such functions do exist, examples are provided by e.g. (almost every) path of a Brownian motion, see [5, Thm. 12.25 & Cor. 12.27]. Since a prospective solution u is to have $u(t) \in C(t)$, it is clear that $u(t)$ can be a very bad function for which no obvious kind of derivative could be defined. To indicate a way out, recall that it would be already sufficient to have u of bounded variation. Moreover, in a general sweeping process the point to be swept will not move (and hence not increase its variation) as long as it stays in the interior of the moving set. This was exactly the problem with $C(t) = \{v(t)\}$, as this set has empty interior.

With some technical efforts one can then prove

Theorem 9. *Let $t \mapsto C(t)$ be continuous in the sense of (69) such that every $C(t) \subset H$ is nonempty, closed, and convex, and let $u_0 \in C(0)$. Assume that*

$$\operatorname{int} C(t) \neq \emptyset, \quad t \in [0, T].$$

*Then (68) has a unique solution which is continuous and of bounded varia-
tion. Here the measure $d\mu$ in Definition 2 can be chosen as $d\mu = |du|$, the
"modulus" of du.*

Proof: See [42, p. 46]. □

We also note that there is a good approximation result for solutions to this
problem by means of the Yosida-Moreau approximations; see [32]. Concrete
examples for such rcbv or continuous sweeping processes can be obtained
from the example in Section 3.2 by corresponding relaxation of the continuity
assumptions on g and c; see hypothesis $(H3)$.

2.) This second remark concerns what has been called "degenerate sweeping
processes", given by

$$-u'(t) \in N_{C(t)}(Au(t)) \quad \text{a.e. in} \quad [0,T], \quad u(0) = u_0 \in D(A), \quad Au_0 \in C(0). \tag{70}$$

Here A is a "maximal monotone" operator with domain of definition $D(A)$.
This concept of a maximal monotone operator can somehow be thought of
being a nonlinear generalization of a selfadjoint linear operator; one may consult [6] for further information. For simplicity we only indicate some results
in this linear case, see [33,34,37] for full details on the nonlinear problems.
Sweeping processes of type (70) are quite general and may fail to have a
solution, even for linear A; therefore they may be considered degenerate.

Example 8. Let $H = \mathbb{R}^2$, $T = 1$, $A = \begin{pmatrix} 1 & 0 \\ 0 & 0 \end{pmatrix}$ and $C(t) = [0,1] \times [t,1]$
for $t \in [0,1]$. Then A is linear, bounded, selfadjoint, and even $\langle Au, u \rangle \geq 0$
for $u \in H$. Also $C(\cdot)$ is d_H-Lipschitz continuous. However, there can be no
solution u of (70) with initial value $u_0 = (0,0) \in D(A)$, since this would
imply $Au(t) = (u_1(t), 0) \in C(t)$ a.e. in $[0,1]$, a contradiction. ◇

If the monotonicity assumption on A is sharpened to

$$\langle Au, u \rangle \geq \beta |u|^2 \quad \text{for} \quad u \in H \tag{71}$$

with some $\beta > 0$, then the situation is better, as (70) will have a unique
solution.

Theorem 10. *Let $A : H \to H$ be linear, bounded, and selfadjoint such that
(71) holds. If (13) is satisfied for $t \mapsto C(t)$ and if $Au_0 \in C(0)$, then (70) has
a unique solution, and this solution is Lipschitz continuous.*

Proof: See [34, Thm. 2]. □

3.) There are also some papers that deal with periodic solutions of sweeping
processes or related models, under various assumptions. In this case one has
to suppose that also $t \mapsto C(t)$ is T-periodic, i.e., $C(0) = C(T)$. See [19–21,31].

4 Second-order problems

In this section, we will analyze a general sweeping process problem of the form

$$-du \in N_{V(q(t))}(u(t)), \quad u(t) \in V(q(t)), \tag{72}$$

where V is a closed convex valued multifunction with a suitable dependence on the state variable q, and u is the first derivative of q. Hence du somehow represents a second derivative of q, thus explaining why such sweeping processes are of second order. Our main concern is to describe a discretization procedure which is similar to those considered in the applications to dynamical problems. For (72), two main situations have been considered so far.

(a) If $q \mapsto V(q)$ is Lipschitz continuous in the sense of Hausdorff distance, then one obtains a Lipschitz continuous solution u.
(b) If $q \mapsto V(q)$ is continuous with nonempty interior, then there exists a continuous solution u with bounded variation.

Below we will investigate a subclass of (b). It should also be noted that in addition Castaing studied (72) either in a finite-dimensional setting or assuming that V is antimonotone (i.e., $-V$ is monotone); see [18], [42, Sect. 5.1], or [22, Thm. 6.1] for three different techniques.

Although the computations to follow look a little awkward, the main ingredients are Euclidean geometry and telescopic sums, so that the reader should not feel discouraged. However, on first reading this technical part can be omitted.

The discretization procedure that we will consider concerns more precisely the problem

$$-du \in N_{V(q(t))}(u(t)), \quad u(t) = \mathrm{Av}(u^-(t), u^+(t)) \in V(q(t)), \tag{73}$$

where $\mathrm{Av}(u^-(t), u^+(t))$ is a kind of linear interpolation, or weighted average, of $u^-(t)$ and $u^+(t)$, the left-limit and right-limit of u at t, respectively; recall Remark 1.) in Section 3.4. For the operator Av our main choice will be

$$\mathrm{Av}(u^-, u^+) = u^+ + k(u^- - u^+), \quad 0 \le k \le 1/2, \tag{74}$$

which has the following properties:

$$u := \mathrm{Av}(u^-, u^+) \in [(u^- + u^+)/2, u^+] \subset [u^-, u^+],$$
$$|u^+ - u| \le k|u^+ - u^-|,$$
$$|u^+ - u| \le k(1-k)^{-1}|u^- - u|,$$
$$|u^+ - u^-| \le 2|u - u^-|; \tag{75}$$

note that here we used the notation $[v, w] = \{\lambda w + (1-\lambda)v : \lambda \in [0,1]\} \subset \mathrm{IR}^d$ for the line joining v and w.

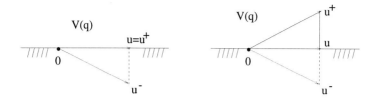

Fig. 6. (a) $k = 0$: inelastic   (b) $k = 1/2$: elastic

In (74), the case $k = 0$, i.e., $u = u^+$, corresponds to purely inelastic collisions, while $k = 1/2$ corresponds to purely elastic collisions; see Fig. 6 (a) and (b).

The relation of k to the usual restitution coefficient e and to Moreau's dissipation index δ will be explained in the next section. The partial "inverse" operator $(u^-, u) \mapsto u^+$ will be denoted Iav, thus $\mathrm{Iav}(u^-, u) = (1-k)^{-1}(u - ku^-)$ for (75). Hence $\mathrm{Iav}(u^-, u) = u$ in the inelastic case and $\mathrm{Iav}(u^-, u) = u^- + 2(u - u^-)$ in the elastic case.

Assume now that $q \mapsto V(q)$ is Hausdorff continuous, with each set $V(q) \subset \mathbb{R}^d$ being closed, convex, and with nonempty interior. The initial data are $q_0 \in \mathbb{R}^d$ and $u_0 \in V(q_0)$. As in the proof of Theorem 2, for each $n \in \mathbb{N}$ we fix a partition

$$0 = t_0^n < t_1^n < \dots < t_n^n = T$$

of the interval $[0, T]$, say, with equal stepsize $h = T/n$. By induction on n we construct finite sequences (q_i^n), (u_i^n), and (w_i^n), for $i = 0, \dots, n$, starting with $q_0^n = q_0$ and $u_0^n = w_0^n = u_0$, as follows. By integrating the constant velocity u_i^n over the time interval $[t_i^n, t_{i+1}^n]$, we define

$$q_{i+1}^n = q_i^n + (t_{i+1}^n - t_i^n)u_i^n. \tag{76}$$

The differential inclusion in (73) will be discretized as

$$-(u_{i+1}^n - u_i^n) \in N_{V(q_{i+1}^n)}(\mathrm{Av}(u_i^n, u_{i+1}^n)). \tag{77}$$

This means that in (77) we update V at q_{i+1}^n, and for the computation of the average we take into account that as $n \to \infty$ the "velocities" u_i^n and u_{i+1}^n should play the role of the left limit u^- and the right limit u^+, respectively. Writing

$$w_{i+1}^n = \mathrm{Av}(u_i^n, u_{i+1}^n), \tag{78}$$

(hence $w_{i+1}^n = u_{i+1}^n$ in the "inelastic" case), we infer from (75) that

$$w_{i+1}^n \in [(u_i^n + u_{i+1}^n)/2, u_{i+1}^n] \subset [u_i^n, u_{i+1}^n],$$
$$|u_{i+1}^n - w_{i+1}^n| \le k|u_{i+1}^n - u_i^n|,$$
$$|u_{i+1}^n - w_{i+1}^n| \le k(1-k)^{-1}|u_i^n - w_{i+1}^n|,$$
$$|u_{i+1}^n - u_i^n| \le 2|w_{i+1}^n - u_i^n|. \tag{79}$$

By (78), $u_{i+1}^n - u_i^n = (1-k)^{-1}(w_{i+1}^n - u_i^n)$, so that (77) is equivalent to

$$-(w_{i+1}^n - u_i^n) \in N_{V(q_{i+1}^n)}(w_{i+1}^n);$$

recall that $N_{V(q_{i+1}^n)}(w_{i+1}^n)$ is a cone and hence not affected by multiplication with a positive number. Thus by the characterization of projections (5)

$$w_{i+1}^n = \operatorname{proj}(u_i^n, V(q_{i+1}^n)) \in V(q_{i+1}^n) \qquad (80)$$

is necessary. We summarize the algorithm as follows. If the values from step i have already been determined, then we

(1.) update the state to q_{i+1}^n by means of (76),

(2.) obtain w_{i+1}^n by (80),

(3.) and introduce the new velocity u_{i+1}^n by

$$u_{i+1}^n = \operatorname{Iav}(u_i^n, w_{i+1}^n) = (1-k)^{-1}(w_{i+1}^n - k u_i^n). \qquad (81)$$

In particular, the estimates from (79) are valid. Notice that when $k \neq 0$ it may happen that $u_{i+1}^n \notin V(q_{i+1}^n)$; see Fig. 7 below. However, if $k = 0$, then $u_{i+1}^n = w_{i+1}^n \in V(q_{i+1}^n)$.

Finally we define the functions $u_n : [0, T] \to \mathbb{R}^d$ and $q_n : [0, T] \to \mathbb{R}^d$ by $u_n(t) = w_i^n$ for $t = t_i^n$,

$$u_n(t) = u_i^n \quad \text{for} \quad t \in]t_i^n, t_{i+1}^n[,$$

and

$$q_n(t) = q_0 + \int_0^t u_n(s)\,ds. \qquad (82)$$

Thus $q_n(t_i^n) = q_i^n$ by (76), and $\dot{q}_n(t) = u_n(t)$ Lebesgue almost everywhere in $[0, T]$.

Next we are going to investigate the convergence of this approximation scheme. Here the assumption that $V(q)$ has nonempty interior will be essential, similarly to what has briefly been mentioned in front of Theorem 9 in Section 3.4. In fact, this hypothesis will allow us to obtain estimates on the total variation of the derivatives (velocities) u_n. To explain this, note first that as $V(q_0)$ has nonempty interior it must contain some ball $\overline{B}_{2r}(a)$. As $q \mapsto V(q)$ is Hausdorff continuous, it then can be shown that for q in a neighbourhood of q_0, $V(q)$ contains at least a ball with the same center $a \in \mathbb{R}^d$ but of a smaller radius, say

$$\overline{B}_r(a) \subset V(q); \qquad (83)$$

see e.g. [42, Lemma 2.4.2] for a proof. With some a priori estimates and working if necessary in a smaller time interval $[0, T_0] \subset [0, T]$, we may suppose that (83) holds for all $q = q_i^n$; hence we may assume that already $T_0 = T$. The following geometrical result, due to Moreau, about projections into convex sets with nonempty interior is crucial for the subsequent estimates.

Lemma 10. *If V is a closed convex set in a Hilbert space H and V contains a ball $\overline{B}_r(a)$, then*

$$|x - \operatorname{proj}(x, V)| \leq \frac{1}{2r}(|x - a|^2 - |\operatorname{proj}(x, V) - a|^2) \quad \text{for all} \quad x \in H.$$

Proof: See [46] and also [42, Lemma 0.4.3] □

Taking $x = u_i^n$ and $V = V(q_{i+1}^n)$, we thus obtain from (83) and (80)

$$|u_i^n - w_{i+1}^n| \leq \frac{1}{2r}(|u_i^n - a|^2 - |w_{i+1}^n - a|^2). \tag{84}$$

The total variation of u_n, recall (10), is given by

$$\operatorname{var}(u_n) = \sum_{i=0}^{n-1} \left(|u_{i+1}^n - w_{i+1}^n| + |w_{i+1}^n - u_i^n| \right) = \sum_{i=0}^{n-1} |u_{i+1}^n - u_i^n|,$$

the latter because the points u_i^n, w_{i+1}^n, and u_{i+1}^n are on a line.

In the simplest case $k = 0$, we have $u_{i+1}^n = w_{i+1}^n$, and an estimate on $\operatorname{var}(u_n)$ is easily obtained by (84) as we get a telescopic sum (Mengoli sum)

$$\operatorname{var}(u_n) \leq \sum_{i=0}^{n-1} \frac{1}{2r}(|u_i^n - a|^2 - |u_{i+1}^n - a|^2)$$

$$= \frac{1}{2r}(|u_0^n - a|^2 - |u_n^n - a|^2) \leq \frac{1}{2r}|u_0 - a|^2. \tag{85}$$

The general "averaged" case $0 < k \leq 1/2$ is much harder. From (79) and (84) we find

$$\operatorname{var}(u_n) \leq \sum_{i=0}^{n-1} 2|w_{i+1}^n - u_i^n| \leq \frac{1}{r} \sum_{i=0}^{n-1} (|u_i^n - a|^2 - |w_{i+1}^n - a|^2). \tag{86}$$

At this point we can invoke an observation of Mabrouk (see [38, Lemma 4]) to rearrange the above expression in order to obtain a telescopic sum.

Lemma 11. *With the above notation, the inequality*

$$|u_{i+1}^n - a|^2 \leq |w_{i+1}^n - a|^2 + |u_{i+1}^n - w_{i+1}^n|^2$$

holds.

Proof: Since

$$|u_{i+1}^n - a|^2 = |w_{i+1}^n - a|^2 + |u_{i+1}^n - w_{i+1}^n|^2 + 2\langle u_{i+1}^n - w_{i+1}^n, w_{i+1}^n - a \rangle,$$

it remains to show that the inner product on the right-hand side is nonpositive. For all $z \in V(q_{i+1}^n)$ we have

$$\langle u_i^n - w_{i+1}^n, z - w_{i+1}^n \rangle \leq 0, \tag{87}$$

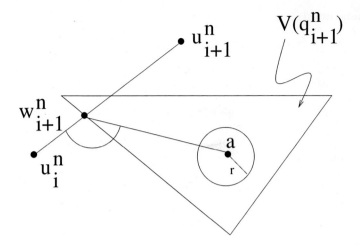

Fig. 7. Geometrical meaning of (88)

recall (80) and (4). Choosing $z = a \in V(q_{i+1}^n)$ yields

$$\langle u_i^n - w_{i+1}^n, a - w_{i+1}^n \rangle \leq 0, \qquad (88)$$

see the Fig. 7: the angle is $\geq \pi/2$.
Furthermore, from (81) it follows that

$$w_{i+1}^n - u_{i+1}^n = k'(u_i^n - w_{i+1}^n), \quad k' = k(1-k)^{-1} \geq 0. \qquad (89)$$

Hence we obtain

$$\langle w_{i+1}^n - u_{i+1}^n, a - w_{i+1}^n \rangle = k' \langle u_i^n - w_{i+1}^n, a - w_{i+1}^n \rangle \leq 0$$

as desired. $\qquad\qquad \Box$

Thus $-|w_{i+1}^n - a|^2 \leq -|u_{i+1}^n - a|^2 + |u_{i+1}^n - w_{i+1}^n|^2$, and so (86) implies

$$\mathrm{var}\,(u_n) \leq \frac{1}{r}\sum_{i=0}^{n-1}(|u_i^n - a|^2 - |u_{i+1}^n - a|^2 + |u_{i+1}^n - w_{i+1}^n|^2)$$

$$\leq \frac{1}{r}|u_0 - a|^2 + \frac{1}{r}\sum_{i=1}^{n}|u_i^n - w_i^n|^2, \qquad (90)$$

since part of the sum again "telescoped". In order to bound the extra terms, as compared with (85) in the case $k = 0$, we have to take into account the variation of $q \mapsto V(q)$. Assume that we already know some a priori estimate on the velocities, say

$$|u_n(t)| \leq M_1, \quad n \in \mathrm{I\!N}, \ t \in [0, T]. \qquad (91)$$

Then by integration

$$|q_n(t)| \leq M_2 := |q_0| + TM_1, \quad n \in \mathbb{N}, \ t \in [0, T].$$

In the compact set $\overline{B}_{M_2}(0) \subset \mathbb{R}^d$ the continuous multifunction $V(\cdot)$ is uniformly continuous. In other words, if we introduce the function

$$\delta(\zeta) := \sup \left\{ d_H(V(q), V(q')) : q, q' \in \overline{B}_{M_2}(0), \ |q - q'| \leq \zeta \right\},$$

then $\delta(\zeta) \to 0$ as $\zeta \to 0^+$. Since $|q_{i+1}^n - q_i^n| \leq M_1(t_{i+1}^n - t_i^n) = M_1 T/n$, we then find

$$d_H(V(q_i^n), V(q_{i+1}^n)) \leq \delta_n := \delta(M_1 T/n) \to 0, \quad n \to \infty.$$

It turns out that when $k > 0$ the continuity assumption on V has to be reinforced if one wants to obtain from (90) a finite upper bound on the total variations.

Definition 3. The multifunction $q \mapsto V(q)$ is called Hölder continuous of exponent $\alpha \in]0, 1]$, if there is $L > 0$ such that

$$d_H(V(q), V(q')) \leq L|q - q'|^\alpha, \quad q, q' \in H. \tag{92}$$

If $\alpha = 1$, this simply means that V is a Lipschitz continuous multifunction. By (92), $\delta(\zeta) \leq L\zeta^\alpha$, and thus in particular

$$\delta_n \leq LM_1^\alpha T^\alpha n^{-\alpha}. \tag{93}$$

Note that although now the function $\delta(\cdot)$ is independent of M_2, the quantities δ_n do depend on M_1, the bound on u_n from (91). To use (93) to estimate the last term in (90), we write $\eta_i^n := |w_i^n - u_i^n|$ and estimate it recursively. With $k' = k(1-k)^{-1}$ we find

$$\eta_{i+1}^n = k'|w_{i+1}^n - u_i^n| = k'\mathrm{dist}\,(u_i^n, V(q_{i+1}^n))$$

from (89). Thus for $i = 0$ we have

$$\eta_1^n = k'\mathrm{dist}\,(u_0, V(q_1^n)) \leq k' d_H(V(q_0), V(q_1^n)) \leq k'\delta_n,$$

since $u_0 \in V(q_0)$. Similarly, as $w_i^n \in V(q_i^n)$,

$$\eta_{i+1}^n = k'\mathrm{dist}\,(u_i^n, V(q_{i+1}^n)) \leq k'\Big(|u_i^n - w_i^n| + \mathrm{dist}\,(w_i^n, V(q_{i+1}^n))\Big)$$
$$\leq k'(\eta_i^n + \delta_n), \tag{94}$$

Notice that if $0 \leq k < 1/2$ (this excludes the "elastic" case $k = 1/2$), then $0 \leq k' < 1$, and by induction on (94)

$$\eta_i^n \leq \delta_n(k' + k'^2 + \ldots + k'^i) \leq \left(\frac{k'}{1-k'}\right)\delta_n = \left(\frac{k}{1-2k}\right)\delta_n. \tag{95}$$

Hence by (93)

$$\sum_{i=1}^{n}|u_i^n - w_i^n|^2 = \sum_{i=1}^{n}(\eta_i^n)^2 \le k^2(1-2k)^{-2}n\delta_n^2 \le \text{const}\, n^{1-2\alpha}, \qquad (96)$$

the constant depending on M_1 and T. Thus if the Hölder exponent satisfies $\alpha \in [1/2, 1]$, then for some $M_3 > 0$ and uniformly in $n \in \mathbb{N}$

$$\sum_{i=0}^{n-1}|u_{i+1}^n - w_{i+1}^n|^2 \le M_3. \qquad (97)$$

Combining this inequality with (90), we finally arrive at the desired uniform bound on the total variations

$$\text{var}\,(u_n) \le M_4, \quad n \in \mathbb{N},$$

for some constant $M_4 > 0$. By Theorem 1 we find subsequences – still denoted by (u_n) and (q_n) – and a function u of bounded variation such that as $n \to \infty$

$$u_n(t) \to u(t) \quad \text{for every} \quad t \in [0, T], \qquad (98)$$

and also

$$q_n(t) \to q(t) \quad \text{uniformly in} \quad t \in [0, T];$$

the uniform convergence is ensured by the Ascoli-Arzelà theorem. Taking limits in (82) we obtain

$$q(t) = q_0 + \int_0^t u(s)\, ds, \quad t \in [0, T]. \qquad (99)$$

These observations can be used to prove the following theorem on the existence of a solution to (72).

Theorem 11. *Let V be a multifunction with closed convex values having nonempty interior in $\mathbb{R}^d$, and assume V is Hölder continuous with exponent $\alpha \in [1/2, 1]$, cf. (92). Define the "averaging" operator $\text{Av}(\cdot, \cdot)$ in (74) with some $0 \le k < 1/2$, and fix initial values $q_0 \in \mathbb{R}^d$ and $u_0 \in V(q_0)$. If the above sequence (u_n) is uniformly bounded, cf. (91), then any pair (u, q) of limit functions on a subsequence of such approximations of the "averaged" problem (73) is a solution of problem (72). More precisely, $u : [0, T] \to \mathbb{R}^d$ is a continuous function of bounded variation, $u(0) = u_0$, (99) holds, we have $u(t) \in V(q(t))$ for all $t \in [0, T]$, and*

$$-\frac{du}{|du|}(t) \in N_{V(q(t))}(u(t)), \quad |du| - a.e., \qquad (100)$$

where du is the Stieltjes measure of u and $|du|$ its measure of total variation.

Remark 6. (a) If $k = 0$, the "bad" terms in the previous estimates vanish and so existence for (100) holds if $q \mapsto V(q)$ is merely continuous. This is the result obtained by Castaing in [18], already mentioned above. What is new here is the study of more general discretization procedures which will be of interest later.

(b) In the applications to dynamics given in the following Section 5, the necessary uniform bound on the velocities can be verified, as a result of some dissipation property inherent to problems describing inelastic collisions or friction. The case of purely elastic collisions, which corresponds to $k = 1/2$, may also be solved, cf. the references below.

(c) For the second order state-dependent problem (72) the Hausdorff continuity of the multifunction is sufficient, as opposed to the first order problems, where Lipschitz continuity was required, see Section 3.3. This is due to the usual smoothing effect of second order problems. $\diamondsuit$

The proof of the existence theorem will be divided into five lemmas, all more or less interesting in their own right. The reader may want to skip the remainder of this section and return to it only later.

From now on we assume (u, q) are limit functions of the sequence (u_n, q_n), which exist by the assumptions and the above arguments.

Lemma 12. *For all $t \in [0, T]$ we have $u(t) \in V(q(t))$.*

Proof: Let $t \in [0, T]$ and $n \in \mathbb{N}$. Then by construction we known $u_n(t)$ equals either u_i^n or w_i^n for some $i = i_n \in \{0, \ldots, n\}$. Hence

$$
\begin{aligned}
\text{dist}\,(u_n(t), V(q(t))) &\le \text{dist}\,(u_i^n, V(q(t))) + |u_i^n - w_i^n| \\
&\le \text{dist}\,(u_i^n, V(q_i^n)) + d_H(V(q_i^n), V(q(t))) + \eta_i^n \\
&\le d_H(V(q_i^n), V(q(t))) + 2\eta_i^n,
\end{aligned}
$$

since $w_i^n \in V(q_i^n)$ by (80). But $\eta_i^n \le \text{const}\,\delta_n \to 0$ as $n \to \infty$ according to (95), uniformly in $i \in \{0, \ldots, n\}$. Moreover, due to $q_i^n = q_n(t_i^n)$, and by (82) and (91),

$$
\begin{aligned}
d_H(V(q_i^n), V(q(t))) &\le L|q_i^n - q(t)|^\alpha \le L\Big(|q_n(t_i^n) - q_n(t)| + |q_n(t) - q(t)|\Big)^\alpha \\
&\le L\Big(\text{const}\,n^{-1} + |q_n - q|_{L^\infty([0,T];\mathbb{R}^d)}\Big)^\alpha \to 0, \quad n \to \infty.
\end{aligned}
$$

Therefore $\text{dist}\,(u_n(t), V(q(t))) \to 0$ as $n \to \infty$, even uniformly in $[0, T]$. By (98), $u(t) = \lim_{n \to \infty} u_n(t)$, and thus taking the limit as $n \to \infty$ we obtain $\text{dist}\,(u(t), V(q(t))) = 0$, i.e., $u(t)$ belongs to the closure of $V(q(t))$ which equals $V(q(t))$. $\square$

Lemma 13. *The limit function u is right-continuous.*

Proof: Fix $t_0 \in [0, T[$ and $\varepsilon > 0$. Since $u(t_0) \in V(q(t_0))$, we find a_0 and $r_0 > 0$ such that $B_{2r_0}(a_0) \subset V(q(t_0))$ and $|u(t_0) - a|^2/r_0 \le \varepsilon$; this is easy, cf. [42, Lemma 2.4.4]. Hence $B_{r_0}(a_0) \subset V(q_n(t))$ for all large $n \in \mathbb{N}$ and all t close to t_0. By using the bound on the total variation of u_n in a small interval $[t_0, t_0 + \sigma]$ instead of $[0, T]$, we obtain

$$
\limsup_{n \to \infty} \operatorname{var} (u_n; [t_0, t_0 + \sigma])
$$

$$
\le \limsup_{n \to \infty} \left(\frac{1}{r_0} |u_n(t_0) - a_0|^2 + \frac{1}{r_0} \sum_{i=p_n}^{s_n} |u_i^n - w_i^n|^2 \right)
$$

$$
\le \varepsilon + \frac{1}{r_0} \left(\frac{k}{1 - 2k} \right)^2 \limsup_{n \to \infty} \delta_n^2 \left(\sum_{i=p_n}^{s_n} 1 \right),
$$

where the number of terms in the sum $\cong (s_n - p_n)$ is of the order of $\sigma/h = \sigma n/T$. Thus, as before in (96),

$$
\limsup_{n \to \infty} \operatorname{var} (u_n; [t_0, t_0 + \sigma]) \le \varepsilon + \limsup_{n \to \infty} \delta_n^2 \left(\frac{k}{1 - 2k} \right)^2 \frac{\sigma n}{r_0 T}
$$

$$
\le \varepsilon + \operatorname{const} \sigma \limsup_{n \to \infty} n^{1-2\alpha} \le \varepsilon + \operatorname{const} \sigma \le 2\varepsilon,
$$

if σ is chosen small enough. For $t \in [t_0, t_0 + \sigma]$, we then have

$$
|u(t) - u(t_0)| = \lim_{n \to \infty} |u_n(t) - u_n(t_0)| \le \limsup_{n \to \infty} \operatorname{var} (u_n; [t_0, t_0 + \sigma]) \le 2\varepsilon.
$$

Thus the lemma is proved. $\qquad\square$

Since u is of bounded variation, recall from Remark 1.) in Section 3.4 that the left limits $u^-(t_0) = \lim_{t \nearrow t_0} u(t)$ do exist.

Lemma 14. *We have $u^-(t_0) \in V(q(t_0))$ for $t_0 \in [0, T[$, in particular u is continuous.*

Proof: 1.) Let $t_n \nearrow t_0$. By Lemma 12, $u(t_n) \in V(q(t_n))$, and we know that $q(t_n) \to q(t_0)$. By continuity of $V(\cdot)$ it then follows that $u^-(t_0) \in V(q(t_0))$.

2.) To verify the second claim, we only have to show u is left-continuous. Fix $\varepsilon > 0$. As in Lemma 13 there is a ball $B_{2r_0}(a_0) \subset V(q(t_0))$ such that $|u^-(t_0) - a|^2/r_0 \le \varepsilon$. Then for sufficiently small $\sigma > 0$ we have analogously to the proof of Lemma 13 that $|u(t_0 - \sigma) - a_0|^2/r_0 \le 2\varepsilon$ and, say, that $\operatorname{var}(u; [t_0 - \sigma, t_0]) \le 3\varepsilon$. $\qquad\square$

By the previous lemmas, to complete the proof of Theorem 11 all that remains is to verify (100). We start with a preliminary observation.

Lemma 15. *If $0 \leq s \leq t \leq T$ and $z \in V(q')$ for all $q' \in \mathbb{R}^d$ in a neighbourhood of $q([s,t]) \subset \mathbb{R}^d$, then*

$$\langle z, u(t) - u(s) \rangle \geq \frac{1}{2}|u(t)|^2 - \frac{1}{2}|u(s)|^2. \tag{101}$$

Proof: Since $q_n \to q$ uniformly on $[0,T]$ and q is Lipschitz continuous, we find $\varepsilon > 0$ such that $z \in V(q_n(\tau))$ for all $n \in \mathbb{N}$ sufficiently large (w.l.o.g. $n \geq 1$) and all $\tau \in [s - \varepsilon, t + \varepsilon]$. Fix $n \geq 1$. Then by definition of u_n we obtain

$$\langle z, u_n(t) - u_n(s) \rangle \geq \langle z, u_l^n - u_j^n \rangle - |z|\Big(|u_l^n - w_l^n| + |u_j^n - w_j^n|\Big)$$

$$\geq \langle z, u_l^n - u_j^n \rangle - \text{const}\, |z|\delta_n, \tag{102}$$

as $\eta_i^n \leq \text{const}\, \delta_n$ for all $i \in \{0, \dots, n\}$; here j and l, both depending on n, are chosen such that $s \in [t_j^n, t_{j+1}^n[$ and $t \in [t_l^n, t_{l+1}^n[$. We may as well assume that $s - \varepsilon \leq t_j^n$ and $t_{l+1}^n \leq t + \varepsilon$. Hence for $j \leq i \leq l$ we find $z \in V(q_n(t_{i+1}^n)) = V(q_{i+1}^n)$. Thus (87) implies $\langle z - w_{i+1}^n, w_{i+1}^n - u_i^n \rangle \geq 0$, and therefore

$$\langle z, u_l^n - u_j^n \rangle = \sum_{i=j}^{l-1} \langle z, u_{i+1}^n - u_i^n \rangle \geq \sum_{i=j}^{l-1} \langle w_{i+1}^n, u_{i+1}^n - u_i^n \rangle.$$

To estimate the right-hand side further, we can write

$$\langle w_{i+1}^n, u_{i+1}^n - u_i^n \rangle$$
$$= \langle u_{i+1}^n, u_{i+1}^n - u_i^n \rangle + \langle w_{i+1}^n - u_{i+1}^n, u_{i+1}^n - u_i^n \rangle$$
$$= \frac{1}{2}|u_{i+1}^n|^2 - \frac{1}{2}|u_i^n|^2 + \frac{1}{2}|u_{i+1}^n - u_i^n|^2 + \langle w_{i+1}^n - u_{i+1}^n, u_{i+1}^n - u_i^n \rangle$$
$$= \frac{1}{2}|u_{i+1}^n|^2 - \frac{1}{2}|u_i^n|^2 + \langle w_{i+1}^n - \frac{u_i^n + u_{i+1}^n}{2}, u_{i+1}^n - u_i^n \rangle.$$

Now note the last inner product is nonnegative, according to $2w_{i+1}^n - u_i^n - u_{i+1}^n = (1 - 2k)(1 - k)^{-1}(w_{i+1}^n - u_i^n)$ and $u_{i+1}^n - u_i^n = (1 - k)^{-1}(w_{i+1}^n - u_i^n)$. It follows that

$$\langle z, u_l^n - u_j^n \rangle \geq \sum_{i=j}^{l-1} \left(\frac{1}{2}|u_{i+1}^n|^2 - \frac{1}{2}|u_i^n|^2 \right) = \frac{1}{2}|u_l^n|^2 - \frac{1}{2}|u_j^n|^2,$$

whence (102) shows

$$\langle z, u_n(t) - u_n(s) \rangle \geq \frac{1}{2}|u_l^n|^2 - \frac{1}{2}|u_j^n|^2 - \text{const}\, |z|\delta_n.$$

Finally, $|u_j^n - u_n(s)| \leq \eta_j^n \leq \text{const}\, \delta_n$ and $|u_l^n - u_n(t)| \leq \eta_l^n \leq \text{const}\, \delta_n$ together with (98) imply in the limit $n \to \infty$ that (101) holds. $\qquad\square$

With the next lemma we finish the proof of Theorem 11. Notice that only this last lemma requires a more specific measure theoretic background. Some material on the subject may be found in [51].

Lemma 16. *For $|du|$-almost every t, one has*

$$-\frac{du}{|du|}(t) \in N_{V(q(t))}(u(t)). \qquad (103)$$

Proof: By a theorem of Jeffery, cf. [42, p. 9], for $|du|$-a.e. t the density $\frac{du}{|du|}(t)$ of the measure du with respect to $|du|$ at t can be calculated as

$$\frac{du}{|du|}(t) = \lim_{\varepsilon \to 0^+} \frac{du([t, t+\varepsilon])}{|du|([t, t+\varepsilon])} = \lim_{\varepsilon \to 0^+} \frac{u(t+\varepsilon) - u(t)}{|du|([t, t+\varepsilon])}.$$

The analogous result holds for the density $\frac{d(|u|^2/2)}{|du|}(t) = \langle u(t), \frac{du}{|du|}(t)\rangle$, where we have used the fact that u is continuous. If z is in the interior of $V(q(t))$, we find $\sigma > 0$ and a neighborhood of $q([t-\sigma, t+\sigma])$ such that $z \in V(q')$ for all q' in this neighborhhood, according to the continuity properties of $V(\cdot)$ and $q(\cdot)$. Hence Lemma 15 can be applied to yield

$$\langle z, u(t_0 + \varepsilon) - u(t)\rangle \geq \frac{1}{2}|u(t+\varepsilon)|^2 - \frac{1}{2}|u(t)|^2$$

for $\varepsilon > 0$ small. Dividing by $|du|([t, t+\varepsilon])$ preserves the inequality, and letting ε tend to zero we obtain

$$\left\langle z, \frac{du}{|du|}(t)\right\rangle \geq \left\langle u(t), \frac{du}{|du|}(t)\right\rangle.$$

By density, this inequality transfers to all $z \in V(q(t))$, and (103) follows. $\square$

5 Applications to the dynamics of unilateral contact

5.1 Frictionless unilateral contact

In the early eighties, Moreau introduced a comprehensive mathematical formulation of the dynamics of a particle or a system under frictionless constraints, [47]. A detailed exposition of what he called "standard inelastic shocks" can be found in [48]. Moreau's formulation leads to quite natural and successful numerical procedures (mainly developed later for problems with friction). The first mathematical proof of existence of a solution to the system was given by Monteiro Marques in [39]; see also [42, Sects. 3.1 & 3.2]. The mathematical formulation and the existence proof recently have been extended by Mabrouk [38] to elastic or partially elastic cases. In this section, we will give an outline of this theory and of the proof of Mabrouk's result, and we also mention other related and/or recent developments. For more details, one may refer to [42] and especially to [8], where nonsmooth mechanical problems are very well introduced to the non-expert reader.

A mechanical system with a finite number of degrees of freedom can be represented by a point in the manifold of configurations, or alternatively using local coordinates, by a point q in some Euclidean space $\mathbb{R}^d$. For brevity, we will write $E = \mathbb{R}^d$. The existence results mentioned above will thus be concerned with motions $t \mapsto q(t) \in E$. These motions are subject to a single smooth constraint, i.e., $q(t)$ has to remain in a fixed set

$$L = \{q \in E : f(q) \le 0\},$$

with $f : E \to \mathbb{R}$ a given smooth function. In other words,

$$f(q(t)) \le 0, \quad t \in I, \tag{104}$$

is required for the solution, where $I \subset \mathbb{R}$ is an interval, say $[0, T]$.

Contact with the boundary $S = \{q \in E : f(q) = 0\}$ may lead to a discontinuity of the velocity. Moreau thus assumed that the velocity $v = \dot{q}$ is a function of bounded variation (in short, a bv function), which is quite natural and advantageous. In fact, recall from Remark 1.) in Section 3.4 that such functions do have left and right limits, and therefore we will be able to consider left and right velocities and contact laws relating them. Moreover, the Stieltjes measure dv of such a (possibly discontinuous) bv velocity will substitute the classical acceleration $\dot{v} = \ddot{q}$ of a smooth motion q. One reason for the usefulness of this approach is that by integration of the "generalized acceleration" dv one obtains right and left limits of the velocity v, see the formulas (67) in Section 3.4. Since any bv function is continuous except at an at most countable set of points, the integration of either v, v^+, or v^- with respect to Lebesgue measure all yields the same result, namely the motion q. Let us explain these points a little further. Classically, one writes

$$q(t) = q_0 + \int_0^t v(s)\, ds, \tag{105}$$

where v is the usual velocity $v(t) = \dot{q}(t) = \lim_{h \to 0}(q(t+h) - q(t))/h$. In smooth dynamics, v would be of class C^1, or at least absolutely continuous, and the acceleration $\dot{v} = \ddot{q}$ would satisfy

$$v(t) - v(s) = \int_s^t \dot{v}(\tau)\, d\tau.$$

Under the constraint (104), this no longer can be expected to hold, since contact with the boundary S usually leads to discontinuities of the velocity. Therefore, we will instead work with velocities of bounded variation, i.e., $v \in \mathrm{bv}(I, E)$. More generally, in (105) we replace the "real" velocity v (meaning the time-derivative of q at each t) by any bv function u which satisfies

$$q(t) = q_0 + \int_0^t u(s)\, ds, \quad t \in I. \tag{106}$$

Since $u \in \text{bv}(I, E)$, u possesses right and left limits u^+ and u^-, respectively, and u has at most a countable number of discontinuity points. Hence $u = u^+ = u^-$ Lebesgue almost everywhere, and (106) implies that also

$$q(t) = q_0 + \int_0^t u^+(s)\, ds = q_0 + \int_0^t u^-(s)\, ds. \qquad (107)$$

Moreover, by a simple computation,

$$\dot{q}^+(t) = \lim_{h \to 0^+} \frac{q(t+h) - q(t)}{h} = \lim_{h \to 0^+} \frac{1}{h} \int_t^{t+h} u(s)\, ds = u^+(t),$$

i.e., the right velocity $\dot{q}^+(t)$ equals the right limit $u^+(t)$; of course the analogous result holds for the left velocity. Thus, (107) means that the motion q can be obtained by integration of either the right or the left velocity. At its continuity points, u coincides with the actual velocity. From all the above, it is thus reasonable to call u a (generalized) velocity of q.

Now we turn to the description of the dynamics, usually governed by

$$M\ddot{q} = p(t, q, \dot{q}) + R.$$

Here $M \in \mathbb{R}^{d \times d}$ is the mass (or inertia) matrix, p is a given force field, say, of gravity, elastic, and viscous forces, and R is the reaction force exerted by the boundary S, i.e., by the constraint. In the sequel we will assume for simplicity that $M = I$, hence we consider

$$R = \ddot{q} - p(t, q, \dot{q}) = \dot{u} - p(t, q, u). \qquad (108)$$

Since we have already admitted that the acceleration $\ddot{q}$ is the "derivative" of a bv function, hence a measure, we thus have to take the next logical step: (108) implies that the reaction must be a measure as well,

$$dR = d\dot{q} - p(t, q, \dot{q})dt = du - p(t, q, u)dt, \qquad (109)$$

where dt is the Lebesgue measure on the interval I.

Away from contact, i.e., as long as $f(q(t)) < 0$, the reaction is zero, and hence $dR = 0$. Without going into more detail, this just means that $d\dot{q} = p(t, q, \dot{q})dt$, or the usual

$$\ddot{q} = p(t, q, \dot{q}), \quad \text{or} \quad \frac{du}{dt} = p(t, q, u).$$

In this case $u^+ = u^-$, and both functions are absolutely continuous (under reasonable assumptions on the forces p). Absolute continuity may also hold for periods of motion with persistent contact, $f(q(t)) \equiv 0$ during some time, usually with nonzero reaction.

When the velocity is discontinuous at an instant t_s, that is, $u^+(t_s) \neq u^-(t_s)$, the situation is different. "Restricted to t_s", du coincides with a Dirac measure, since $du|_{\{t_s\}} = (u^+(t_s) - u^-(t_s))\delta_{t_s}$ because of

$$du(\{t_s\}) = \int_{\{t_s\}} du = u^+(t_s) - u^-(t_s).$$

As the Lebesgue measure of $\{t_s\}$ is zero, we have $\int_{\{t_s\}} p(t, q(t), u(t)) dt = 0$, and therefore (109) implies that

$$dR(\{t_s\}) = \int_{\{t_s\}} dR = \int_{\{t_s\}} du = u^+(t_s) - u^-(t_s). \tag{110}$$

Consequently, the impact or contact laws relating a pre-impact velocity u^- to a post-impact velocity u^+ can be expressed in terms of the reaction dR. To determine what are the conditions the left and right velocities have to satisfy, let $f(q(t_s)) = 0$. Since $f(q(t)) \leq 0$ for all t, $t = t_s$ is a maximum of that function. Taking right and left derivatives yields

$$\langle u^+(t_s), \nabla f(q(t_s)) \rangle \leq 0 \quad \text{and} \quad \langle u^-(t_s), \nabla f(q(t_s)) \rangle \geq 0. \tag{111}$$

Thus for all $t \in I$ we find

$$u^+(t) \in V(q(t)) \quad \text{and} \quad -u^-(t) \in V(q(t)),$$

where

$$V(q) = \begin{cases} \{w \in E : \quad \langle w, \nabla f(q) \rangle \leq 0\} & : \quad f(q) \geq 0 \\ E & : \quad f(q) < 0 \end{cases} \tag{112}$$

is called the "tangent cone to L at the point q", although in fact $V(q)$ is defined for all $q \in E$ which is more convenient, especially in view of numerical applications. Therefore $V(q)$ is either the whole space or a halfspace, so that for all $q \in E$

$$\text{int } V(q) \neq \emptyset. \tag{113}$$

If $f(q(t_s)) = 0$ and $\langle u^-(t_s), \nabla f(q(t_s)) \rangle > 0$, it is clear from (111) that $u^+(t_s) \neq u^-(t_s)$: a collision or impact occurs. For perfect (frictionless) unilateral contacts, the reaction force exerted by the boundary $S = \partial L$ is normal and pointing inwards (no adhesion). Hence there exists $\lambda \geq 0$ such that

$$R = -\lambda \nabla f(q).$$

If we "integrate for the infinitesimally small duration of the impact", and assuming that the reaction does not change direction, we obtain by (110)

$$u^+(t_s) - u^-(t_s) = -\alpha \nabla f(q(t_s)) \tag{114}$$

for some $\alpha > 0$, the so-called "liaison percussion"; see Fig. 8.

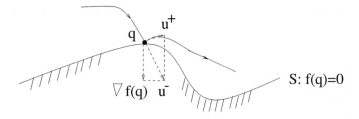

Fig. 8. A partially elastic collision at q

When $L = \{q : f(q) \leq 0\}$ is convex, this can be reformulated as dR being inward normal to L at q, or

$$-dR \in N_L(q), \tag{115}$$

since the outward normal cone is given by $N_L(q) = \{0\}$ if $q \in \text{int } L$, and $N_L(q) = \{\alpha \nabla f(q) : \alpha \geq 0\}$ if $f(q) = 0$. (This might be extended to more general sets L, by using the appropriate notion of normal cone). Notice that (114) or (115) are not enough to determine $u^+ = u^+(t_s)$ given $u^- = u^-(t_s)$, as there is a range of possible choices for u^+ due to the fact that α is unknown.

To describe a **purely inelastic impact**, we just require that the post-impact velocity u^+ be tangential, i.e.,

$$u^+ \in T(q) = \{w \in E : \langle w, \nabla f(q) \rangle = 0\},$$

where $f(q) = 0$. Then the outward normal cone to $V(q)$ at $u = u^+$ is again

$$N_{V(q)}(u) = \{\alpha \nabla f(q) : \alpha \geq 0\},$$

so that, with $u = u^+$

$$-dR \in N_{V(q)}(u). \tag{116}$$

A **purely elastic impact** corresponds to conservation of the kinetic energy, $\mathcal{E}(\dot{q}^+) = \mathcal{E}(\dot{q}^-)$ with $\mathcal{E}(\dot{q}) = \frac{1}{2}|\dot{q}|^2$. Thus a purely elastic impact is characterized by

$$|u^+| = |u^-|. \tag{117}$$

This can simply be expressed as $\frac{1}{2}(u^+ + u^-) \in T(q) \subset V(q)$, and then

$$-dR \in N_{V(q)}\left(\frac{u^+ + u^-}{2}\right).$$

Besides these two classical impact laws, there are more general choices. In order to consider partially elastic, partially inelastic collisions, it suffices to assume $u = u(t_s)$ is a convex combination of $u^- = u^-(t_s)$ and of $u^+ = u^+(t_s)$ and then require that (116) holds. In fact, (116) forces u to belong to $V(q)$. Moreover, u cannot lie in the interior of $V(q)$, since otherwise the normal cone at u would reduce to zero and the reaction would be zero, which is false. Even away from contact (116) has to hold, because then the reaction must be zero and in fact $V(q) = E$ implies that the corresponding outward normal cone is zero.

The general approach described above was adopted by Mabrouk [38], while Schatzman and Paoli [58] studied the problem in the form (115) and (117) (the purely elastic case was treated much earlier by Schatzman [66]). Here the convex combination is taken of the form

$$u = \frac{1}{1+e} u^+ + \frac{e}{1+e} u^-, \tag{118}$$

where $e \in [0, 1]$ is the so-called restitution coefficient. Notice that $e = 0$ leads to $u = u^+$, i.e., purely inelastic impacts, while $e = 1$ leads to $u = \frac{1}{2}(u^+ + u^-)$, that is, the purely elastic case; cf. Fig. 9, and also Fig. 6 above.

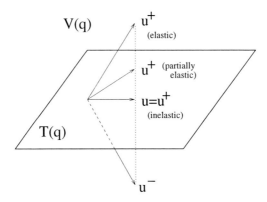

Fig. 9. Elastic, partially elastic, and inelastic impact

In the notation from (74) in Section 4, (118) reads as

$$u = \mathrm{Av}(u^-, u^+) \quad \text{with} \quad k = \frac{e}{1+e}.$$

Let us note in passing that therefore $k' = k(1-k)^{-1} = e$. Moreau [50, Remark 10.2] also proposed, at least informally, an extension of his model of standard inelastic impacts for the purpose of numerical computations. The extension consisted in the introduction of a "dissipation index" δ and the requirement that

$$u = \frac{1+\delta}{2} u^+ + \frac{1-\delta}{2} u^-;$$

hence with

$$e = \frac{1-\delta}{1+\delta}, \quad \delta = \frac{1-e}{1+e}, \quad k = \frac{1-\delta}{2}$$

we are back to (118).

Turning now to the more mathematical part of the problem, we say a function $u : I \to E$ of bounded variation is e-averaged, if it satisfies (118), and we write $u \in \text{bvem}(I, E)$. This "normalization" only concerns the at most countable discontinuity points of u. Hence an e-averaged function u obtained from u^+ and u^- satisfying (107) will also satisfy (106), and its right and left limits are again u^+ and u^-, respectively. Moreover, since $u \in [u^-, u^+]$, the Stieltjes measure is not affected, i.e., $du = du^+ = du^-$; see [53, Section 6] for some information on such "aligned jumps". Notice that for $e = 0$, $\text{bvem}(I, E)$ reduces to the standard space of right-continuous bv functions.

Summarizing the above discussion, we want to solve the following problem.

Problem P: *Given $q_0 \in L = \{q : f(q) \leq 0\}$, $u_0 \in V(q_0)$ (cf. (112)), and a force field $p : I \times E \times E \to E$, try to find $u \in \text{bvem}(I, E)$ with $u(0) = u_0$ such that, putting $q(t) = q_0 + \int_0^t u(s)ds$, we have $u(t) \in V(q(t))$ for $t \in I$, and moreover*

$$-dR = p(t, q(t), u(t))dt - du \in N_{V(q(t))}(u(t)), \tag{119}$$

in the following sense: There exists a nonnegative measure $d\mu$ on the interval I relative to which both the Lebesgue measure dt and the Stieltjes measure du possess densities $t'_\mu \in L^1(I, d\mu; \mathbb{R})$ and $u'_\mu \in L^1(I, d\mu; E)$, respectively, such that the differential inclusion

$$p(t, q(t), u(t))t'_\mu(t) - u'_\mu(t) \in N_{V(q(t))}(u(t)) \tag{120}$$

holds for $d\mu$–almost every $t \in I$.

Remark 7. (a) We can also request that (120) holds for every $t \in I$, by appropriately modifying the (representatives) of the densities in a set of measure zero.

(b) This formulation is essentially also valid for a more general set of constraints

$$L = \{q \in E : f_1(q) \leq 0, \ldots, f_m(q) \leq 0\} \tag{121}$$

for $q \in \partial L$, where the tangent cone now is

$$V(q) = \{v \in E : \langle v, \nabla f_i(q) \rangle \leq 0 \text{ if } f_i(q) = 0\}. \tag{122}$$

(c) Mabrouk has shown in [38, Prop. 1] that any solution of Problem P has several expected properties, such as the generalized versions of the theorem of kinetic energy ($\langle u, dR \rangle = \langle u, pt'_\mu - u'_\mu \rangle = 0$) or of dissipation ($|u^+| \leq |u^-|$). More precisely, by [38, (16)] the dissipation is given by

$$\frac{1}{2}|u^+|^2 - \frac{1}{2}|u^-|^2 = -\frac{1}{2}\left(\frac{1-e}{1+e}\right)|u^+ - u^-|^2 = -\frac{\delta}{2}|u^+ - u^-|^2,$$

or $\mathcal{E}(u^+) - \mathcal{E}(u^-) = -\delta\mathcal{E}(u^+ - u^-) \leq 0$. This justifies that δ is called the dissipation index. ◇

According to the similarity of Problem P with the "averaged" second order sweeping process (73) from the previous section, it is natural to use a discretization procedure of the same type. This yields the following existence result, [38, Theorem 2]; recall $E = \mathbb{R}^d$.

Theorem 12. *Let the constraint function f belong to $C^{1,\alpha}(E)$, with $\frac{1}{2} < \alpha \le 1$, and let $\nabla f(q) \ne 0$ for all $q \in E$. Assume that the force field $p = p(t,q)$ is continuous and bounded. Then:*

(i) *The sequence (u_n, q_n) to be constructed below has a subsequence which converges pointwise×uniformly to a limit (u, q), where $u \in \mathrm{bv}(I, E)$ and (106) holds.*

(ii) $\mathrm{Av}(u^-, u^+)$ *is a solution of Problem P.*

Before giving a sketch of the proof, let us make several comments.

Remark 8. (a) When the given forces p are allowed to depend also on the velocity u, then it is necessary to suppose that p be Lipschitz continuous with respect to the u-variable; see [38, Section 4.7]. If the global boundedness assumption on p is weakened, the procedure still yields local solutions of Problem P.

(b) Uniqueness of solution has been shown when the data (f, p) are real analytic, by Schatzman [67] in the one-dimensional case with one constraint. For elastic collisions inside a convex set, there is a specific result on uniqueness in [66]. Other results have been obtained by Percivale [60,61] and by Carriero and Pascali [16]. More recently, Ballard [3] made a significant extension of the uniqueness result to the situation mentioned in the next Remark (c). Nonuniqueness examples have been given e.g. by Schatzman [65] in the elastic case and by Ballard [3] in the inelastic case with a C^∞ force.

(c) The existence of a solution in the general case of several constraints, cf. (121) and (122), with real analytic data and a non-constant mass matrix has recently been shown by Ballard [3].

(d) The existence of a solution in the situation of several constraints and more general (nonanalytic) data remains open. In the case of orthogonal constraints, there are strong indications (the last one being Ballard [3, Sect. 6 & 7]) that there is a solution. Notice that the use of a fixed point argument for some sweeping processes could only establish the existence of solution for a related simpler problem, which may be called "externally induced inelastic collisions"; see [42, Ch. 4].

(e) The elastic problem, say (115), (117), has attracted a lot of attention on its own; a part of the relevant references are mentioned in [42, Ch. 3]. Let us add here just Buttazzo and Percivale [14,15] and some of the work of Schatzman and Paoli. Existence for elastic collisions inside a convex set L has been shown by Schatzman [65,66]. Paoli [57] proceeded to study dissipative collisions, still inside a convex set L. Recently, these results have been generalized to nonconvex L with C^3 boundary, the mass matrix being allowed to vary, [59]. Some further insight might be gained by comparing this to Mabrouk's result, where $\partial L \in C^{1,\alpha}$ with $\frac{1}{2} < \alpha \leq 1$. $\diamond$

Let us now start with a sketch of the proof of Theorem 12 by describing the discretization procedure. As in the previous section, the interval $I = [0,T]$ is partitioned by points t_i^n, and the approximating functions u_n and q_n are defined with the help of the specific values $q_i^n = q_n(t_i^n)$ and $u_i^n = u_n^+(t_i^n) = u_n^-(t_{i+1}^n)$. We discretize (118), (119) in the form of the following difference inclusion, cf. with (77):

$$(t_{i+1}^n - t_i^n)p(t_{i+1}^n, q_{i+1}^n) - (u_{i+1}^n - u_i^n) \in N_{V(q_{i+1}^n)}(\text{Av}(u_i^n, u_{i+1}^n)).$$

Writing $p_{i+1}^n = p(t_{i+1}^n, q_{i+1}^n)$, $h = t_{i+1}^n - t_i^n$, and $w_{i+1}^n = \text{Av}(u_i^n, u_{i+1}^n) = ku_i^n + (1-k)u_{i+1}^n$, we also have

$$(1-k)hp_{i+1}^n - (1-k)(u_{i+1}^n - u_i^n) \in N_{V(q_{i+1}^n)}(w_{i+1}^n),$$

where $k = e(1+e)^{-1}$. By (5), this is equivalent to

$$w_{i+1}^n = \text{proj}\left(u_i^n + (1-k)hp_{i+1}^n, V(q_{i+1}^n)\right), \tag{123}$$

while by definition of w_{i+1}^n and of the "inverse averaging function" Iav we have

$$u_{i+1}^n = \text{Iav}(u_i^n, w_{i+1}^n).$$

The discretization procedure thus is quite similar to the one used in the previous section. The additional terms are of the order of the stepsize h, since

$$|(1-k)hp_{i+1}^n| \leq Mh,$$

where M is a bound for the norm of the force p. Hence the estimates for the present problem, although being more complicated, resemble those we have been dealing with before. To further explain the analogy, note the multifunction $q \mapsto V(q)$ (the tangent cone from (112)) satisfies (113), but it is not Hausdorff continuous in the whole space as it has "jumps" at the boundary points of L. Nevertheless, V is what is called lower semicontinuous, and this still ensures that $V(q)$ contains a fixed ball $B_r(a)$ for all q in a neighbourhood of the initial position q_0. This fact is a crucial ingredient for estimating the total variations of the approximate velocities u_n. But first we have to obtain the required a priori estimate on the norm of the u_n, cf. (91). Using the projections on the $V(q)$'s and on their normal cones as well as Moreau's lemma of the two cones, one obtains, see [38, Prop. 2], that

$$|u_i^n| \leq |u_0| + ihM \leq |u_0| + TM, \quad \forall n, i.$$

Then the total variations can be bounded as in (90), but with some extra terms due to the contributions $(1-k)hp_i^n$ of the force. More precisely, recalling that $|u_{i+1}^n - w_{i+1}^n| \leq k|u_{i+1}^n - u_i^n|$ by (79), we obtain

$$\text{var}\,(u_n) \leq c_1|u_0 - a|^2 + c_2 \sum_{i=0}^{n-1} |u_{i+1}^n - u_i^n|^2 + c_3 \sum_{i=0}^{n-1} h + c_4 \sum_{i=0}^{n-1} h^2,$$

whence

$$\text{var}\,(u_n) \leq c_5 + c_2 \sum_{i=0}^{n-1} |u_{i+1}^n - u_i^n|^2. \tag{124}$$

To deal with the sum, it could be shown that

$$|u_{i+1}^n - u_i^n| \leq c_6 h + c_7 |\nu_{i+1}^n - \nu_i^n|, \tag{125}$$

see [38, p.829], where ν_i^n denotes the unit outward normal vector to the set $\{q : f(q) \leq f(q_i^n)\}$ at q_i^n. Moreover, if $f \in C^{1,1/2}$ (i.e., if ∇f is Hölder continuous with exponent $\alpha = 1/2$), then there is $c_8 \geq 0$ such that

$$\sum_{i=0}^{n-1} |\nu_{i+1}^n - \nu_i^n|^2 \leq c_8. \tag{126}$$

This is not surprising since ν_i^n is related to $\nabla f(q_i^n)$. One could also observe that if one intersects the halfspaces $V(q)$ with a fixed ball B, then the resulting multifunction $q \mapsto V(q) \cap B$ is 1/2-Hölder continuous. As shown in Section 4, cf. e.g. (97), this yields an adequate uniform estimate for the sum of the squared norms. To summarize, inequalities (124)-(126) imply an upper bound for the total variations of the velocities. Hence, there exist limit functions u and q such that, for subsequences still denoted (u_n) and (q_n), one has $u_n(t) \to u(t)$ pointwise and $q_n(t) \to q(t)$ uniformly for $t \in I$ as $n \to \infty$. Define $w = \text{Av}(u^-, u^+)$, or more explicitly

$$w(t) = \frac{1}{1+e} u^+(t) + \frac{e}{1+e} u^-(t) = (1-k)u^+(t) + ku^-(t),$$

so that $w = u$ at continuity points and $q(t) = q_0 + \int_0^t w(s)\,ds$.

Then Mabrouk [38] proceeded to prove that w is a solution of Problem P in the following steps.

Step 1.) It is shown that $f(q_n(t)) \leq c_9 h$ for some constant c_9, so that in the limit $f(q(t)) \leq 0$, i.e., $q(t) \in L$.

Step 2.) This immediately implies that

$$\dot{q}^+(t) = w^+(t) = u^+(t) \in V(q(t)),$$

i.e., the right velocity is kinetically admissible.

Step 3.) Similarly to Lemma 15, one can show next, see [38, Lemma 6], that for $f \in C^{1,1/2}$ and if $z \in V(q')$ for all q' in a neighbourhood of the range $q([s,t])$, then

$$\langle z, u(t) - u(s) \rangle \geq \frac{1}{2}|u(t)|^2 - \frac{1}{2}|u(s)|^2 + \int_s^t \langle z - u(\tau), p(\tau, q(\tau)) \rangle \, d\tau.$$

Step 4.) Now let $d\mu = |du| + dt$. Then the required densities do exist, and the previous inequality implies that for $d\mu$-almost every continuity point t of w (or equivalently, for $d\mu$-almost every continuity point of u),

$$\langle z, w'_\mu(t) \rangle \geq \langle w(t), w'_\mu(t) \rangle + \langle z - w(t), p(t, q(t)) \rangle,$$

from which it follows that

$$p(t, q(t)) t'_\mu(t) - u'_\mu(t) \in N_{V(q(t))}(u(t)).$$

To obtain this result, one uses again the theorem of Jeffery, cf. the proof of Lemma 16 or [53], and the fact that w and u differ only at discontinuity points.

Step 5.) To deal with the discontinuities, the assumption $f \in C^{1,\alpha}(E)$ for $\frac{1}{2} < \alpha \leq 1$ is used in its full strength for the first time; observe that up to now $f \in C^{1,1/2}(E)$ was sufficient. The arguments refine those in [39]. First, one checks that any discontinuity instant t_s must correspond to a left velocity that is not admissible as a right velocity, i.e., $w^-(t_s) \notin V(q(t_s))$; this can be obtained similarly to Lemmas 13 and 14. Therefore one has to consider $q_s = q(t_s)$ with $f(q_s) = 0$ and $w^- = w^-(t_s) \notin V(q_s)$, i.e., $\langle w^-, \nabla f(q_s) \rangle > 0$; notice that $w^- = u^-(t_s)$. It is then not difficult to see that the condition to be satisfied at t_s is

$$w(t_s) = \text{proj}\,(w^-(t_s), V(q(t_s))) = \text{proj}\,(w^-, V(q_s));$$

this may be compared with (123) with step $h = 0$ and u_i^n standing for the left-limit w^-.

Step 6.) Since $w^-(t_s) \neq w^+(t_s)$, i.e., $u^-(t_s) \neq u^+(t_s)$, we find $\eta > 0$ with $|u(t_s + \eta) - u(t_s - \eta)| \geq 2\delta$, whence from the pointwise convergence it follows that $|u_n(t_s + \eta) - u_n(t_s - \eta)| \geq \delta > 0$ for all large $n \in \mathbb{N}$ and some $\delta > 0$. If all the $t_i^n \in]t_s - \eta, t_s + \eta[$ were such that $f(q_n(t_i^n)) = f(q_i^n) < 0$, then $V(q_i^n) = E$ and by (123), $w_i^n = u_i^n + (1-k)hp_i^n$. This would imply that the variation of u_n or of w_n (with w_n being defined through the w_i^n) in the above subinterval is bounded by its length 2η multiplied by the bound M on the force p. With an appropriate choice of η and δ, this would lead to $|u_n(t_s + \eta) - u_n(t_s - \eta)| < \delta$, a contradiction. We conclude that, for large n, all the approximations q_n must have $f(q_n(t_i^n)) \geq 0$ for some $t_i^n \in]t_s - \eta, t_s + \eta[$.

Step 7.) Given the first such t_i^n, the above argument implies that the approximate velocities u_n (hence also w_n) do not vary much before that instant. At

t_i^n, the definition of w_i^n from (123) means that the required law is satisfied by the approximate functions. The harder part of the proof consists in showing that in the remaining subinterval $]t_i^n, t_s + \eta[$ the variations of u_n and w_n are small enough. This is done in [38, Lemma 9], where the stronger assumption on f is used. By taking limits as $n \to \infty$ and $\eta \to 0$, the desired impact law follows. □

The extension of this existence result to the case of several constraints is certainly difficult, because a discontinuous dependence on the data can occur. Consider e.g. only two constraints $f_1(q) \leq 0$ and $f_2(q) \leq 0$. In the situation described in the last Step 7.) above, one can show similarly that for some t_i^n either $f_1(q_i^n) \geq 0$ or $f_2(q_i^n) \geq 0$. However, limit functions of (q_n) may depend on which of the two constraints is satisfied or violated first. A simple example of this behaviour is that of a ball in a purely inelastic billiard, moving towards a corner with an opening angle larger than $\pi/2$: it can go either way or be stopped at the corner. Nevertheless, this theoretical handicap does not preclude the successful application of discretization procedures based on generalizations of the above algorithm, often with appropriate modifications for better convergence and practicality; see for instance [49, Sect. 15].

5.2 Unilateral contact with friction

If friction is added to the dynamical model problem as described in the last section, then there is a vast literature (see e.g. [8]). In this final subsection we will give a very brief account on how the previous ideas apply to isotropic Coulomb friction. The formulation given here follows [42, p. 78-80]. It concerns only the simplest case considered already in the pioneering works of Moreau [49,50]; see also [30].

Let us assume that the classical isotropic Coulomb's law of friction holds when there is contact with the boundary $S = \{q \in E : f(q) = 0\}$. The friction coefficient ν may depend on the boundary point $q \in S$, i.e., $\nu = \nu(q)$. To each $q \in S$ we associate the friction cone $C(q)$, a (revolution) cone about the inward normal vector $n = n(q) = -|\nabla f(q)|^{-1}\nabla f(q)$, with opening angle $\alpha = \alpha(q)$ such that $\nu(q) = \tan\alpha(q)$. To be precise,

$$C = C(q) = \{v \in E : \langle v, n(q)\rangle \geq |v|\cos\alpha(q)\}.$$

In [42] it is assumed that $\alpha(q) \in]0, \pi/2[$, i.e., that friction is finite in every direction, see Fig. 10.

The reaction force r has to obey several restrictions: it belongs to the friction cone,

$$r \in C;$$

it vanishes when there is no contact,

$$f(q) < 0 \Rightarrow r = 0,$$

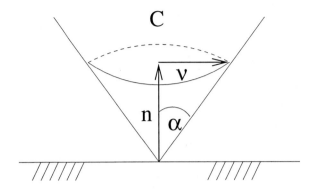

Fig. 10. The friction cone

or when there is contact and the right velocity u points inward, that is

$$[f(q) = 0, \langle u, n \rangle > 0] \Rightarrow r = 0.$$

Finally, in the nontrivial case where the right velocity $u = \dot{q}^+$ is tangential, that is, when u belongs to the tangent hyperplane $T = T(q)$, the classical form of the friction law can be equivalently expressed as

$$[f(q) = 0, \langle u, n \rangle = 0] \Rightarrow -u \in \text{proj}_T\left(N_C(r)\right),$$

cf. [49, (11.15), p. 75-76]. That is to say $-u$ belongs to the orthogonal projection on T of the outward normal cone to the friction cone C at r. In the spirit of the previous subsection we introduce the reaction measure $dR = du - p(t, q(t))dt$, and we formulate the mathematical problem by replacing r with the density

$$r'_\mu = \frac{dR}{d\mu} \in L^1(I, d\mu; E),$$

where $d\mu$ is any positive measure with respect to which dR is absolutely continuous. Then we have the following result.

Theorem 13. *Let the constraint function $f \in C^{1,1}(E)$ satisfy $\nabla f(q) \neq 0$ for all $q \in E$. Assume that the force field $p = p(t, q) : I \times E \to E$ is continuous and bounded and that the angle of friction $\alpha(q)$ is a continuous function of $q \in E$ with values in $]0, \pi/2[$. Then, given $q_0 \in L = \{q : f(q) \leq 0\}$ and $u_0 \in V(q_0)$, there exists a right continuous bv function $u : I \to E$ (the right-velocity) defining the motion $q(t) = q_0 + \int_0^t u(s)\, ds$ for $t \in I$ such that $q(0) = q_0$, $u(0) = u_0$, $q(t) \in L$, and $u(t) \in V(q(t))$ for all $t \in I$. Moreover, the following implications hold $d\mu$-almost everywhere, where $d\mu$ and r'_μ are* *as above:*

$$f(q(t)) < 0 \Rightarrow r'_\mu(t) = 0;$$
$$[f(q(t)) = 0, \langle u(t), \nabla f(q(t)) \rangle < 0] \Rightarrow r'_\mu(t) = 0;$$
$$[f(q(t)) = 0, \langle u(t), \nabla f(q(t)) \rangle = 0] \Rightarrow -u(t) \in \mathrm{proj}_{T(q(t))}\left(N_{C(q(t))}(r'_\mu(t))\right).$$
$$(127)$$

This existence theorem was first given in [41], cf. also [42, Ch. 3]. It is based on a discretization technique similar to the one used above for frictionless contact problems. The algorithm is suggested by an equivalent form of (127) at a discontinuity point t_s: If $u^-(t_s) \neq u(t_s) = u^+(t_s)$, then t_s is an atom of the measure dR, with $dR(\{t_s\}) = u(t_s) - u^-(t_s)$ and, a fortiori, an atom of $d\mu$. It follows that $r'_\mu(t_s) = \lambda(u(t_s) - u^-(t_s))$ for some $\lambda > 0$ and thus the inclusion in (127) is equivalent to

$$-u(t_s) \in \mathrm{proj}_{T(q(t_s))}\left(N_{C(q(t_s))}(u(t_s) - u^-(t_s))\right).$$

The latter was shown by Moreau [49, p.78-79] to be equivalent to

$$u(t_s) = \mathrm{proj}\left(0, [u^-(t_s) + C(q(t_s))] \cap T(q(t_s))\right); \qquad (128)$$

see Fig. 11 below.

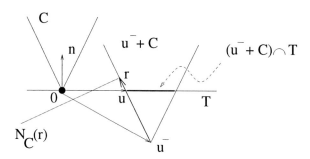

Fig. 11. Illustration of (128)

In the algorithm, this is translated roughly as follows; see [42, (3.3.8)]:

$$u^n_{i+1} = \mathrm{proj}\left(0, [u^n_i + hp^n_{i+1} + C(q^n_{i+1})] \cap T(q^n_{i+1})\right),$$

using the previous notation. The proof of existence, although more involved, follows the general scheme as in the frictionless case.

It should finally be mentioned that there is currently much interest in the study of nonsmooth dynamics with friction within the mathematical framework of measure differential inclusions (that is, differential inclusions involving bv functions and their Stieltjes measures). A full description of all the recent developments exceeds the scope of these notes, hence we will only mention a few of them.

Glocker [29] also used a measure equality to define the reaction. He considered so-called hemivariational inequalities, developed by Panagiotopoulos [55], which allow to include nonconvex potentials.

Frémond [28] invented a theory of rigid body collisions which may be formulated by means of a measure differential inclusion. The existence of solutions with bv velocities was shown by Cholet in [24].

Stewart [68] extended the existence theorem above, by combining the approach of measure differential inclusions with that of complementarity problems (for the latter see e.g. [56] or [69]). Stewart's remarkable existence result for the one contact case allows for more general friction laws, for a varying mass matrix, and a relaxed restriction on the friction cone. This restriction is satisfied in the celebrated Painlevé problem, which is thus shown to have a solution exhibiting shocks (impulsive contact forces without collisions); a numerical hint for that phenomenon can be found in [50, Sect. 15]. Stewart's approach is also numerically efficient in the case of multiple contacts. However, the corresponding existence theory is naturally not so complete. In that situation, even the formulation of the friction law is a matter of continuing research.

References

1. Apostol T. (1969) Mathematical Analysis. Addison-Wesley, Reading MA
2. Baiocchi C., Capelo A. (1984) Variational and Quasivariational Inequalities. John Wiley, New York
3. Ballard P. (1999) The Dynamics of Discrete Mechanical Systems with Perfect Unilateral Constraints. Preprint
4. Ballard P. (1999) A Counter-Example to Uniqueness in Quasi-Static Elastic Contact Problems with Small Friction. Int J Engineering Sci 37: 163-178
5. Breiman L. (1968) Probability. Addison-Wesley, Reading MA
6. Brezis H. (1973) Opérateurs Maximaux Monotones. North Holland, Amsterdam London
7. Brocks W., Hao S., Steglich D. Micromechanical Modelling of the Damage and Toughness Behaviour of Nodular Cast Iron Materials. In: Proc Euromech-Mecamat, Fontainebleau 1996. To appear
8. Brogliato B. (1999) Nonsmooth Mechanics. 2nd edition. Springer, Berlin Heidelberg New York
9. Brokate M. (1998) Elastoplastic Constitutive Laws of Nonlinear Kinematic Hardening Type. In: Brokate M., Siddiqi A.H. (Eds.) Functional Analysis with Current Applications in Science, Technology and Industry. Longman, Harlow, 238-272
10. Brokate M. (1999) Manuscript
11. Brokate M., Krejčí P. (1998) Wellposedness Of Nonlinear Kinematic Hardening Models in Elastoplasticity. Math Modelling Numer Anal 32: 177-209

12. Brokate M., Krejčí P. On the Wellposedness of the Chaboche Model. In: Proc Vorau Conf on Control and Estimation of Distributed Parameter Systems, 1996. Birkhäuser, Basel Boston Berlin. To appear

13. Brokate M., Krejčí P. (1997) Maximum Norm Wellposedness of Nonlinear Kinematic Hardening Models. Preprint

14. Buttazzo G., Percivale D. (1981) The Bounce Problem on n-Dimensional Riemannian Manifolds. Atti Accad Naz Lincei, Cl Sci Fis Mat Natur 70: 246-250

15. Buttazzo G., Percivale D. (1983) On the Approximation of the Elastic Bounce Problem on Riemannian Manifolds. J Differential Equations 47: 227-245

16. Carriero M., Pascali E. (1982) Uniqueness of the One-Dimensional Bounce Problem as a Generic Property in $L^1([0,T],\mathbb{R})$. Boll Unione Mat Ital A 5: 87-91

17. Carstensen C., Mielke A. (1997) A Formulation of Finite Plasticity and an Existence Proof for One Dimensional Problems. Preprint

18. Castaing C. (1988) Quelques Problèmes d'Évolution du Second Ordre. Séminaire d'Analyse Convexe Montpellier 18: Exposé 5

19. Castaing C., Monteiro Marques M.D.P. (1995) Periodic Solutions of Evolution Problems Associated with a Moving Convex Set. C R Acad Sci Paris, Sér I Math 321: 531-536

20. Castaing C., Monteiro Marques M.D.P. (1995) BV Periodic Solutions of an Evolution Problem Associated with Continuous Moving Convex Sets. Set-Valued Anal 3: 381-399

21. Castaing C., Monteiro Marques M.D.P. (1997) Topological Properties of Solution Sets for Sweeping Processes with Delay. Portugal Math 54: 485-507

22. Castaing C., Truong Xuan Duc Ha, Valadier M. (1993) Evolution Equations Governed by the Sweeping Process. Set-Valued Anal 1: 109-139

23. Castaing C., Valadier M. (1977) Convex Analysis and Measurable Multifunctions. Lecture Notes in Mathematics Vol 580. Springer, Berlin Heidelberg New York

24. Cholet C. (1998) Chocs de Solides Rigides. Ph D Thesis, Université de Paris VI, Paris

25. Chraibi Kaadoud M. (1987) Étude Théorique et Numérique de Problèmes d'Évolution en Présence de Liaisons Unilatérales et de Frottement. Ph D Thesis, Université de Montpellier, Montpellier

26. Deimling K. (1985) Nonlinear Functional Analysis. Springer, Berlin Heidelberg New York

27. Deimling K. (1992) Multivalued Differential Equations. de Gruyter, Berlin New York

28. Frémond M. (1995) Rigid Bodies Collisions. Physics Letters A 204: 33-41

29. Glocker Ch. (1998) The Principles of d'Alembert, Jourdain and Gauss in Nonsmooth Dynamics, Part I: Scleronomic Multibody Systems. Z Angew Math Mech 78: 21-37

30. Jean M., Moreau J.J. (1985) Dynamics in the Presence of Unilateral Contacts and Dry Friction: A Numerical Approach. U.S.T.L. Montpellier, Preprint 85-5

31. Kunze M. (1999) Periodic Solutions of Non-Linear Kinematic Hardening Models. Math Meth Appl Sci 22: 515-529

32. Kunze M., Monteiro Marques M.D.P. (1996) Yosida-Moreau Regularization of Sweeping Processes with Unbounded Variation. J Differential Equations 130: 292-306

33. Kunze M., Monteiro Marques M.D.P. (1997) Existence of Solutions for Degenerate Sweeping Processes. J Convex Anal 4: 165-176
34. Kunze M., Monteiro Marques M.D.P. (1998) On the Discretization of Degenerate Sweeping Processes. Portugal Math 55: 219-232
35. Kunze M., Monteiro Marques M.D.P. (1998) On Parabolic Quasi-Variational Inequalities and State-Dependent Sweeping Processes. Topol Methods Nonlinear Anal 12: 179-191
36. Kunze M., Monteiro Marques M.D.P. (1999) A Note on Lipschitz Continuous Solutions of a Parabolic Quasi-Variational Inequalitiy. In: Li Ta-tsien, Lin Longwei, Rodrigues J.F. (Eds.) Luso-Chinese Symposium on Nonlinear Evolution Equations and Their Applications, Macau 1998. World Scientific, Singapore New Jersey London, 109-115
37. Kunze M., Monteiro Marques M.D.P. (1999) Degenerate Sweeping Processes. In: Argoul P., Frémond M., Nguyen Q.S. (Eds.) Proc IUTAM Symposium on Variations of Domains and Free-Boundary Problems in Solid Mechanics, Paris 1997. Kluwer Acad Press, Dordrecht, 301-307
38. Mabrouk M. (1998) A Unified Variational Model for the Dynamics of Perfect Unilateral Constraints. Eur J Mech A/Solids 17: 819-842
39. Monteiro Marques M.D.P. (1985), Chocs Inélastiques Standards: Un Résultat d'Existence. Séminaire d'Analyse Convexe Montpellier 15: Exposé 4
40. Monteiro Marques M.D.P. (1987) Regularization and Graph Approximation of a Discontinuous Evolution Problem. J Differential Equations 67: 145-164
41. Monteiro Marques M.D.P. (1988) Inclusões Diferenciais e Choques Inelásticos. Ph D Thesis, Universidade de Lisboa, Lisbon
42. Monteiro Marques M.D.P. (1993) Differential Inclusions in Nonsmooth Mechanical Problems–Shocks and Dry Friction. Birkhäuser, Basel Boston Berlin
43. Moreau J.J. (1974) On Unilateral Constraints, Friction and Plasticity. In: Capriz G., Stampacchia G. (Eds.) New Variational Techniques in Mathematical Physics, CIME ciclo Bressanone 1973. Edizioni Cremonese, Rome, 171-322
44. Moreau J.J. (1976) Application of Convex Analysis to the Treatment of Elastoplastic Systems. In: Germain P., Nayroles B. (Eds.) Applications of Methods of Functional Analysis to Problems in Mechanics. Lecture Notes in Mathematics Vol 503. Springer, Berlin Heidelberg New York, 56-89
45. Moreau J.J. (1977) Evolution Problem Associated with a Moving Convex Set in a Hilbert Space. J Differential Equations 26: 347-374
46. Moreau J.J. (1978) Un Cas de Convergence des Itérées d'une Contraction d'un Espace Hilbertien. C R Acad Sci Paris, Sér A-B 286: A143-A144
47. Moreau J.J. (1983) Liaisons Unilatérales sans Frottement et Chocs Inélastiques. C R Acad Sci Paris, Sér II 296: 1473-1476
48. Moreau J.J. (1985) Standard Inelastic Shocks and the Dynamics of Unilateral Constraints. In: del Piero G., Maceri F. (Eds.) Unilateral Problems in Structural Analysis. Springer, Berlin Heidelberg New York, 173-221
49. Moreau J.J. (1985) Dynamique de Systèmes à Liaisons Unilatérales avec Frottement Sec Éventuel: Essais Numériques. U.S.T.L. Montpellier, Technical Report 85-1
50. Moreau J.J. (1988) Unilateral Contact and Dry Friction in Finite Freedom Dynamics. In: Moreau J.J., Panagiotopoulos P.D. (Eds.) Nonsmooth Mechanics and Applications, CISM Courses. Springer, Berlin Heidelberg New York, 1-82

51. Moreau J.J. (1988) Bounded Variation in Time. In: Moreau J.J., Panagiotopoulos P.D., Strang G. (Eds.) Topics in Non-Smooth Mechanics. Birkhäuser, Basel Boston Berlin, 1-74

52. Moreau J.J. (1998) Numerical Aspects of the Sweeping Process. In: Martins J.T., Klarbring A. (Eds.) Computer Methods in Applied Mechanics and Engineering. Special issue of: Computational Modeling of Contact and Friction. To appear

53. Moreau J.J., Valadier M. (1984) Quelques Résultats sur les Fonctions Vectorielles à Variation Bornée d'une Variable Réelle. Séminaire d'Analyse Convexe Montpellier 14: Exposé 16

54. Natanson I.P. (1955) Theory of Functions of a Real Variable. Frederick Ungar Publ Co, New York

55. Panagiotopoulos P.D. (1985) Inequality Problems in Mechanics and Applications. Convex and Nonconvex Energy Functions. Birkhäuser, Basel Boston Berlin

56. Pang J.-S., Trinkle J.C. (1996) Complementarity Formulations and Existence of Solutions of Dynamic Multi-rigid-body Contact Problems with Coulomb Friction. Math Programming 73: 199-226

57. Paoli L. (1993) Analyse Numérique de Vibrations avec Contraintes Unilatérales. Ph D Thesis, Université Claude Bernard, Lyon

58. Paoli L., Schatzman M. (1993) Mouvement a un Nombre Fini de Degrés de Liberté avec Contraintes Unilatérales: Cas avec Perte d'Énergie. RAIRO Modél Math Anal Numér 27: 673-717

59. Paoli L., Schatzman M. (1999) A Numerical Scheme for Impact Problems. Preprint 300, UMR 5585, Lyon

60. Percivale D. (1985) Uniqueness in the Elastic Bounce Problem I. J Differential Equations 56: 206-215

61. Percivale D. (1991) Uniqueness in the Elastic Bounce Problem II. J Differential Equations 90: 304-315

62. Prigozhin L. (1996) Variational Model of Sandpiles Growth. European J Appl Math 7: 225-235

63. Rockafellar R.T. (1971) Convex Integral Functionals and Duality. In: E.H. Zarantonello (Ed.) Contributions to Nonlinear Functional Analysis. Academic Press, New York, 215-236

64. Rudin W. (1973) Functional Analysis. McGraw-Hill, New Delhi

65. Schatzman M. (1977) Le Système Différentiel $(du^2/dt^2) + \partial\phi(u) \ni f$ avec Conditions Initiales. C R Acad Sci Paris, Sér A 284: 603-606

66. Schatzman M. (1978) A Class of Nonlinear Differential Equations of Second Order in Time. Nonlinear Anal 2: 355-373

67. Schatzman M. (1998) Uniqueness and Continuous Dependence on Data for One Dimensional Impact Problems. Math and Computational Modelling 28: 1-18

68. Stewart D.E. (1998) Convergence of a Time-Stepping Scheme for Rigid-Body Dynamics and Resolution of Painlevé's problem, Arch Rat Mech Anal 145: 215-260

69. Stewart D.E., Trinkle J.C. (1996) An Implicit Time-Stepping Scheme for Rigid Body Dynamics with Inelastic Collisions and Coulomb Friction. Intern J Num Meths Eng 39: 2673-2691

70. Studt H. Ph D Thesis, Universität Kiel, Kiel. In preparation

71. Yosida K. (1995) Functional Analysis. Reprint. Springer, Berlin Heidelberg New York

Dynamic Simulation of Rigid Bodies: Modelling of Frictional Contact

Michel Abadie

Schneider Electric, Research Center A2, Rue Volta, F-38050 Grenoble cedex 9, France
michel_abadie@mail.schneider.fr

Abstract. The object of this work is to deal with frictional contact problems in the mechanisms. The classical approach consist in modelling the contact by using penalized methods. The contact forces are then considered as external forces defined by some regularized contact laws. The main problems of this approach are:
- stiff system of differential equations;
- oscillations of the contact forces and of the accelerations;
- sticking is not correctly modelled because a regularized Coulomb law is used.

In order to avoid these problems, we have chosen to implement a frictional contact model based on unilateral contact theory. This work details the implementation of this approach, while trying to highlight its advantages and its drawbacks.

1   Introduction

The object of this chapter is to present the work accomplished in the Schneider Electric company in order to deal with frictional contact problems in mechanisms. By mechanism we mean, a set of perfectly rigid bodies which are in relation via classical joints (like pins or sliders) or via frictional contacts. The calculation of mechanisms is a tool which is extremely helpful in the design process, because it intervenes at all the stages of a project:

- At the beginning of a project, when we have a very simple idea of the principal functionalities of the mechanism we must design, we can run some kinematic studies starting from a skeleton mechanism. We just want to test at this stage the general principle of the mechanism and also to select the best solution among the various options of design which arise. Here, the shape of the parts can be outlined and one can already be interested in problems of interferences.
- A little later in the project, when the structure of the mechanism is defined, the model is enriched and we can start static calculations. We affect masses and inertias for each body and the characteristics of the actuators are coarsely defined (springs, drivers…).
- Lastly, we reach the dynamic analysis, the most complex analysis, which must give us an idea of the real behaviour of our mechanism. The masses and inertias of the bodies and the characteristics of the actuators are refined in order to optimize numerically the behaviour of the mechanism.

The gains which we can draw from the use of this kind of tool are clear. The two principal objectives are:

- give to the designers an effective tool to simulate the behaviour of their mechanisms at the earliest stage of the project;

- postpone to a maximum the realization of prototypes.

We have started to do some mechanism simulations ten years ago at Schneider Electric. But the mechanisms which we design are very particular and very delicate to model, because we have to deal with a great number of contacts. Let us consider for example one of our circuit breakers, C60, which is a domestic circuit breaker low voltage that you can find in your electrical equipment box. The mechanical part of this system comprises only seven moving bodies, but all the difficulty is, that in the course of operation, twelve contacts come into play (for a complete description see paragraph 5). In addition, this circuit breaker, in closed position, is in equilibrium only thanks to the friction and thanks to the sticking of one of the contacts.

The modelling of frictional contact (and especially phenomena of sticking) is thus an essential point for the applications we wish to treat.

There exists commercial softwares of mechanism simulation which today are widely diffused. Let's quote here for example the three principal actors of this market, Pro-Mechanica Motion (PTC), Adams (MDI) and Dads (LMS). These softwares are perfectly equivalent in continuous motions, and even if there are some differences in computing time, the results obtained are almost identical and perfectly satisfactory. On the other hand, when we have to deal with frictional contact problems, it becomes more delicate. All these softwares have adopted the same contact model, that we will call penalized model. Let us describe here briefly this model and underline its principal drawbacks. With this approach, the normal and tangential contact forces are calculated as follows:

1.1 Normal Contact Force – Penalized Method

The normal contact force is represented by a non-linear spring to which a damping term is added:

$$F_{normal} = F_{stiffness} + F_{damping} \qquad (1)$$

Classically, the term of stiffness is calculated using Hertz theory [1]:

$$F_{stiffness} = \frac{4E_1E_2r^{1/2}}{3(E_1(1-v_2^2)+E_2(1-v_1^2))}y^{3/2} = Ky^{3/2} \qquad (2)$$

Where y represents the penetration of the contacting bodies,
(E_1, E_2, v_1, v_2) the material properties of the contacting bodies,
r is a function of the curvature radius of the contacting bodies.

If everyone agrees on the definition of this term, one finds for the damping various definitions. We quote here two of them:

$$\begin{cases} F_{damping} = \left(2e_r m \sqrt{\dfrac{\overline{K}}{m}}\right)\dot{y} \quad \text{with} \quad \overline{K} = \dfrac{3}{2}\sqrt[3]{mgK^2} \\ F_{damping} = \left(e_r K y^{3/2}\right)\dot{y} \end{cases} \qquad (3)$$

Where $\dot{y} = dy/dt$ is the penetration velocity of the bodies in contact,

e_r represents a restitution coefficient,

m the equivalent mass of the contacting bodies (which is not an easy parameter to estimate).

Note: The first formula is used in the Dads software (LMS) and the second in the Mechanica Motion software (PTC).

The normal contact force can be represented as a function of the penetration by the following figure:

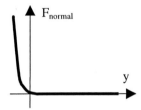

Fig.1. Regularized normal contact la

It is important to note that a light penetration is tolerated between the contacting bodies. We have to note that the calculation of the stiffness K has been defined b Hertz with the hypothesis of a quasistatic phenomenon and a central contact (the gravity centers of the contacting bodies are aligned with the contact point). The extensions of this theory to the problem of shocks, far from being quasistatic, and to the non-central contacts are strong assumptions which are seldom underlined and which do not appear easily justifiable to us. Nevertheless, we will not enlarge on this discussion, because we will further see that it is not the most disturbing point.

1.2 Tangential Contact Force – Penalized Method

The tangential contact force is calculated using a regularized Coulomb law, which can be represented in 2D by the following figure:

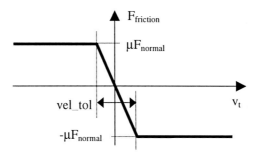

Fig.2. Tangential regularized contact la

Where μ is the friction coefficient,
v_t represents the slip velocity,
vel_tol is a parameter fixed by the user.

Thus, here is the "traditional" penalized method. This modelling of the frictional contact has the following advantages:

- simplicity of implementation (the contact forces are regarded as external forces);
- we can ensure the existence and the uniqueness of the solution calculated by this method;
- the contact does not influence the degree of hyperstaticity of the system.

For these reasons, this approach is extremely widespread. However, we can raise the following objections:

- The introduction of stiffness, generally very large, in the dynamic equations leads to stiff differential equations difficult to integrate. Although implicit algorithms, extremely robust, are used (as Dassl [2], for example, that one finds in Mechanica Motion, Adams and Dads), the results remain very sensitive to the tuning of the algorithm.
- When we deal with contact problem, even without friction, we often observe some oscillations in the case of permanent contact. These oscillations are particularly visible on the efforts and the accelerations.
- The sticking does not respect the law of Coulomb, but a regularized law. Thus, if the sliding velocity is null, the tangential contact force is always null.

The objective of this work is to develop a frictional contact model which makes possible to avoid all the problems that we have highlighted. Taking into account the importance of friction in the Schneider Electric mechanisms, we will particularly take care to respect the Coulomb law (and thus to model the sticking problems correctly).

In order to do this, we have chosen an approach based on the concept of unilateral joints, that is to say non-permanent joints. In a very diagrammatic way, we will replace the preceding regularized problem by a strongly non-linear problem. We can represent this operation by the following figure (in 2D):

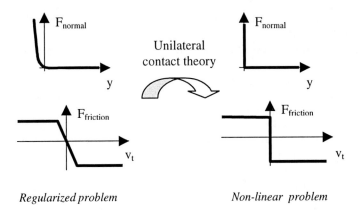

Regularized problem   *Non-linear problem*

Fig.3. Regularized contact law versus non-linear contact laws

The algorithms, which we have developed, have been implemented in the SDS software of the Solid Dynamics company. This software is widely used within Schneider Electric, it thus has been a privileged platform of development. When we have begun our collaboration, SDS treated only arborescent systems with a first order integration scheme (and a constant step of time). The aim of this article is to report how we have deeply modified the structure of this program in order to implement our frictional contact model. We will here present a certain number of principles, without going into all the details, in order to arouse the reader's interest and to raise a certain number of important problems. We will thus approach all the aspects of the mechanisms simulation.

This chapter is composed of four sections:

In the first section, we will see how the dynamics equations are formulated for arborescent systems, then we will be interested in the closed-loop systems. We will see how to integrate these equations using Runge-Kutta methods. We will be interested particularly in methods of polynomial interpolation which make it possible to estimate certain variables on the step of time. Lastly, we will present constraints stabilization methods which guarantee that the constraints are verified at each step of time. The tools introduced here will find their utility in the treatment of frictional contact.

In the second section, we will be interested in the frictional contact problem. We will start by giving a definition of the normal and tangential contact laws. We will then present a naive approach to highlight the difficulties of the problem we have to solve. We will examine how this problem is treated in the literature. We will justif the choice of the Moreau method [3] which we have adopted and we will discuss the advantages and the disadvantages of velocity and acceleration formulations. We will be concerned with the existence and uniqueness of the solution, and we will state some results of convergence. We will then see how we have modified the Moreau method to adapt it to our needs (formulation in acceleration, driven joints, bilateral or permanent joint). Lastly, we will examine two extremely important problems (inconsistency and indeterminacy), of which it is advisable to be informed, and we will see how we can face them.

The third section of this document will be devoted to the study of the shocks. We will start establishing the dynamic equations as a function of the velocity. We will be interested then in the definition of a shock law, for that, we will introduce the concept of restitution coefficient. We will see that there are several definitions of this parameter and how to implement a coefficient of restitution in our algorithms. We will then insist on certain problems which appear important to us, and speak successively about the propagation of the impulses, of the interpretation of the impulses in forces and of the problems of capture. To finish this chapter, we will outline the structure of our program

In order to illustrate our work, we will present an industrial example. That will be the subject of our fourth section.

2 Continuous Motion

The aim of this paragraph is to briefly describe the setting and the integration of the dynamic equations in the case of continuous motions. We just want here to give some basic elements to the reader and to raise certain important problems.

2.1 Setting of the Dynamic Equations.

Let us consider a mechanism (Σ) made up of (N) bodies ($S_1,\ldots,S_N$), perfectly rigid, connected by (N+P) perfect joints. To simplify, we will suppose that all these connections are pins or sliders. Subsequently to this document, we will work in relative coordinates. We will not discuss the choice of these parameters, because they are imposed by SDS.

2.1.1 Open Systems

For a system with a tree structure (P=0), we thus obtain a minimum number of independent parameters. The number of relative parameters then represents the number of degrees of freedom (N) of the system. Let us note q the set of these (N) generalized coordinates:

$$q = \left[q_1,\ldots,q_N\right]^T \quad \text{si} \quad P = 0 \tag{4}$$

The dynamic equations in relative coordinates are obtained using the algorithm of Newton-Euler [4]. We thus obtain a system of (N) equations in state space form:

$$H(q)\ddot{q} - U(q,\dot{q},t) + C(q,\dot{q},t) = 0 \tag{5}$$

Where H is the mass matrix, symmetric definite positive,
    C represents the external efforts and the inertia forces,
    U the efforts applied by the user along the axes of the joints.

2.1.2 Closed loop system

To parameterize a mechanism containing (P) closed chains, we have recourse to a traditional artifice, which consists in «removing» (P) joints in order to obtain an associated open system. For this new system, we define a set of generalized coordinates q of dimension (N). We have then to take into account the constraints of (P) the connections which have been «removed». We will call these connections, closed loop joints or additional joints. The other connections will be called principal joints. For a system with closed chains, we then obtain a minimal set of dependent parameters q. The position and the orientation of the system Σ) is entirely determined by the vector q of dimension (N)

$$q = \left[q_1, ..., q_N \right]^T, \tag{6}$$

subjected to the (5P) geometrical constraints of the form

$$\Phi^L(q) = \left[\Phi^{L_1}(q), ..., \Phi^{L_P}(q) \right]^T = 0, \tag{7}$$

where $\Phi^{L_i} = \left[\Phi_1^{L_i}, \Phi_2^{L_i}, \Phi_3^{L_i}, \Phi_4^{L_i}, \Phi_5^{L_i} \right]^T$ represents the five geometrical constraints associated to the connection (L_i).

The principle of the Lagrange multipliers enables us to write the (N) dynamic equations in the following form:

$$H(q)\ddot{q} - U(q, \dot{q}, t) + C(q, \dot{q}, t) - (\Phi^L)^T_{,q} \lambda^L = 0 \tag{8}$$

Where $(\Phi_L)_{,q}$ is the matrix of kinematic constraints,
λ^L are the Lagrange multipliers.

The Lagrange multipliers λ^L represent the efforts in the closed loop joints.
Note: The interested reader will find in classical text books (see for example [5] or [6]) the detailed calculation of the geometrical and kinematic constraints for a certain number of traditional connections and also a precise interpretation of the Lagrange multipliers.
These (N) equations must be solved by taking account the (5P) geometrical constraints equations:

$$\Phi^L(q) = 0 \tag{9}$$

Here is briefly for the setting of the dynamic equations. Obviously other sets of parameters and other formulations of the dynamic equations exist, for a complete presentation, we return the reader interested in the work of [6]. Let us see now how to integrate this system of equations.

2.2   Integration schem

We will quickly present here the basic principles of an integration scheme. We will give a general description of the Runge-Kutta methods we have chosen to use. We will then present a very interesting tool which will enable us to interpolate the results between two Runge-Kutta steps.

2.2.1   Method of Integration

Let us consider the resolution of the following differential problem:

Problem : *Let f be a continuous function on* $\mathbf{R} \times \mathbf{R}^N$ *with values in* $\mathbf{R}^N$ *and* y_o *an element of* $\mathbf{R}^N$. *It is necessary to find a function y, continuous and derivable on* $\mathbf{R}$, *with values in* $\mathbf{R}^N$, *which satisfies:*

$$\begin{cases} \forall t \in \mathbf{R}, \quad y' = \dfrac{d}{dt} y(t) = f(t, y(t)) \\ t_o \in \mathbf{R}, \quad y(t_o) = y_o \end{cases} \tag{10}$$

If the function f is continuous and locally lipschitzian, the existence and the uniqueness of the solution (y) are given by the Cauchy-Péano and Cauchy-Lipschitz theorems [7]. We will admit them here.

In our particular case, the parameter t represents time. We will seek the solution of the problem (10) on an interval of time $I = [t_o, t_f]$ of $\mathbf{R}^+$. Let $[t_n, t_{n+1}]$ be a subinterval of I. We define the step of time by the following relation:

$$h_n = t_{n+1} - t_n \qquad h = \max_{o \le n < f} (h_n) \tag{11}$$

If we consider the interval $[t_n, t_{n+1}]$, our problem consists in finding an estimate of the exact solution:

$$y(t_{n+1}) \cong y(t_n + h_n) \tag{12}$$

that we note y_{n+1}. Once this operation is carried out, as we know the function f, we are then able to calculate its derivative using equation (10). Thus, gradually, we solve a discretized problem.

We can distinguish two types of methods which allow the estimate of the equation (12):

- ▪   One-step Methods (Runge-Kutta).

  $$y_{n+1} = y_n + h_n F(t_n, h_n, y_n)$$

- Multi-step Methods (Adams, Backward differentiation methods).

$$y_{n+1} = y_n + h_n G(t_n, h_n, y_n, y_{n-1}, ..., y_{n-s})$$

The multi-step methods are particularly interesting, because we can obtain $_{n+1}$ without additional evaluation of the function f. However, we have chosen to use one-step methods for two essential reasons:

- In the multi-step methods the integration order and the step of time both vary. In the literature, their descriptions are usually given with a constant step of time [5], but in practice, these methods are very heavy to implement (see [7][2] for example).
- Moreover, although we considered here only continuous motions, we should not lose sight of the fact that our aim is to treat discontinuous movements. But after each discontinuity, the order of the multi-step methods falls to one, we then need several steps to be again effective.

Considering these two remarks, we have naturally chosen to use one-step methods. We now will give a general description of the Runge-Kutta methods.

2.2.2 Runge-Kutta Methods

We will calculate an approximate value y_{n+1} of $y(t_{n+1})$ starting from the $_n$ estimation of $y(t_n)$. Our problem can be written:

$$\begin{cases} y_{n+1} = y_n + \int_{t_n}^{t_{n+1}} f(t, y)dt = y_n + h_n F(t_n, y_n, h_n) \\ y_0 = y(t_0) \end{cases} \tag{13}$$

We consider a set of (s) real numbers $(c_1, c_2, ..., c_s)$ which satisfy : $0 \leq c_i \leq 1$ for i=1..s. Let us note $T_{ni} = t_n + c_i h_n$ the intermediate instants between t_n and t_{n+1}. A Runge-Kutta method with (s) stages can be written in the form:

$$\begin{cases} Y_{ni} = y_n + \int_{t_n}^{T_{nj}} f(t, y)dt = y_n + h_n \sum_{j=1..s} a_{ij} f(T_{nj}, Y_{nj}) \qquad i = 1..s \\ y_{n+1} = y_n + \int_{t_n}^{t_{n+1}} f(t, y)dt = y_n + h_n \sum_{j=1..s} b_j f(T_{nj}, Y_{nj}) \end{cases} \tag{14}$$

The method is completely defined when the coefficients (a_{ij}), (b_i) and (c_i) for i,j=1..s are known. The following table is thus a characteristic of the method:

c_1	a_{11}	a_{12}	$\cdots$	a_{1s}
c_2	a_{21}	a_{22}		a_{2s}
$\vdots$	$\vdots$	$\vdots$		$\vdots$
c_s	a_{s1}	a_{s2}	$\cdots$	a_{ss}
	b_1	b_2	$\cdots$	b_s

Among all the available Runge-Kutta methods, we have chosen to implement the Dormand&Prince algorithm which is order 5.

For reasons of clearness, we will not describe here the step of time management we have used, although it is a little particular since it relies on a PI control. We will not consider either the stability problems of the Runge-Kutta methods (stability domains, stiff problems, automatic stiffness detection…). The reader interested by these problems is invited to refer to the book [8] in which these principles are described.

In the continuation of our work, we will deal with discontinuous problems. So that the Runge-Kutta methods are applicable, we have to locate the discontinuit instants. Once this operation is carried out, we treat a piecewise continuous problem. The determination of these transition instants is realized thanks to a polynomial interpolation which is described below.

2.2.3 Interpolation of the Results

Let us consider that we have solved the problem (10) on the interval $[t_n, t_{n+1}]$, we thus know y_n, y'_n, y_{n+1} and y'_{n+1}. Our aim is to calculate an approximate value of y on the whole interval $[t_n, t_{n+1}]$. We will use a polynomial interpolation. We consider again the formula (14), where the coefficients b are replaced by polynomials $b(\theta)$, where $\theta \in [0,1]$. We can then write:

$$y(t_n + \theta h_n) = y_n + h_n \sum_{j=1..s'} b_j(\theta) f(T_{nj}, Y_{nj}) \qquad (15)$$

We consider here only the interpolation methods which check (s'=s), where (s) is the number of steps of the Runge-Kutta method. In this case, no additional evaluation of the function f is necessary. The calculation cost of this process is thus very small.
The simplest approach consists in using an Hermite interpolation:

$$y(t_n + \theta h_n) = (1-\theta)y_n + \theta y_{n+1} +$$
$$\theta(\theta - 1)((1 - 2\theta)(y_{n+1} - y_n) + (1-\theta)h_n y'_n + \theta h_n y'_{n+1}) \qquad (16)$$

If the order of the method is equal to three or higher, we thus obtain an interpolation of order three. But, if we wish an interpolation of order higher than three, calculations become extremely complex. The polynomials $b(\theta)$ must then satisfy a certain number of relations, necessary and sufficient, in order to ensure the result order. Once again, the interested reader is returned to the work of [8] which presents this calculation using graphs. It is interesting to note, that for a given method of Runge-Kutta, there are several interpolations of the same order.
Thereafter, we will use an algorithm of order 5. For this algorithm, it would be damaging to be limited to an interpolation of order 3. We will thus consider an interpolation of order 4, of which we will admit the form without calculation.
This procedure of interpolation is a very important tool of our algorithm. Without this procedure, the research of the transition instants would be extremely expensive.

2.3   Application to the Dynamic Equations

The application of the Runge-Kutta method to the movement equations is direct for arborescent systems. In this case, we just have to integrate a traditional differential system. For the closed loop systems, the things are more complicated. We must integrate a differential-algebraic system, that is to say a differential system subjected to equality constraints, which can be written:

$$\begin{cases} H(q)\ddot{q} - U(q,\dot{q},t) + C(q,\dot{q},t) - (\Phi^L)_{,q}^T \lambda^L = 0 \\ \Phi^L = 0 \end{cases} \tag{17}$$

Let us see now why the direct integration of this differential-algebraic system is not possible with the Runge-Kutta methods and how to modify this system in order to treat this problem.

2.3.1 Integration of the Differential-Algebraic Systems by the Runge-Kutta Methods

The differential-algebraic systems are characterized by their index:

> ***Definition [2]:*** *The index of a differential-algebraic system represents the number of times that is necessary to derive the algebraic equations, in order to get a differential system of equations.*

The preceding problem, that we will call (P3)

$$(P3) \quad \begin{cases} H(q)\ddot{q} - U(q,\dot{q},t) + C(q,\dot{q},t) - (\Phi^L)_{,q}^T \lambda^L = 0 \\ \Phi^L = 0 \end{cases}$$

is of index 3. It is very delicate to show the convergence of a Runge-Kutta method for a differential-algebraic problem of index 3. This demonstration does not appear in the literature which deals with differential-algebraic systems [8][2].

Taking into account this difficulty, which is for the moment insuperable, the only possibility we have is to decrease the index of the problem (P3). We are thus going to formulate a system of index 2. For that, we have to derive the geometrical constraint equations (9) to obtain velocity constraint equations:

$$(\Phi^L)_{,q} \dot{q} = 0 \tag{18}$$

As before, we will call (P2) the problem of index 2 constituted by the dynamic equations (16) and the velocity constraint equations (18):

$$(P2) \quad \begin{cases} H(q)\ddot{q} - U(q,\dot{q},t) + C(q,\dot{q},t) - (\Phi^L)_{,q}^T \lambda^L = 0 \\ (\Phi^L)_{,q}\dot{q} = 0 \end{cases}$$

The system (P2) ca be written in the form:

$$\begin{cases} y' = f(t,y,z) \\ 0 = g(t,y) \end{cases} \quad \text{avec} \quad \begin{cases} y = \begin{bmatrix} q \\ \dot{q} \end{bmatrix} \\ z = \lambda \end{cases} \tag{19}$$

We can find in [8] a convergence demonstration for the index 2 systems. But we have to note that for a method of Runge-Kutta of order (p), the variable z is calculated with an order lower than (p). This order reduction does not appear acceptable to us.

Note : No phenomenon of this type exists for the multi-step methods of retrograde differentiation [9].

For the reason which have been just evoked, we give up the formulations (P3) and (P2). This can be done without any regret, because even if we do not consider the convergence problems, it is easy to notice that the nature of the problem (P3) or (P2) equations is very different. We thus suspect that their numerical treatment leads to complex methods.

Finally, we are going to formulate an index 1 problem, for which the convergence of the Runge-Kutta methods is ensured without any order reduction [8]. We then derive the geometrical constraint equations (9) twice to obtain acceleration constraint equations:

$$(\Phi^L)_{,q}\ddot{q} + ((\Phi^L)_{,q}\dot{q})_{,q}\dot{q} = (\Phi^L)_{,q}\ddot{q} + \eta^L \tag{20}$$

We thus obtain the problem (P1):

$$(P1) \quad \begin{cases} H(q)\ddot{q} - U(q,\dot{q},t) + C(q,\dot{q},t) - (\Phi^L)_{,q}^T \lambda^L = 0 \\ (\Phi^L)_{,q}\ddot{q} + \eta^L = 0 \end{cases}$$

To obtain a problem of index 1, we write (P1) in the following form:

$$(P1) \quad \begin{cases} \begin{bmatrix} q' \\ u' \end{bmatrix} = \begin{bmatrix} u \\ \omega \end{bmatrix} \\ \begin{bmatrix} H(q) & -(\Phi^L)_{,q}^T \\ (\Phi^L)_{,q} & 0 \end{bmatrix} \begin{bmatrix} \omega \\ \lambda^L \end{bmatrix} = \begin{bmatrix} U(q,u,t) - C(q,u,t) \\ -\eta^L \end{bmatrix} \end{cases} \tag{21}$$

It is clear that this system is equivalent to:

$$\begin{cases} y' = f(t, y, z) \\ 0 = g(t, y) \end{cases} \quad avec \quad y = \begin{bmatrix} q \\ u \end{bmatrix} \quad z = \begin{bmatrix} \omega \\ \lambda \end{bmatrix} \tag{22}$$

The system (P1) now presents a form which is integrable by the Runge-Kutta methods. The existence and the uniqueness of the solution of the problem (P1) can be shown simply [10]. The following traditional results are found:

- If $\operatorname{rank}(\Phi^L)_{,q} = 5P$ and $5P \leq N$. We can ensure that the solution of the problem (21) is unique if the rank of the kinematic constraint matrix is maximum, and if the number of constraints is lower or equal to the number of parameters. Our mechanism is then isostatic or hypostatic.

- If $\operatorname{rank}(\Phi^L)_q < 5P$ or $5P > N$. We can show that the existence and the uniqueness of the acceleration vector ω, but only the existence of the Lagrange multipliers λ. We are then in a traditional case of a hyperstatic system.

When we deal with hyperstatic systems, we have recourse to the same artifices in order to ensure the method convergence and to solve the system (21):

- If $\operatorname{rank}(\Phi^L)_q < 5P$ and $5P \leq N$, when we solve the system (21) with a Gauss procedure, one at least of the pivots $(r_i)_{i=1..5P}$ is null. We force the value of the Lagrange multiplier associated to this (i)-th constraint to zero. So, in the course of the Gauss procedure, we set to the (i)-th line of the matrix and the result vector to zero, then we set the (i)-th matrix diagonal element to one.

- If $5P > N$, (N) independent constraints are selected in the course of the Gauss procedure. The other constraints are ignored and the associated Lagrange multipliers are set to zero using the preceding method.

In fact, in these two cases, we choose one set of Lagrange multipliers. This procedure is completely correct, but the user has to be informed that he must not trust the calculated efforts. Only the kinematic parameters (position, velocity, acceleration) of the mechanism are analysable.

With such a procedure, the convergence of the Runge-Kutta methods can be demonstrated [10] even for a hyperstatic system. Subsequently, we thus suppose that $\operatorname{rank}(\Phi^L)_{,q} = 5P$ and $5P \leq N$. We thus have all the necessary elements for the resolution of the problem (P1).

But before going further, it is advisable to make the following remark: when we solve the problem (P1), we check only the acceleration constraints. But because of the integration errors, we then observe a drift of the geometrical and velocity constraints. The following result can be shown easily:

Theorem [8]: *If we use a method of order (p) in order to solve the problem (P1), and if at time $t_o = 0$, the initial conditions check the geometrical and the velocity constraint equations, then at time t_n, we have:*

$$\left\|\Phi^L(q_n)\right\| \le h^P(At_n + Bt_n^2) \qquad \left\|(\Phi^L)_{,q} u_n\right\| \le h^P(Ct_n)$$

$$\text{avec} \quad h = \max_{i=o..n}(h_i)$$

In practice, this problem is very important, because it leads to aberrant results. We will see how to correct this phenomenon.

2.3.2 Stabilization of the Constraints

There are mainly two types of methods which make it possible to stabilize the geometrical and the velocity constraints:

Limitation of the drift methods

These methods just slow down the drift of the constraints, but we cannot guarantee that at the end of a certain number of steps the constraints are always checked with an acceptable precision. Among these methods, we will quote the Baumgarte method [11] and an original penalized method presented by Garcia de Jalon [6]. We can note that the parameters used in these methods are obtained in an empirical way and that the quality of the results is very sensitive to their adjustment. For certain applications, where the speed of treatment is an essential element (real time), these methods should be interesting. However, these methods must not be used if precise results are wanted.

Methods of correction of the drift

With these methods, we can check the geometrical and velocity constraints with a precision fixed by the user. Haug [5] proposes a method which consists in determining a set of independent parameters. This method is based on the analysis of the kinematic constraint matrix $(\Phi^L)_{,q}$ of dimension (5P,N). Applying a Gauss procedure to this matrix, we obtain a matrix of the form:

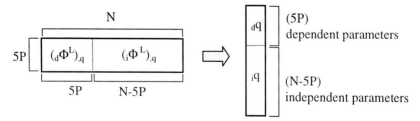

Fig.4. Decomposition of the kinematic constraint matrix

The analysis of this matrix, enables us to define a set of independent and dependent parameters. We are going to use this partition in the following way:

1. Statement.

Let us suppose that we solve the problem (P1) (21) on the interval $[t_n, t_{n+1}]$. We then know $_{n+1}$, and u_{n+1}. The problem is that q_{n+1}, and u_{n+1} do not check the geometrical and velocity constraint equations:

$$\Phi^L(q_{n+1}) \neq 0 \qquad (\Phi^L)_{,q} u_{n+1} \neq 0 \tag{23}$$

The method principle consists in correcting the dependent parameters, by leaving the independent parameters invariant.

2. Correction of the geometry.

We solve the non-linear problem of (5×P) equations with (5×P) unknowns:

$$\Phi^L(_d q_{n+1}) = 0 \tag{24}$$

For that, we use a Newton or quasi-Newton method which can be written in the following form:

$$\begin{cases} (\Phi^L)_{,q}(\Delta_d q) = -\Phi^L(_d \tilde{q},_i q) \\ _d\tilde{q} =_d q + \Delta_d q \end{cases} \tag{25}$$

We thus obtain a vector $\tilde{q}_{n+1}$ which verifies:

$$\left| \Phi^L(\tilde{q}_{n+1}) \right| \leq \varepsilon \tag{26}$$

Where ε is a parameter specified by the user.

3. Correction of the velocity.

In the same spirit, we correct the velocity u_{n+1} by solving the linear system:

$$(_d\Phi^L)_{,q}(_d u_{n+1}) = -(_i\Phi^L)_{,q}(_i u_{n+1}) \tag{27}$$

It is advisable to re-evaluate the matrix $(\Phi^L)_{,q}$ with the geometry $\tilde{q}_{n+1}$. A new velocity $\tilde{u}_{n+1}$ is thus obtained.

4. Calculation of the acceleration and of the Lagrange multipliers.

We then calculate the parameters ω_{n+1} and λ_{n+1} with these new geometry and velocity. one solves:

$$\begin{bmatrix} H(\tilde{q}_{n+1}) & (\Phi^L)_{,q}^T \\ (\Phi^L)_{,q} & 0 \end{bmatrix} \begin{bmatrix} \omega_{n+1} \\ \lambda_{n+1}^L \end{bmatrix} = \begin{bmatrix} U(\tilde{q}_{n+1}, \tilde{u}_{n+1}, t_{n+1}) - C(\tilde{q}_{n+1}, \tilde{u}_{n+1}, t_{n+1}) \\ -\eta^L \end{bmatrix} \tag{28}$$

This method enables us to ensure that the geometrical constraints are checked with a precision ε. It is often said that the parameter ε represents the assembly precision of the mechanism. This method is called the Global Coordinate Partitioning [5], and is used in the Dads software (LMS).

We can note, that in continuous motion, we determine a set of dependent and independent parameters only once, at the beginning of movement. This partition remains valid as long as we do not reach singular positions. These positions are located very simply while controlling, at each step of time, the pivot values used during the decomposition of the matrix $(\Phi^L)_{,q}$. If a pivot becomes very small, a new partition is carried out.

This method is applicable even in the case of hyperstatic systems. It enables us to determine the hyperstaticity degree DH of the mechanism by the following formula:

$$DH=(5P)-dependent\_parameter\_number \qquad (29)$$

It is a very powerful method which we have implemented in SDS (Solid Dynamics). As we said previously, our aim is to deal with discontinuous motions. So it is important to note that after each discontinuity, a new set of dependent and independent parameters will have to be evaluated.

Of course other stabilization methods exist, we can quote for example the projection methods [8]. These methods avoid the determination of dependent and independent parameters. However, we have to note that the equation system which is solved is more important than with the preceding method. For this reason, we have chosen to keep the Global Coordinate Partitioning approach, without however being able to affirm that a method is superior to another.

Note: One has the right to wonder whether these stabilization methods influence the method order. We will admit that they do not modify the order of the results.

3   Modelling of Frictional Contact

We are now going to see how to take into account frictional contacts. This section is composed of four principal paragraphs. We will be first interested in the definition of the unilateral constraints and specify the concept of contact law. We will see how to apply these principles through a «First Approach», voluntarily naive, which will enable us to highlight the difficulties of the method. In a second paragraph, we will analyze how this problem is treated in the literature and justify the choice of the method we have adopted. In a third paragraph, we will examine existence and uniqueness of the solution, and state certain results of convergence. In order to end this chapter, we will give a pseudo code of the method. We will then describe the algorithm modifications we have made in order to take into account bilateral joints and to model driven joints. We will consider two cases, particularly interesting, of indeterminacy and inconsistency, and see how our algorithm is able to detect them.

We consider here a system (Σ) made of (N) principal connections, (P) closed loop connections and (Q) potential frictional contacts.

3.1 Principle of the Method

We chose a model of contact which is based on the concept of unilateral constraints. We will specify here this concept and will see how to define the contact between two solids (S_i) and (S_j). We will propose then a first approach, very intuitive, which will enable us to emphasize the difficulties of the method.

3.1.1 Definition of a Unilateral Connection

Let (S_i) and (S_j) be two solids of the system (Σ). Suppose that (S_i) and (S_j) are in contact, we note (C) this contact. Let C_i and C_j be the points which move on the surface of (S_i) and (S_j) and who coincide, at the moment t_n of our analysis, with the point of contact (C). One notes $(\Re_i)=(C_i, x_i, y_i, z_i)$, $(\Re_j)= (C_j, x_j, y_j, z_j)$ two coordinate systems attached to the points C_i and C_j. Let (n_i) be the contact normal at the point C_i. It is supposed that $(z_i= n_i)$ and $(z_j= -n_i)$ and that the vectors $(t_i)=(x_i, y_i)$ define the tangent plan to the contact. The system can be represented by the following figure:

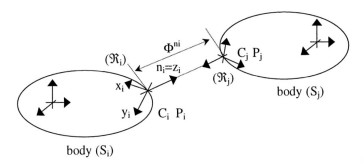

body (S_i)

Fig.5. Bodies in contact

Let us consider now two points P_i and P_j which have the same positions, at the moment tn of our analysis, than the points C_i and C_j. We will suppose that these points belong respectively to the solids (S_i) and (S_j).

Note : We do not discuss here the cases where the normal and the tangent plane to the contact cannot be defined. That can occur, for example, when two corners of the bodies (S_i) and (S_j) are in contact.

We define the distance between the solids (S_i) and (S_j) by:

$$\Phi^{n_i} = P_iP_j.n_i \tag{30}$$

We are going to associate to the contact (C) a set of three kinematic constraints. We thus define a normal constraint and two tangential constraints of the form:

$$\begin{cases} {}^{n_i}v = \left(\Phi^{n_i}\right)_q \dot{q} = 0 \\ {}^{t_i}v = \begin{bmatrix} {}^{x_i}v \\ {}^{y_i}v \end{bmatrix} = \begin{bmatrix} \left(\Phi^{x_i}\right)_q \dot{q} \\ \left(\Phi^{y_i}\right)_q \dot{q} \end{bmatrix} = \left(\Phi^{t_i}\right)_q \dot{q} = 0 \end{cases} \tag{31}$$

Where the vector $[{}^{x_i}v, {}^{y_i}v, {}^{n_i}v]$ represents the relative velocity of the point P_j with respect to P_i expressed in the coordinate system $(\mathfrak{R}_i)$.

It is important to note that in the case of a three-dimensional contact, we cannot define the tangential geometric constraints, because the non-holonomic kinematic constraints (31) can not be integrated. This is not very important, because only the normal geometric constraints are necessary to our analysis.

We are going to take into account the kinematics constraints (31) by means of Lagrange multipliers, exactly as we did in the paragraph (2.1.2.). For that reason, it is necessary to derive the equation (31), in order to obtain constraints in acceleration. Which gives us:

$$\begin{cases} {}^{n_i}a = \left(\Phi^{n_i}\right)_q \ddot{q} + \eta^{n_i} = 0 \\ {}^{t_i}a = \begin{bmatrix} {}^{x_i}a \\ {}^{y_i}a \end{bmatrix} = \begin{bmatrix} \left(\Phi^{x_i}\right)_q \ddot{q} + \eta^{x_i} \\ \left(\Phi^{y_i}\right)_q \ddot{q} + \eta^{y_i} \end{bmatrix} = \left(\Phi^{t_i}\right)_q \ddot{q} + \eta^{t_i} = 0 \end{cases} \tag{32}$$

Where the vector $[{}^{x_i}a, {}^{y_i}a, {}^{n_i}a]$ represents the derivative of the vector $[{}^{x_i}v, {}^{y_i}v, {}^{n_i}v]$.

But the vector $[{}^{x_i}a, {}^{y_i}a, {}^{n_i}a]$ does not represent the relative acceleration of the point P_j with respect to P_i. However, and by a language abuse, we will speak about normal and tangential relative accelerations.

We note $[\lambda^{t_i}, \lambda^{n_i}] = [\lambda^{x_i}, \lambda^{y_i}, \lambda^{n_i}]$ the Lagrange multipliers associated with the acceleration constraints (32). These parameters respectively represent the contact forces (normal and tangential) which exerts the body (S_i) on (S_j). These efforts are expressed in the coordinate system $(\mathfrak{R}_i)$.

Very schematically, when we have to deal with a problem with bilateral joints and a frictional contact, we must solve a problem (P1) (see paragraph 2.3.1.) of the form (in the case of sticking contacts):

$$\begin{bmatrix} H & -(\Phi^L)_{,q}^T & -(\Phi^{n_i})_{,q}^T & -(\Phi^{t_i})_{,q}^T \\ (\Phi^L)_{,q} & & & \\ (\Phi^{n_i})_{,q} & & 0 & \\ (\Phi^{t_i})_{,q} & & & \end{bmatrix} \begin{bmatrix} \omega \\ \lambda^L \\ \lambda^{n_i} \\ \lambda^{t_i} \end{bmatrix} = \begin{bmatrix} U - C \\ -\eta^L \\ -\eta^{n_i} \\ -\eta^{t_i} \end{bmatrix} \tag{33}$$

In fact, we have just added to the differential-algebraic system (21), defined for a system with closed loop joints, the acceleration constraint equations (32). But unlike the bilateral constraints, the unilateral constraints are not permanent. Thus, the principle of the method consists in adding or suppressing the constraints of our equations system according to certain principles. These principles are dictated by a contact law which we now will define.

3.1.2 Contact Law

We will make here the distinction between normal contact law and tangential contact law. It is just a convenience of presentation, but we should not lose sight of the fact that these two laws are dependent.

Normal Contact Law

It is first of all necessary to define a contact normal law, for that we adopt the Signorini conditions which are stated in the following way:

Signorini conditions *(see for example [12]):*
Let (S_i) and (S_j) be two bodies which may be in contact, the three following conditions must be checked:

- *Impenetrability:*
$$\Phi^{n_i} \geq 0$$
- *In the case of contact, the bodies do not present adhesion (no attraction):*
$$\Phi^{n_i} = 0 \quad \Rightarrow \quad \lambda^{n_i} \geq 0$$
- *If the contact is not active, the normal contact force is null:*
$$\Phi^{n_i} > 0 \quad \Rightarrow \quad \lambda^{n_i} = 0$$

These three relations are equivalent to the complementarity conditions:
$$\Phi^{n_i} \geq 0 \quad \lambda^{n_i} \geq 0 \quad \Phi^{n_i}\lambda^{n_i} = 0$$

This law is expressed as a function of the body positions. In order to be able to decide if a constraint of the problem (P1) must be added or released, it is necessary to express this law in terms of accelerations. We will then admit the two following proposals.
Let I_n be the set of the active contacts defined by:

$$I_n = \{ \quad i \in (1,2,..,Q) \quad / \quad \Phi^{n_i} = 0 \quad \}$$

Using this definition, it can be shown [13] that the Signorini conditions imply:

$$\textbf{Proposal:} \begin{cases} \text{if} \quad i \in I_n \quad {}^{n_i}v \geq 0 \quad \lambda^{n_i} \geq 0 \quad ({}^{n_i}v)(\lambda^{n_i}) = 0 \\ \text{else} \quad \lambda^{n_i} = 0 \end{cases}$$

Moreover, if we consider the I_{nn} set, define as follows:

$$I_{nn} = \{ \quad i \in I_n \quad / \quad {}^{n_i}v = 0 \quad \},$$

the preceding conditions imply:

$$\textbf{Proposal:} \begin{cases} \text{if} \quad i \in I_{nn} \quad {}^{n_i}a \geq 0 \quad \lambda^{n_i} \geq 0 \quad ({}^{n_i}a)(\lambda^{n_i}) = 0 \\ \text{else} \quad \lambda^{n_i} = 0 \end{cases}$$

A more compact formulation of these complementarity conditions can be obtained using the tools of convex analysis. We return the reader to [14] and [13] for more details on these formulations.

The principles which we have just stated constitute a "complete" law of contact (see [15] for example), when there is no friction. We will now describe the tangential contact law.

Tangential Contact Law

We will adopt here the Coulomb law. In order to simplify this presentation, we do not consider rolling friction, nor swivelling friction. Moreover, we consider an isotropic friction and we do not make distinction between coefficients of static and dynamic friction.
Let I_s and I_r be the two sets:

$$\begin{cases} I_r = \{ i \in I_{nn} / \left\| {}^{t_i}v \right\| = 0 \} & \equiv \quad \text{sticking} \\ I_s = \{ i \in I_{nn} / \left\| {}^{t_i}v \right\| \neq 0 \} & \equiv \quad \text{sliding} \end{cases}$$

According to the hypotheses and the preceding notations, we state the following principle:

Definition (see for example [16]): *The Coulomb law is given by the relations:*

$$
\begin{cases}
\text{if} \quad i \in I_r \quad \left\| \lambda^{t_i} \right\| = \sqrt{(\lambda^{x_i})^2 + (\lambda^{y_i})^2} \leq \mu \lambda^{n_i} \\[2em]
\text{if} \quad i \in I_s \quad \lambda^{t_i} = \begin{bmatrix} \lambda^{x_i} \\ \lambda^{y_i} \end{bmatrix} = \begin{bmatrix} -\dfrac{x_i v}{\left\| t_i v \right\|} \mu \lambda^{n_i} \\[1.5em] -\dfrac{y_i v}{\left\| t_i v \right\|} \mu \lambda^{n_i} \end{bmatrix} = -\dfrac{t_i v}{\left\| t_i v \right\|} \mu \lambda^{n_i} = \tilde{\mu} \lambda^{n_i}
\end{cases}
\tag{34}
$$

We still seek for expressing a law in acceleration. Contrary to the Signorini conditions, it is clear that the Coulomb law cannot be formulated directly in acceleration. Indeed, in the sliding phases, relative tangential velocity and acceleration do not have necessarily the same direction. We thus need to make the distinction between sliding and sticking contacts.

Sliding
The tangential contacts forces are perfectly defined as a function of the tangential relative velocities and of the normal contact force. No particular test on accelerations is necessary. The result is thus direct.

Sticking
The things are appreciably more complex. We must distinguish two cases:

$$
i \in I_r \begin{cases} \left\| {}^{t_i} a \right\| = 0 & \equiv \quad \text{sticking continues} \\[1em] \left\| {}^{t_i} a \right\| \neq 0 & \equiv \quad \text{stick} - \text{slip transition} \end{cases}
$$

We will suppose that the law of Coulomb can be written in the form:

Proposal [16]:

$$
\text{if} \quad i \in I_r \quad \begin{cases} \left\| \lambda^{t_i} \right\|^2 \leq \mu^2 \left(\lambda^{n_i} \right)^2 \\[1em] ({}^{t_i} a)^T (\lambda^{t_i}) \leq 0 \\[1em] \left(\left\| \lambda^{t_i} \right\|^2 - \mu^2 \left(\lambda^{n_i} \right)^2 \right) \left(\left\| {}^{t_i} a \right\|^2 \right) = 0 \end{cases}
\tag{35}
$$

In fact, if we consider the two preceding cases:

- If the contact preserves a status of sticking: ${}^{t_i} a = [{}^{x_i} a, {}^{y_i} a]^T = 0$ and the preceding conditions are verified.

- If there is a transition to sliding: $\left\|\lambda^{t_i}\right\|^2 = (\lambda^{x_i})^2 + (\lambda^{y_i})^2 = \mu^2(\lambda^{n_i})^2$ and the third equality is automatically checked. The problem comes here from the second inequality. The Coulomb law imposes that the friction force is directly opposed to the sliding velocity, but nothing is specified about acceleration. Thus we just can say that the friction force and tangential relative acceleration must have opposite signs (One should not lose sight of the fact that here tangential relative velocity is null). In the case of 2D frictional contacts, this indeterminacy vanishes. For 3D frictional contacts, there are two manners to treat this problem:
  ⇒ The first consists in completely prolonging the Coulomb law in acceleration, while advancing that the friction force and the relative tangential acceleration must be directly opposite [16].
  ⇒ The second consists in directing the effort of friction in the direction of the last sticking force [17][18].
  The first approach seems more coherent to us, in the sense that, just after the transition, relative tangential velocity and acceleration will have the same direction. We thus adopt the point of view of [16]. In fact, we extend the law Coulomb in acceleration if the tangential relative velocity is null.

The Coulomb law is thus formulated in acceleration.

Let us see now how we will use these contact laws to control the constraints addition/deletion of the differential-algebraic system (33).

3.1.3 «A First Approach»

At time t_{n-1}, we suppose known the positions, velocities and accelerations of our mechanism, as well as the contact status. We understand by the term status, the fact of knowing if a contact is active or passive, and if its state is sliding or sticking. Let us admit that all the parameters of the system (Σ) are compatible with the law of contact at time t_{n-1}. Our aim is to determine new parameters at time t_n. For that, we will assume that the contact statuses do not vary until time t_n. The unilateral constraints, associated with each contact, can then be regarded as locally bilateral constraints. We thus integrate, on the interval $[t_{n-1}, t_n]$, a system of differential-algebraic equations, whose form will be specified later. Once this calculation is carried out, it is necessary to check if the results obtained are compatible with the contact law. If it is not the case, we seek the first time $t_n{}^* \in [t_{n-1}, t_n]$, where the contact law is violated. This operation is carried out using the interpolation methods described in the paragraph (2.2.3.). At this time, it is necessary to determine new initial conditions. For that, we need a decision procedure which is able to find new contact statuses.

The object of this paragraph is to describe an excessively simple decision procedure, which we call «First Approach ». As we will see, this approach led to erroneous conclusions. However, it has seemed important to us to present this method, because its presentation will allow the not-informed reader to become familiar with certain important concepts: status tests, structure of the differential-algebraic systems, events finding...

We will first consider a problem in which we have only one contact (C) and no friction. We will consider all the system states, and see how the structure of our system of equations is modified as a function of the contact laws. This paragraph will enable us to describe the research of the various events which punctuate the simulation (lift-off, shock...). We will then consider the same procedure by adding friction and see to what point this approach can be prolonged to several contacts.

We place ourselves at the time t_n of simulation, the unknown factors of the problem are the accelerations and the Lagrange multipliers (which represent the efforts in the closed loop joints and the contact forces).

A Contact without Friction

Only the normal contact law is to be considered. Two cases can occur:

1. $\Phi^{n_i} \leq 0$

If $\Phi^{n_i} = 0$ and $(^{n_i}v) = 0$,
the contact (C) is active, we then solve a problem (P1) which can be written in the form:

$$
\begin{bmatrix}
H(q) & -(\Phi^L)^T_{,q} & -(\Phi^{n_i})^T_{,q} \\
(\Phi^L)_{,q} & & \\
(\Phi^{n_i})_{,q} & & 0
\end{bmatrix}
\begin{bmatrix}
\omega \\
\lambda^L \\
\lambda^{n_i}
\end{bmatrix}
=
\begin{bmatrix}
U - C \\
-\eta^L \\
-\eta^{n_i}
\end{bmatrix}
\tag{36}
$$

Following that, if the value of the normal contact force λ^{n_i} is positive, we will say that the contact (C) remains active. We preserve then the structure of the system (36) to continue the resolution until time t_{n+1}. On the other hand if λ^{n_i} is negative, the contact (C) is broken. Using the interpolation procedures, described in the paragraph (2.2.3.), we will seek the moment $t_n{}^*$ for which this value cancels. A this moment, we will remove the normal constraint, the problem (P1) will then be written in the form:

$$
\begin{bmatrix}
H(q) & -(\Phi^L)^T_{,q} \\
(\Phi^L)_{,q} & 0
\end{bmatrix}
\begin{bmatrix}
\omega \\
\lambda^L
\end{bmatrix}
=
\begin{bmatrix}
U - C \\
-\eta^L
\end{bmatrix}
\tag{37}
$$

The contact (C) cease and solids (S_i) and (S_j) separate. We thus continue integration with a system of the form (37).

If $\Phi^{n_i} \leq 0$ and $(^{n_i}v) < 0$,
There is an impact between bodies (S_i) and (S_j). The problem cannot be dealt with any more in acceleration. Our first work consists in evaluating the time $t_n{}^*$ for which $\Phi^{n_i} = 0$. At this moment, we determine the velocity jump:

$$\Delta^{n_i} v = (^{n_i} v)^+ - (^{n_i} v)^- \tag{38}$$

we return its calculation to the Chapter 4. We are only interested here in the value $(^{n_i} v)^+$:

> If $(^{n_i} v)^+ = 0$,

the contact (C) is maintained and we continue integration with a system of the form (36).

> If $(^{n_i} v)^+ > 0$,

the contact (C) is broken, we continue the resolution with a system of the form (37).

2. $\Phi^{n_i} > 0$

The contact (C) is not active, the problem (P1) is then identical to problem (37).

Let us now complicate the problem by adding friction.

A Contact with Friction

We consider again the various preceding stages.

1. $\Phi^{n_i} \le 0$

> If $\Phi^{n_i} = 0$ and $(^{n_i} v) = 0$,

the contact (C) is active. The Coulomb law in acceleration requires that we make the distinction between sliding and sticking. We consider the two cases:

> If $\left\| ^{t_i} v \right\| \ge \varepsilon$,

where ε is a parameter specified by the user (generally we set its value to the absolute integration precision Atol).
In the case of sliding, the problem (P1) can be written in the form:

$$
\begin{bmatrix}
H & -(\Phi^L)_{,q}^T & -(\Phi^{n_i})_{,q}^T & -(\Phi^{t_i})_{,q}^T \\
(\Phi^L)_{,q} & 0 & & \\
(\Phi^{n_i})_{,q} & & 0 & \\
0 & & -\tilde{\mu} & I
\end{bmatrix}
\begin{bmatrix}
\omega \\
\lambda^L \\
\lambda^{n_i} \\
\lambda^{t_i}
\end{bmatrix}
=
\begin{bmatrix}
U(q,u,t) - C(q,u,t) \\
-\eta^L \\
-\eta^{n_i} \\
0
\end{bmatrix}
\tag{39}
$$

$$\text{Where } \tilde{\mu} = -\frac{^{t_i} v}{\left\| ^{t_i} v \right\|} \mu .$$

Once this linear system is solved, we control the sign of λ^{n_i} and we the decide if the contact (C) must be maintained or released. The detection of a possible slip-stick transition is done before the resolution of this system. Consequently, one tests the sign of the scalar product between tangential relative velocity at times t_{n-1} and t_n. If this sign is positive, sliding continues. If it is negative, we seek the time t_n* where the scalar product vanishes, we then consider that we have a possible slip-stick transition. When we have to deal with 3D contact problems, we add to this test the additional condition:

$$\left\| {}^{t_i} v(t_n *) \right\| < \varepsilon \qquad (40)$$

In the case of a possible slip-stick transition, we move to the next step.

If $\left\| {}^{t_i} v \right\| < \varepsilon$,

the problem (P1) is written:

$$
\begin{bmatrix}
H(q) & -(\Phi^L)^T_{,q} & -(\Phi^{n_i})^T_{,q} & -(\Phi^{t_i})^T_{,q} \\
(\Phi^L)_{,q} & & & \\
(\Phi^{n_i})_{,q} & & 0 & \\
(\Phi^{t_i})_{,q} & & &
\end{bmatrix}
\begin{bmatrix}
\omega \\
\lambda^L \\
\lambda^{n_i} \\
\lambda^{t_i}
\end{bmatrix}
=
\begin{bmatrix}
U(q,u,t) - C(q,u,t) \\
-\eta^L \\
-\eta^{n_i} \\
-\eta^{t_i}
\end{bmatrix}
\qquad (41)
$$

Once this linear system is solved, we control, once more, the sign of λ^{n_i} and we decide whether the contact (C) must be maintained or released. If this contact is maintained, it is advisable to test the value of $\left\| \lambda^{t_i} \right\|^2$.

If $\left\| \lambda^{t_i} \right\|^2 \leq \mu^2 \left(\lambda^{n_i} \right)^2$,

Sticking continues. We keep on working with a system of the form (41).

If $\left\| \lambda^{t_i} \right\|^2 > \mu^2 \left(\lambda^{n_i} \right)^2$,

we seek the moment t_n* where these two quantities are equal, and we then solve a system of the form (39) with:

$$\tilde{\mu} = -\frac{{}^{t_i} a}{\left\| {}^{t_i} a \right\|} \mu \qquad (42)$$

We have a stick-slip transition.

If $\Phi^{n_i} \leq 0$ and $(^{n_i}v) < 0$,

the course of operations is exactly the same. We consider again the following phase of tests:

If $(^{n_i}v)^+ = 0$,

the contact (C) is maintained.

If $\left\| (^{t_i}v)^+ \right\| = 0$,

we continue integration with a system of the form (41).

If $\left\| (^{t_i}v)^+ \right\| > 0$,

we continue integration with a system of the form (39).

If $(^{n_i}v)^+ > 0$,

the contact (C) is broken, we continue the resolution with a system of the form (38).

2. $\Phi^{n_i} > 0$

The contact (C) is not active, the problem (P1) is then identical to the proble (37).

As a conclusion to this "First approach", we have to deal here with four types of transition: **lift-off , slip-stick, stick-slip, shock.**

All these problems can be treated with an acceleration formulation except the shocks, we return their study to the Chapter 4. In this chapter, we will consider only the first three types of transition. In this context, our problem can be written as follows: At time t $_n$, we suppose known the positions and the velocities of our mechanism and we seek the accelerations, the Lagrange multipliers and the status of the contact points.

In this paragraph, we have considered all the possible evolutions of our system for a single frictional contact. We now will try to extend our «First Approach » to the case of multiple contacts.

Generalization

We will show in this paragraph, that the approach that we have just introduced cannot be generalized to the case of multiple dependent contacts. Indeed, we will present a very simple example, without friction, for which the preceding method fails.

We propose here an example [19] which is, according to us, even clearer than the traditional Delassus example [20]. We consider two identical blocks, (S_1) and (S_2), of mass m, the position of which is pictured in the figure below. We suppose that the contacts are point-to-point contacts, and we apply forces F_1 and F_2 on each of these blocks. We note λ^{n1} and λ^{n2}, the normal contact forces (ground-body1) and (body1-body2) respectively. The system can evolve towards one of the four following states:

- Case n°1: *The system stays in equilibrium.*
- Case n°2: *(S_1) remains in contact with the ground and (S_2) takes off.*

- Case n°3: *(S₁) and (S₂) take off but remain in contact.*
- Case n°4: *(S₁) and (S₂) take off separately.*

Let us suppose that (m=1kg), (F₁=12N), (F₂=11N), and we fix the gravity along a downward vertical axis to (g=10m/s²).

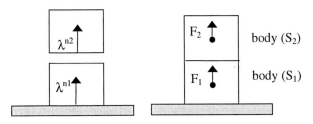

Fig.6. Two blocks example

Let us suppose that at the time t_n, the two contacts are active. We must then solve a system of the form (36), which gives us:

$$\lambda^{n1} = -3 \qquad \lambda^{n2} = -1$$

In accordance with what was stated in the preceding paragraph, we conclude that the two contacts must be released. However it is obvious that the bodies (S_1) and (S_2) will remain in contact because $F_1 > F_2 > mg$.

It is clear that the preceding strategy is not sufficient and leads, even in an excessively simple case, to erroneous results. The problem is that, when the contacts are dependant, a contact status modification has effects on the other contact status. Thus, the contacts which correspond to negative normal contact forces must not necessarily be released.

How to solve this problem? A first approach consists in applying the preceding method until the contact laws are satisfied. If we take again the example of [19], we then have to consider the cases (1,2,3,4) successively, until we obtain a valid solution. We have to deal with a combinative exercise which can become very heavy when the number of contacts becomes large (2^N possibilities for (N) active contacts). The things quickly become inextricable when, in addition, friction is considered. It has been the only method available for a long time. Indeed we find in [20] the following sentence:

«*Practically, a problem with unilateral connections is treated by successive tests of the various hypotheses, tests which finish when a satisfactory solution was obtained.* »

This method is still used for example by [18-21-22], [23] or more recently [24]. But, we have today at our disposal more effective tools.

Moreau [25][26] was the first to give to the Delassus objections a comprehensible mathematical form. His approach has consisted in formulating the contact problem, without friction, in the form of a Quadratic Programming problem (QP), subjected to constraint inequalities. This approach has been extended to the problem of 2D frictional contact by [27][28], then generalized in 3D by a great number of authors.

Following the Moreau idea, we will say that we have to formulate a problem of Optimization under Constraints in order to treat frictional contact problem. Furthermore, we will see that in certain cases the frictional contact problem can be written in a well defined mathematical form (LCP,QP,NCP,MCP). Let us now detail this approach, it is the object of the following paragraph.

3.2 Formulation and Resolution of a Problem of Optimization under Constraints

First, we will see how to write our problem in the form of a problem of Optimization under Constraints. This will enable us to stress certain modelling difficulties. We will examine how certain authors formulate this problem in the literature, we will underline their hypothesis, and we will mention the resolution methods they use. Lastly, we will present the approach we have adopted.

3.2.1 Formulation of a Problem of Optimization under Constraints

We consider the system (Σ) with (N) principal joints, (P) closed loop joints and (Q) contacts $(C_i)_{i=1..Q}$ active at time t_n. We suppose moreover that these (Q) contacts $(C_i)_{i=1..Q}$ verify:

$$i \in I_{nn} = \{ j \in I_n \ / \ ^{n_j}v = 0 \}$$

Where $I_n = \{ j \in (1,...,Q) \ / \ \Phi^{n_j} = 0 \}$

This hypothesis is fundamental, because it enables us to write the contact laws in acceleration.

We note a_n and a_t the normal and tangential relative accelerations defined by:

$$\begin{cases} a_n = [^{n_1}a,..., ^{n_Q}a]^T \\ a_t = [^{t_1}a,..., ^{t_Q}a]^T = [^{x_1}a,..., ^{x_Q}a, ^{y_1}a,..., ^{y_Q}a]^T \end{cases} \tag{43}$$

In the same manner, we define the contact forces:

$$\begin{cases} \lambda^n = [\lambda^{n_1},...,\lambda^{n_Q}]^T \\ \lambda^t = [\lambda^{t_1},...,\lambda^{t_Q}]^T = [\lambda^{x_1},...,\lambda^{x_Q},\lambda^{y_1},...,\lambda^{y_Q}]^T \end{cases} \tag{44}$$

And matrices of kinematics constraints:

$$\begin{cases} (\Phi^n)_{,q} = [(\Phi^{n_1})_{,q},...,(\Phi^{n_Q})_{,q}]^T \\ (\Phi^t)_{,q} = [(\Phi^{t_1})_{,q},...,(\Phi^{t_Q})_{,q}]^T = [(\Phi^{x_1})_{,q},...,(\Phi^{x_Q})_{,q},(\Phi^{y_1})_{,q},...,(\Phi^{y_Q})_{,q}]^T \end{cases} \quad (45)$$

For the clearness of our presentation, we will first consider the contact without friction, then the contact with friction. Moreover, we suppose that we do not have any bilateral constraint (P=0), we will see, in the paragraph (3.4.2.), how to very simply take them into account.

Contact without Friction

Since we have (Q) active contacts at the time t_n, the dynamics equations can be written in the following form:

$$H\ddot{q} - U + C - (\Phi^n)^T_{,q} \lambda^n = 0 \quad (46)$$

These (N) equations are subjected to (Q) the complementarity conditions, imposed b the contact law:

$$a_n \geq 0 \quad \lambda^n \geq 0 \quad (a_n)^T (\lambda^n) = 0 \quad (47)$$

With $\quad a_n = (\Phi^n)_{,q} \ddot{q} + \eta^n$

In order to formulate an optimization problem under constraints, we are going to express the normal relative acceleration a_n as a function of the normal contact forces λ_N. We therefore multiply the equation (46) by H^{-1}, which gives us:

$$\ddot{q} = H^{-1}(U - C) + H^{-1}(\Phi^n)^T_{,q} \lambda^n \quad (48)$$

This operation is completely legitimate because H is definite positive, therefore invertible. If we defer this expression in the expression of a_n, we obtain:

$$a_n = \left((\Phi^n)_{,q} H^{-1}(\Phi^n)^T_{,q}\right) \lambda^n + (\Phi^n)_{,q} H^{-1}(U - C) + \eta^n = A\lambda^n + b \quad (49)$$

By convention, matrix A is called the Delassus operator [3].
Once this operation is carried out, our problem can be written in the three different forms. The equation (49) enables us to formulate a Linear Complementarity problem. We seek λ_N such as:

$$(A\lambda^n + b)^T (\lambda^n) = 0 \quad A\lambda^n + b \geq 0 \quad \lambda^n \geq 0 \quad (LCP) \quad (50)$$

Another manner of posing the problem consists in formulating a Quadratic Programming problem. For that, we seek λ_N checking:

$$\min_{\lambda^n}\left((\lambda^n)^T A\lambda^n + (\lambda^n)^T b\right) \quad A\lambda^n + b \geq 0 \quad \lambda^n \geq 0 \qquad \text{(QP1)} \qquad (51)$$

What we can also write:

$$\min_{\lambda^n}\left(\frac{1}{2}(\lambda^n)^T A\lambda^n + (\lambda^n)^T b\right) \quad \lambda^n \geq 0 \qquad \text{(QP2)} \qquad (52)$$

The optimality conditions of Karush, Kuhn and Tucker (KKT) ensure equivalence between the problems (QP1) and (QP2) [29].

It is very important to make the difference between (LCP) and (QP) because these two formulations lead to different methods of resolution.

Let us see now see how to write the problem when friction is introduced.

Contact with Friction

Once more, we will make the distinction between sliding and sticking. In these two cases, the dynamics equations are put in the form:

$$H\ddot{q} - U + C - (\Phi^n)^T_{,q}\lambda^n - (\Phi^t)^T_{,q}\lambda^t = 0 \qquad (53)$$

Sliding

In the sliding case, we have seen that the tangential contact force can be expressed as a function of the normal contact force. We have the relation:

$$\lambda^t = \tilde{\mu}\lambda^n \qquad (54)$$

The equations (53) and (54) enable us to write:

$$H\ddot{q} - U + C - \left((\Phi^n)^T_{,q} + \tilde{\mu}(\Phi^t)^T_{,q}\right)\lambda^n = 0 \qquad (55)$$

We thus have eliminated the tangential contact force and we then treat only the normal contact problem. Using the preceding procedure, we can write a Linear Complementarity problem of the form:

$$(A_g\lambda^n + b)^T(\lambda^n) = 0 \quad A_g\lambda^n + b \geq 0 \quad \lambda^n \geq 0 \qquad \text{(LCP}_g) \qquad (56)$$

Where $A_g = (\Phi^n)_{,q}H^{-1}\left((\Phi^n)^T_{,q} + \tilde{\mu}(\Phi^t)^T_{,q}\right) = A + (\Phi^n)_{,q}H^{-1}\left(\tilde{\mu}(\Phi^t)^T_{,q}\right)$ is the modified Delassus operator.

In the same manner, we can write two Quadratic Programming problems:

$$\min_{\lambda^n}\left((\lambda^n)^T A_g \lambda^n + (\lambda^n)^T b\right) \quad A_g \lambda^n + b \geq 0 \quad \lambda^n \geq 0 \qquad (QP1_g) \qquad (57)$$

$$\min_{\lambda^n}\left(\frac{1}{2}(\lambda^n)^T A_g \lambda^n + (\lambda^n)^T b\right) \quad \lambda^n \geq 0 \qquad (QP2_g) \qquad (58)$$

Sticking

We are going to express normal and tangential relative accelerations as a function of contact forces. So, we proceed in the same way than previously and we thus obtain:

$$
\begin{bmatrix} a_n \\ a_t \end{bmatrix} = \begin{bmatrix} A & (\Phi^n)_{,q} H^{-1}(\Phi^t)^T_{,q} \\ (\Phi^t)_{,q} H^{-1}(\Phi^n)^T_{,q} & (\Phi^t)_{,q} H^{-1}(\Phi^t)^T_{,q} \end{bmatrix} \begin{bmatrix} \lambda^n \\ \lambda^t \end{bmatrix} + \begin{bmatrix} b \\ (\Phi^t)_{,q} H^{-1}(U-C)+\eta^t \end{bmatrix}
$$

$$= A_r \begin{bmatrix} \lambda^n \\ \lambda^t \end{bmatrix} + b_r$$

$$(59)$$

This relation is accompanied by the following complementarity conditions:

$$
\left\{
\begin{array}{l}
a_n \geq 0 \quad \lambda^n \geq 0 \quad (a_n)^T(\lambda^n) = 0 \\[4pt]
\left\|\lambda^{t_i}\right\|^2 \leq \mu^2 \left(\lambda^{n_i}\right)^2 \quad (\lambda^{t_i})^T (^{t_i} a) \leq 0 \\[4pt]
\left(\left\|\lambda^{t_i}\right\|^2 - \mu^2 \left(\lambda^{n_i}\right)^2\right)\left(\left\|^{t_i} a\right\|^2\right) = 0 \quad \left((\lambda^{t_i}) \wedge (^{t_i} a)\right) = 0
\end{array}
\right\} \qquad \text{for} \quad i \in I_r
\qquad (60)
$$

Note: In accordance with our choice of the paragraph (3.1.2.), we have added the equation $\left((\lambda^{t_i}) \wedge (^{t_i} a)\right) = 0$ to the Coulomb law.

These constraints relations are non-linear, even when we have to deal with a 2D frictional contact problem. We cannot thus formulate a Linear Complementarit problem in the case of sticking. In this precise case, we cannot either write a quadratic functional minimization (we cannot formulate a quadratic functional). It is thus impossible to deal with the problem in the form of Quadratic Programming problem.

We will say, very generally, that we have to solve two coupled optimization problems under constraints, without more precise details.

Let's see know how this problem is treated in the literature, by examining onl the formulation of the problem in the case of sticking, since sliding does not pose major problems.

3.2.2 Bibliographical Analysis

Our aim here is to present different approaches of the frictional contact problem which were developed by a certain number of researchers. We will outline their approach while trying to emphasize the forces and the weaknesses of their formulation. We do not respect here a chronological order, but a logical order of increasing difficulty. We will finish this paragraph by drawing up an assessment of these various approaches. Thus we will justify the choice of the method we adopted.

Different formulations of the sticking proble

Pfeiffer, GLocker [19][13]
The problem is formulated in acceleration and is limited to the study of 2D friction. However, we quote here these authors work, because it describes very well the necessary efforts to write the sticking problem in the form of a Linear Complementarity problem. Pfeiffer and Glocker use an artifice of calculation which, as we will see it, is rather heavy.
For 2D sticking, the tangential contact law can be put in the form:

$$\left|\lambda^{t_i}\right| \le \mu\lambda^{n_i} \qquad (\lambda^{t_i})(^{t_i}a) \le 0 \qquad \left(\left|\lambda^{t_i}\right| - \mu\lambda^{n_i}\right)(^{t_i}a) = 0 \qquad \text{for} \quad i \in I_r \quad (61)$$

The absolute value of the friction force is highly disturbing. The uneasy question is how to modify this relation in order to obtain a complementarity condition exploitable by a LCP. The answer is traditional in optimization, it is necessary to introduce new variables, that we call "signed variables". In our case, the authors set:

$$\begin{cases} \lambda^{t_i} = \lambda_+^{t_i} - \lambda_-^{t_i} \\ {}^{t_i}a = {}^{t_i}z^+ - {}^{t_i}a^- = {}^{t_i}a^+ - {}^{t_i}z^- \end{cases} \quad \text{with} \quad \begin{cases} {}^0\lambda_+^{t_i} = \mu\lambda^{n_i} - \lambda_-^{t_i} \\ {}^0\lambda_-^{t_i} = \mu\lambda^{n_i} - \lambda_+^{t_i} \end{cases} \quad \text{for} \quad i \in I_r \quad (62)$$

With:

$$\begin{cases} (\lambda_+^{t_i}, \lambda_-^{t_i}, {}^0\lambda_+^{t_i}, {}^0\lambda_-^{t_i}) \ge 0 \quad (\lambda_+^{t_i})(\lambda_-^{t_i}) = 0 \\ (^{t_i}a^+, {}^{t_i}a^-, {}^{t_i}z^+, {}^{t_i}z^-) \ge 0 \quad (^{t_i}a^+{}^{t_i})(z^-) = 0 \quad (^{t_i}z^+)(^{t_i}a^-) = 0 \end{cases} \quad \text{for} \quad i \in I_r$$

The Coulomb graph is thus broken up into four, as shown below:

93

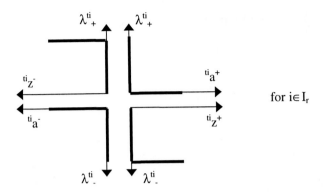

for $i \in I_r$

Fig.7. Decomposition of the Coulomb law in 2D

Once this decomposition is carried out, a LCP is formulated without any difficulty as if we have four corner laws (like Signorini law). We thus obtain a problem of the form:

$$
\begin{bmatrix} \lambda^{n_i} \\ \lambda^{t_i}_+ \\ \lambda^{t_i}_- \\ {}^0\lambda^{t_i}_+ \\ {}^0\lambda^{t_i}_- \end{bmatrix} \geq 0 \quad
\begin{bmatrix} {}^{n_i}a \\ {}^{t_i}a^+ \\ {}^{t_i}a^- \\ {}^{t_i}z^+ \\ {}^{t_i}z^- \end{bmatrix} \geq 0 \quad
\begin{bmatrix} \lambda^{n_i} \\ \lambda^{t_i}_+ \\ \lambda^{t_i}_- \\ {}^0\lambda^{t_i}_+ \\ {}^0\lambda^{t_i}_- \end{bmatrix}^T
\begin{bmatrix} {}^{n_i}a \\ {}^{t_i}a^+ \\ {}^{t_i}a^- \\ {}^{t_i}z^+ \\ {}^{t_i}z^- \end{bmatrix} = 0 \quad \text{for} \quad i \in I_r \quad (63)
$$

We measure here all the heaviness of this method which obliges us to introduce eight variables and four complementarity conditions, for each 2D sticking contacts.
This method thus makes it possible to formulate sliding and sticking in the form of a LCP. The authors solve this problem by using the Lemke algorithm, which is an alternative of the Simplex method. We will not give details of this algorithm, for more information the reader should consult for example the work of Ciarlet [29].
    Let us see now what happens in the case of 3D sticking contacts.

Trinkle, Pang [16][30][31]
Here again, the problem is formulated in acceleration. In 3D, we have to deal with the following sticking constraints:

$$\begin{cases} a_n \geq 0 \quad \lambda^n \geq 0 \quad (a_n)^T(\lambda^n) = 0 \\ \left\| \lambda^{t_i} \right\|^2 \leq \mu^2 \big(\lambda^{n_i}\big)^2 \quad (\lambda^{t_i})^T({}^{t_i}a) \leq 0 \\ \Big(\left\| \lambda^{t_i}\right\|^2 - \mu^2\big(\lambda^{n_i}\big)^2\Big)\Big(\left\|{}^{t_i}a\right\|^2\Big) = 0 \quad \big((\lambda^{t_i}) \wedge ({}^{t_i}a)\big) = 0 \end{cases} \quad \text{for} \quad i \in I_r \tag{64}$$

These authors also seek to write a Linear Complementarity problem. In order to simplify the sticking conditions, these authors put the problem in the form:

$$\begin{cases} \left| \lambda^{x_i} \right| \leq \mu\lambda^{n_i} \quad (\lambda^{x_i})({}^{x_i}a) \leq 0 \quad \big(\left|\lambda^{x_i}\right| - \mu\lambda^{n_i}\big)({}^{x_i}a) = 0 \\ \left| \lambda^{y_i} \right| \leq \mu\lambda^{n_i} \quad (\lambda^{y_i})({}^{y_i}a) \leq 0 \quad \big(\left|\lambda^{y_i}\right| - \mu\lambda^{n_i}\big)({}^{y_i}a) = 0 \end{cases} \quad \text{for} \quad i \in I_r \tag{65}$$

Which is equivalent to transform 3D frictional problem into two problems of 2D friction. We can see that differently by noticing that quadratic Coulomb cone is replaced by a polyhedral cone. Indeed, this operation can be pictured as follows:

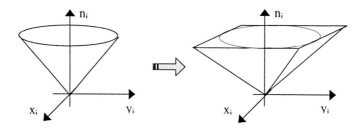

Fig.8. Decomposition of the Coulomb law in 3D

We can notice that the traditional Coulomb cone is completely included in the pyramid. Trinkle and Pang then use the same type of decomposition as Pfeiffer and Glocker to put the problem (65) in a LCP form. This operation is carried out in a slightly more elegant way, and led to the introduction of eight variables in three dimensions (the same number as Pfeiffer in two dimensions). These authors set:

$$\begin{cases} s_+^{x_i} = \mu\lambda^{n_i} + \lambda^{x_i} \quad s_-^{x_i} = \mu\lambda^{n_i} - \lambda^{x_i} \\ s_+^{y_i} = \mu\lambda^{n_i} + \lambda^{y_i} \quad s_-^{y_i} = \mu\lambda^{n_i} - \lambda^{y_i} \\ {}^{x_i}a = {}^{x_i}a^+ - {}^{x_i}a^- \\ {}^{y_i}a = {}^{y_i}a^+ - {}^{y_i}a^- \end{cases} \quad \text{for} \quad i \in I_r \tag{66}$$

The variable s represents the distance to the polyhedral Coulomb cone (s is often called saturation variable) as pictured in the following figure in two dimensions:

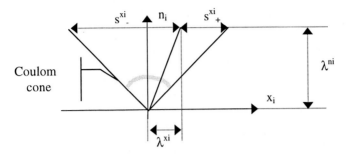

Fig.9. Signification of the s variables

We can then write a problem of the form:

$$
\begin{bmatrix} \lambda^{n_i} \\ s_+^{x_i} \\ s_-^{x_i} \\ s_+^{y_i} \\ s_-^{y_i} \end{bmatrix} \geq 0 \qquad
\begin{bmatrix} {}^{n_i}a \\ {}^{x_i}a^+ \\ {}^{x_i}a^- \\ {}^{y_i}a^+ \\ {}^{y_i}a^- \end{bmatrix} \geq 0 \qquad
\begin{bmatrix} \lambda^{n_i} \\ s_+^{x_i} \\ s_-^{x_i} \\ s_+^{y_i} \\ s_-^{y_i} \end{bmatrix}^T
\begin{bmatrix} {}^{n_i}a \\ {}^{x_i}a^+ \\ {}^{x_i}a^- \\ {}^{y_i}a^+ \\ {}^{y_i}a^- \end{bmatrix} = 0 \qquad \text{for} \quad i \in I_r \qquad (67)
$$

This LCP is once again treated with the Lemke algorithm, but we can notice that these authors have also used Interior Point methods to solve this problem.
This polyhedral approximation of the cone of Coulomb raises two remarks:
- First, in the case of persistent sticking, we cannot guarantee that the contact force is contained in the quadratic Coulomb cone.
- Secondly, in the case of a stick-slip transition, the friction force will be opposed to acceleration, but inevitably directed towards one of the corners of the pyramid:

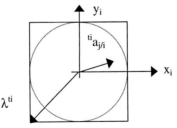

Fig.10. Direction of the friction force
in the case of a stick-slip transition

The second remark hardly sets a problem, because as we have mentioned it, we do not know, in this case, the direction of the friction force. The first remark is more disturbing.

More recently, Glocker have generalized this approach by proposing the following formulation:

Glocker [32]

The principle proposed by Glocker consists in using a polyhedral Coulomb cone approximation, but without being limited to a pyramid. The equivalent cone can have here as many facets as the user wishes. Once more, the problem is formulated in acceleration. This author describes a cone, external to the Coulomb cone, which in 3D is perfectly equivalent to that of Pang and Trinkle, and which in 2D allows one to introduce only 4 variables (thus replacing the first model we have presented). The tangential contact law is written

$$-^{ti} a = \sum_{i=1..a} e_i k_i \quad \sigma_i = \mu \lambda^{n_i} - e_i^T \lambda^{t_i} \geq 0 \quad k_i \geq 0 \quad \sigma_i k_i = 0 \quad i \in I_r, \quad (68)$$

for a cone of (a) facets, whose principal vectors $(e_i)_{i=1..a}$ are represented in the following figure:

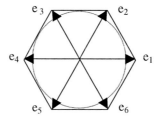

Fig.11. Exterior polyhedral approximation of the Coulomb cone

Caring for clarity, we will not report the way in which the author modifies this tangential contact law to formulate a LCP of dimension (a+1) for a sticking contact. We simply leave to the reader the care to imagine the number and the complexity of the complementarity conditions which result from it. It is obvious that the more the number of facets increases, the more we approach the quadratic Coulomb cone, but there is then a dramatic increase of the number of the variables to treat. Moreover, as we noticed it for the model of Pang and Trinkle, certain directions are privileged in the case of a stick-slip transition.

In order to avoid these problems, Glocker proposes (in the same article) a different approach which makes it possible to respect the quadratic Coulomb cone perfectly. This method consists in formulating a Non-linear Complementarity problem (NCP: $x \geq 0$ $F(x) \geq 0$ $x.F(x) = 0$), by using the properties of convex sets intersections. Glocker indeed considers the intersection of the two following cones:

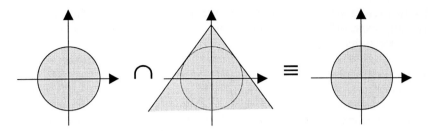

Fig.12. Intersection of convex sets

which is the cone of quadratic Coulomb cone. Using the same technique which is used to formulate a LCP, Glocker sets a NCP of dimension (5) for a sticking contact.

It is important to note that it is one of the first times that the sticking problem is written in a standard mathematical form with no approximation of the Coulomb law. This new formulation is very promising, but it has unfortunately not yet been tested as these lines are written. The reader interested by resolution methods of NCP should refer to [33].

In the same spirit, Trinkle and Stewart were interested in a velocit formulation of the sticking contact problem [34][35][36]. Anitescu and Potra use exactly the same approach, we will describe it now.

Anitescu, Potra, Stewart, Trinkle [34][35][36][37][38]

These authors use a velocity formulation. The problem is set by discretizing the dynamics equations using a first order integration scheme. Their work is also based on a polyhedral approximation of the Coulomb cone, but this time it is about a cone interior to the Coulomb cone. We have to note that the maximum principle of dissipation is used here in order to formulate a tangential contact law. This principle simply states that the friction forces must be calculated in order to maximize the energy loss. Another time, these authors seek to write a LCP, whose size is slightl higher than the LCP of Glocker (size (a+2) for a contact). Just for information, here is a view of the polyhedral cone that these authors use:

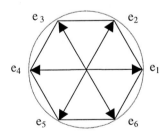

Fig.13. Interior polyhedral approximation of the Coulomb cone

Contrary to the preceding approaches, these authors use an approximation of the Coulomb cone, even for sliding contacts (because of the velocity formulation). Under these conditions, the friction force is not opposed any more to the tangential relative velocity, even in the case of sliding, but it points towards one of the corners of the polyhedral cone. The principal directions are thus more privileged than with an acceleration formulation.

Because of this approach disadvantages, Stewart and Pang [36] have proposed a formulation which makes it possible to take into account the quadratic Coulomb cone (see also [38]). Their approach can be outlined as follows. Always using a velocity formulation, these authors use the maximum principle of dissipation which is transformed, using an argument of Fritz John, in complementarity relations (the same type of argument is used in the case of a polyhedral cone). The problem is then written in the form of a highly non-linear mixed complementarity problem (MCP). This work constitutes another new approach to formulate the frictional contact problem in a standard mathematical form. These authors advise the use of homotopy methods to solve this problem, but unfortunately no numerical example is presented.

Wosle, Pfeiffer [39][40]
Wosle and Pfeiffer propose two formulations in acceleration in order to correctl model the Coulomb cone. The first approach [39] is inspired from concepts presented by Klarbring [41] for deformable bodies. These concepts are modified in order to apply to rigid bodies mechanisms. The normal and tangential contact laws are written in the form of variational inequalities, and then transformed into equality using the projection operator:

$$\begin{cases} \lambda^{n_i} = \text{proj}_{C_{n_i}} \, (\lambda^{n_i} - b_n{}^{n_i} a) = \Pi_{n_i} \, (\lambda^{n_i}, \ddot{q}) \\ \lambda^{t_i} = \text{proj}_{C_{t_i}(\lambda^{n_i})} \, (\lambda^{t_i} - b_t{}^{t_i} a) = \Pi_{t_i} \, (\lambda^{n_i}, \lambda^{t_i}, \ddot{q}) \end{cases} \quad \text{for } i \in I_{nn} \qquad (69)$$

Where $\begin{cases} C_{n_i} = \left\{ \lambda^{n_i} / \lambda^{n_i} \geq 0 \right\} \\ C_{t_i}(\lambda^{t_i}) = \left\{ \lambda^{t_i} / \left\| \lambda^{t_i} \right\| \leq \mu \lambda^{n_i} \right\} \end{cases}$ and $(b_n, b_t) > 0$

The normal and tangential contact forces are then replaced in the dynamics equations. A non-linear problem is thus obtained of which the principal unknowns are $(\ddot{q}, \lambda^{n_i}, \lambda^{t_i})$. The solution of this problem is obtained in two stages:

1. The contact forces are kept constant and we solve the non-linear problem as a function of $\ddot{q}$:

$$H(q)\ddot{q} - (\Phi^n)_{,q}^T \Pi_n (\lambda^n, \ddot{q}) - (\Phi^t)_{,q}^T \Pi_t (\lambda^n, \lambda^t, \ddot{q}) = U - C \qquad (70)$$

using a method of Newton.

2. The contact forces are then corrected using the new accelerations:

$$\begin{cases} \lambda^{n_i} = \Pi_{n_i}(\lambda^{n_i}, \ddot{q}) \\ \lambda^{t_i} = \Pi_{t_i}(\lambda^{n_i}, \lambda^{t_i}, \ddot{q}) \end{cases} \quad \text{for } i \in I_{nn} \tag{71}$$

These two stages are repeated until convergence of the method. This solving algorithm is in fact a Newton method encased in a fixed point algorithm. A second method of resolution of this problem is presented by Klarbring [41], but it is not presented by Wolse and Pfeiffer.

A second approach suggested by Wosle and Pfeiffer consists in introducing additional parameters in order to formulate the tangential contact law. These authors write:

$$\begin{cases} \lambda^{t_i} = \begin{bmatrix} \cos\varphi_i \\ \sin\varphi_i \end{bmatrix} \left\| \lambda^{t_i} \right\| = \begin{bmatrix} \cos\varphi_i \\ \sin\varphi_i \end{bmatrix} (\mu\lambda^{n_i} - \lambda^{t_i0}) \\ \lambda^{t_i0} = \mu\lambda^{n_i} - \left\| \lambda^{t_i} \right\| \\ {}^{t_i}a = -\begin{bmatrix} \cos\varphi_i \\ \sin\varphi_i \end{bmatrix} k_i \end{cases} \quad \text{for } i \in I_r \tag{72}$$

λ^{t_i0} can be seen as the saturation variable of the tangential contact force. This decomposition allows Wosle and Pfeiffer to formulate complementarity conditions between (k_i, λ^{t_i0}). These complementarity conditions are then transformed into equalities by using a Mangasarian argument. The problem thus reduces to the resolution of a non-linear system of the form:

$$F({}^n a, \lambda^n, k, \varphi, \lambda^{t0}, \lambda^t, \ddot{q}) = 0 \tag{73}$$

This system is then solved using a homotopy method which seems to give good results in a 3D example.

Moreau [3][14][42][43]
Moreau has been the first one to use a velocity formulation in order to solve frictional contact problem. For reasons of clearness, we place ourselves within the framework of the standard inelastic shocks (null restitution coefficients). Moreau's approach can be summarized as follows. It is still supposed that we have Q active contacts.
The equations of dynamics are discretized with an algorithm of order «1-1/2»:

$$\left\{ q_{1/2} = q_{n-1} + \frac{h_n}{2}\dot{q}_{n-1} \quad \dot{q}_{1/2} = \dot{q}_{n-1} \quad \ddot{q}_{1/2} = \frac{\dot{q}_n - \dot{q}_{n-1}}{h_n} \right\} \tag{74}$$

A system of the form is obtained:

$$H(\dot{q}_n - \dot{q}_{n-1}) = h_n(U-C) + h_n(\Phi^n)_{,q}^T \lambda^n + h_n(\Phi^t)_{,q}^T \lambda^t \tag{75}$$

Where all the vectors and matrices are calculated as a function of $q_{1/2}$ and $\dot{q}_{1/2}$. This system can be written:

$$\dot{q}_n - \dot{q}_{n-1} = h_n H^{-1}(U-C) + H^{-1}\left[(\Phi^n)_{,q}^T \quad (\Phi^t)_{,q}^T\right]\begin{bmatrix} S^n \\ S^t \end{bmatrix}$$

$$= h_n H^{-1}(U-C) + H^{-1}(\Phi)_{,q}^T S \tag{76}$$

Where the vector S represents the contact impulses.
Finally, we obtains:

$$(\Phi)_q(\dot{q}_n - \dot{q}_{n-1}) = h_n(\Phi)_{,q}H^{-1}(U-C) + (\Phi)_{,q}H^{-1}(\Phi)_{,q}^T S \tag{77}$$

That we note:

$$\begin{bmatrix} ^n v_F \\ ^t v_F \end{bmatrix} = \begin{bmatrix} ^n v_L \\ ^t v_L \end{bmatrix} + A_r \begin{bmatrix} S^n \\ S^t \end{bmatrix} \tag{78}$$

Where $(^n v_F, {}^t v_F)$ represent respectively the normal and tangential relative velocities at time tn (F for final).
$(^n v_L, {}^t v_L)$ are the normal and tangential relative velocities obtained when no contact is active (L for Libre=Free in French, i.e. free-motion here).
Ar is the Delassus operator.

The system (78) is subject to the conditions imposed by the contact law:

$$\begin{cases} \text{Signorini}(^n v_F, S^n) \\ \\ \text{Coulomb}(^t v_F, S^t) \end{cases} \tag{79}$$

To deal with this problem, Moreau uses a relaxation method which is similar to the Gauss-Siedel method for the linear systems. This method consists in solving the problem contact by contact, that is to say to treat a set of Q simple contacts. Let us consider the (i)-th contact (C_i). It is supposed here that the problem is solved for all the contacts $(C_j)_{j\neq i}$, we can thus write:

$$\begin{cases} S^{n_i} = \dfrac{1}{A_{r(i,i)}}{}^{n_i}v_F + \dfrac{1}{A_{r(i,i)}}\left(-{}^{n_i}v_L + \displaystyle\sum_{\substack{j=1..3Q \\ j\neq i}} A_{r(i,j)} S^j \right) \\[4ex] S^{x_i} = \dfrac{1}{A_{r(i+Q,i+Q)}}{}^{x_i}v_F + \dfrac{1}{A_{r(i+Q,i+Q)}}\left(-{}^{x_i}v_L + \displaystyle\sum_{\substack{j=1..3Q \\ j\neq i+Q}} A_{r(i,j)} S^j \right) \qquad i=1..Q \quad (80) \\[4ex] S^{y_i} = \dfrac{1}{A_{r(i+2Q,i+2Q)}}{}^{y_i}v_F + \dfrac{1}{A_{r(i+2Q,i+2Q)}}\left(-{}^{y_i}v_L + \displaystyle\sum_{\substack{j=1..3Q \\ j\neq i+2Q}} A_{r(i,j)} S^j \right) \end{cases}$$

Note : *If the contact* (C_i) *is active, the terms* $A_{r(i,i)}$, $A_{r(i+Q,i+Q)}$ *and* $A_{r(i+2Q,i+2Q)}$ *are strictly positive, because the* A_r *operator is semi-definite positive.*
We can write these relations in the form of three linear equations:

$$\begin{cases} S^{n_i} = c_{n_i}{}^{n_i}v_F + b_{n_i} \\[2ex] S^{x_i} = c_{x_i}{}^{x_i}v_F + b_{x_i} \qquad i=1..Q \\[2ex] S^{y_i} = c_{y_i}{}^{y_i}v_F + b_{y_i} \end{cases} \qquad (81)$$

These three relations are subjected to the constraints:

$$\begin{cases} \text{Signorini}({}^{n_i}v_F, S^{n_i}) \\[2ex] \text{Coulomb}({}^{x_i}v_F, {}^{y_i}v_F, S^{x_i}, S^{y_i}) \end{cases} \qquad i=1..Q \qquad (82)$$

Let us see how to solve this problem.

Signorini($ {}^{n_i}v_F, S^{n_i}$)

We are brought to study the intersection between the Signorini graph and the line:

$$S^{n_i} = c_{n_i}{}^{n_i}v_F + b_{n_i} \qquad (83)$$

Taking into account the preceding remark, we can ensure that the slope of this line is strictly positive. We thus obtain two cases (figure 14) :

$$\begin{cases} \text{if} \quad b_{n_i} \geq 0 \quad \text{then} \quad S^{n_i} = b_{n_i} \qquad {}^{n_i}v_F = 0 \\ \text{if} \quad b_{n_i} < 0 \quad \text{then} \quad S^{n_i} = 0 \qquad {}^{n_i}v_F = -b_{n_i}/c_{n_i} \end{cases} \tag{84}$$

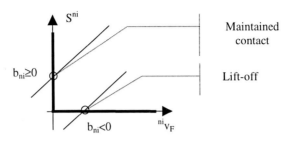

Fig.14. Normal contact problem for the i-th contact

Coulomb$(\,^{x_i}v_F\,,^{y_i}v_F\,,S^{\,x_i}\,,S^{\,y_i}\,)$

We have:

$$\begin{cases} \text{if} \quad S^{n_i} = 0 \quad \text{then} \quad S^{x_i} = S^{y_i} = 0 \\ \\ \text{if} \quad S^{n_i} \neq 0 \quad \text{then} \begin{cases} \text{if} \quad \left\| b_{t_i} \right\| < \mu S^{n_i} \begin{cases} S^{x_i} = b_{x_i} \\ S^{y_i} = b_{y_i} \end{cases} \\ \\ \text{if} \quad \left\| b_{t_i} \right\| \geq \mu S^{n_i} \begin{cases} S^{x_i} = \mu S^{n_i} \dfrac{b_{x_i}}{\left\| b_{t_i} \right\|^2} \\ \\ S^{y_i} = \mu S^{n_i} \dfrac{b_{y_i}}{\left\| b_{t_i} \right\|^2} \end{cases} \end{cases} \end{cases} \tag{85}$$

With $\left\| b_{t_i} \right\| = \sqrt{\left(b_{x_i} \right)^2 + \left(b_{y_i} \right)^2}$

This operation corresponds to a projection on the cone of quadratic Coulomb cone. We can represent this process in 2D by the following figure:

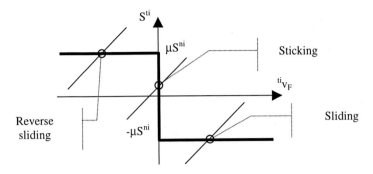

Fig.15. Tangential contact problem for the i-th contact

We reiterates this process until a certain criterion of convergence is checked. The algorithm is presented in the following way:

```
{ Procedure J.J. Moreau }
{ Estimation of the invariant vectors and matrix }
calcul_q1/2
calcul_vL
calcul_Ar
{ Initialization of the contact impulses }
for i=1 to Q
     Sⁿ=0
     Sˣ=0
     Sʸ=0
endfor
{ Relaxation method }
while a==0
    k=k+1
    for i=1 to Q
        { Geometric detection of the contacts }
        if Φⁿⁱ≤0
            calcul_bi
            calcul_ci
            if bni>0
                { Normal contact }
                Sⁿⁱ=bni
                { Tangential contact }
                if (bxi²+byi²)<μ²(Sⁿⁱ)²
                    { Sticking }
                    Sˣⁱ=bxi
                    Sʸⁱ=byi
                else
                    { Sliding }
                    Sˣⁱ =μ Sⁿⁱ bxi /(bxi²+byi²)^(1/2)
                    Sʸⁱ =μ Sⁿⁱ byi /(bxi²+byi²)^(1/2)
                endif
            else
```

```
        { Lift-off }
        S^{ni}=0
        S^{xi}=0
        S^{yi}=0
      endif
    endif
  endfor
  { Convergence test }
  if norm(Θ(k,k+1,S^{ni},S^{xi},S^{yi}))<ε
    a=1
  endif
endwhile
```

This method proposes an implicit treatment of the relative velocities and of the contact impulses. No freedom is taken with respect to the Coulomb law. But the most interesting with this approach is that the velocity formulation makes it possible to treat, at the same time, all the transitions, slip-stick, stick-slip, separation and even the shocks. This type of method is called "time-stepping". Moreau has applied this algorithm to the modelling of granular mediums (simulating the behaviour of samples containing several thousands of grains (balls and polygons) of different sizes) and also to the modelling of monumental buildings.

Baraff [17][44][45]
The work of this author is also particularly interesting. The problem is formulated in acceleration. After having used during a long time the traditional method (LCP formulation, Lemke algorithm) [44][45], we can find in [17] the description of a new algorithm inspired from the Dantzig method. It is a pivotation method which is used to solve linear complementarity and quadratic programming problems. The advantage of this method is that, as in the case of the Moreau algorithm, the problem is treated contact by contact. This algorithm makes it possible to handle friction correctly, without polyhedral approximation of the Coulomb cone. We return the reader to the article [17] for a complete description of the Dantzig algorithm.

Note : This author has used the Lemke algorithm for a long time, whereas he was perfectly conscious of the existence of the relaxation methods. Indeed, he presented a Jacobi method [45], very close to the Moreau algorithm. But this author didn't judged the convergence of the method good enough to be used as principal method. Consequently, the role of this algorithm was simply to predict the sign of the tangential contact forces. This operation made possible the removal of the absolute values of the expressions (61) or (65). Then, the problem was formulated on a LCP form and its resolution was entrusted to the algorithm of Lemke.

This ends our literature survey. We do not claim that this bibliographical review is exhaustive, but we think that the principal proceedings were presented. Let us see now the strategy we have chosen.

3.2.3 Determination of our Strategy

First of all, it seems clear that the treatment of the frictional contact problem in the form of a linear complementarity problem is not satisfactory for two essential reasons:
- The polyhedral approximation of the Coulomb cone which is necessary in the sticking case.
- The introduction of new variables.

We will not give our opinion on the formulations NCP of Glocker and MCP of Pang and Stewart, because these approaches are still too recent to allow a sound judgement of their effectiveness on examples. If the resolution algorithms of these problems are not too complex and if they present interesting convergence properties, it will be necessary in the future to be interested in these approaches.

At the moment, we prefer to limit our choices to the methods of Wosle, Moreau and Baraff:

Wosle treats all the contacts at the same time, i.e. all the contact status are modified during the same iteration. The first method of Wosle proposes a tempting formulation, but the resolution algorithm appears rather heavy (with each iteration of the fixed point, one solves a non-linear system with a method of Newton). In his second formulation, he introduces additional variables which increase unpleasantly the size of the problem (12 variables for a sticking contact in 3D). Moreover, the homotopy method which is used for the resolution appears, once again, quite heavy to implement.

For reasons of simplicity, we prefer here to adopt Moreau and Baraff attitude, which consists in using algorithms which deal with the problem contact by contact. Between these two algorithms, we have naturally adopted Moreau's and that for various reasons. It is clear that the fact that we have worked in the same laborator was an important factor, for we did benefit from well-advised counsels. But, there are also more scientific reasons for this choice. First, the Gauss-Seidel method rests on a simple principle. The clearness of this algorithm enables very simple modification in order, for example, to take into account permanent constraints, to define joint drivers or to introduce a restitution coefficient. This method is much easier to implement and to handle than the Dantzig algorithm or than all the other algorithms which we have quoted. Admittedly, the convergence of the Gauss-Seidel method is probably lower than that of the Dantzig method, but the calculations carried out in the Gauss-Seidel procedure are much less heavy. With this method, we avoid the resolution of linear systems in the course of iterations, whereas at each step of the Dantzig method such a calculation is necessary. Furthermore, Moreau has extensively used his algorithm for at least ten years, so we can really rely on the robustness of this procedure.

So, for reasons of simplicity and of robustness, we have chosen to adopt the decision procedure of Moreau.

Note : Whichever method is adopted, it is necessary to check if this method is able to treat hyperstatic systems (see remark in [32] for example).

A second point which seems important is: should we formulate the problem in velocity or in acceleration? To answer that, we deliver to the reader the following reflections:

The acceleration formulation raises some remarks:

- It is necessary to seek an approximation of each transition time (where an event occurs). We will say that we have a method of the type «Event Driven ». It is important to note that this research of the events is perfectl obligatory, because the contact laws which we have formulated are onl valid under certain velocity conditions. This aspect is very seldo approached, for example [39] does not evoke this subject at all. As a consequence, every transition times can be reported to the user. That allows a precise analysis of the results and guarantees that no important event will be neglected.
- Before running the decision procedure in acceleration, we must be able to carry out a first classification of our contacts. This classification is necessary because, as we have seen, the Coulomb law can't be expressed directly in acceleration. We proceed in the following way:

$$\begin{cases} \left\| \mathbf{^{t_i}v} \right\| \geq \varepsilon & \equiv \quad \text{sliding} \\ \left\| \mathbf{^{t_i}v} \right\| < \varepsilon & \equiv \quad \text{sticking} \end{cases} \quad \text{for} \quad i = 1,...,Q$$

The danger here is to make a mistake in the classification because of the rounding errors. If we give a sticking status to a sliding contact, it is not a problem because the decision procedure can carry out a stick-slip transition. On the other hand, if a contact, whose real status is sticking, is classified as sliding contact, the algorithm will not be able to catch up with that. It is thus advisable to stabilize the constraints with the greatest care in order to avoid errors due to simple drifts (see paragraph 2.3.2.).
- The shocks are treated with an additional algorithm (see chapter 4).

From this point of view, the velocity formulation of Moreau presents obvious advantages:
- No status classifying.
- The shocks are treated in the same footing than the other events.
- No research procedure of the transition times is necessary.

On the other hand, the Moreau algorithm uses an integration scheme of order «1-1/2» with a constant step of time. It is conceived to face several thousands of simultaneous contacts. Our objective is much more modest, and we can estimate that forty contacts is a reasonable limit for the applications which we intend to treat. Furthermore, we will make the hypothesis that the number of events, which punctuate simulations, is not too large. With this assumption, we can then use high order integration scheme, in order to move as fast as possible between two consecutive events. We could be tempted to use a velocity formulation with high integration scheme, but, that would probably give a process of integration slower than if we worked in acceleration (because of the discretization errors of the dynamic equations), with all the same the obligation to locate all the events. For these reasons, we have chosen to formulate the problem in acceleration.

We will now focus on the existence and the uniqueness of the results as well as on the convergence of the Gauss-Seidel algorithm.

3.3   Existence, Uniqueness, Convergence

Let us see the results of existence and uniqueness which are at our disposal.

3.3.1 Existence, Uniqueness

Once again, we will make the distinction between contact with and without friction. We still consider that we do not have bilateral constraints (P=0). We can note that all the results we will state remain valid when (P≠0), because a bilateral constraint can always be expressed in the form of two unilateral constraints. In this paragraph, we will state several theorems and return the reader to the references quoted for their demonstration. All the results we will present, are valid for the quadratic and polyhedral Coulomb.

We still consider a system (Σ) with (N) principal joints and (Q) contacts $(C_i)_{i=1..Q}$ with $i \in I_{nn}$.

Contact without Friction

We must deal with a complementarity problem of the form:

$$ a_n \geq 0 \quad \lambda^n \geq 0 \quad (a_n)^T (\lambda^n) = 0 \qquad (86) $$

With $a_n = \left((\Phi^n)_{,q} H^{-1} (\Phi^n)_{,q}^T \right) \lambda^n + (\Phi^n)_{,q} H^{-1} (U - C) + \eta^n = A\lambda^n + b$

The following theorem is stated:

> ***Theorem [27]:***
>
> *There is always a single solution* $\left(a_n, \left(\Phi^n \right)_{,q}^T \lambda^n \right)$ *to the problem (86).*
>
> *If moreover, rang=min(N,Q),* $\left(\Phi^n \right)_{,q}^T$ *then the solution* $\left(a_n, \lambda^n \right)$ *is unique.*

We can note that Moreau [25] has already proposed an equivalent theorem.

It shows that we always have the existence and uniqueness of accelerations and of the contact forces, when the Delassus operator A is definite positive, but only the existence of the contact forces when A is semi-definite positive. This theorem is valid for the three formulations (LCP), (QP1) and (QP2).

We find the same result as in the case of the bilateral constraints. Remember that we had been brought, paragraph (2.3.1.), to modify the kinematic constraint matrix, for a hyperstatic system, in order to ensure the convergence of the Runge-Kutta method. No operation of this kind is necessary here, because the optimization problem is solved only at the end of the step of time.

Let us now add friction to our problem.

Contact with friction

The distinction between sticking and sliding contacts has to be made.

Sticking
We must solve the problem:

$$
\left.
\begin{cases}
a_n \geq 0 \quad \lambda^n \geq 0 \quad (a_n)^T(\lambda^n) = 0 \\[2mm]
\left\| \lambda^{t_i} \right\|^2 \leq \mu^2 \left(\lambda^{n_i}\right)^2 \quad (\lambda^{t_i})^T ({}^{t_i} a) \leq 0 \\[2mm]
\left(\left\| \lambda^{t_i} \right\|^2 - \mu^2 \left(\lambda^{n_i}\right)^2 \right)\left(\left\| {}^{t_i} a \right\|^2 \right) = 0 \quad \left((\lambda^{t_i}) \wedge ({}^{t_i} a) \right) = 0
\end{cases}
\right\} \quad \text{for} \quad i \in I_r
$$

(87)

With:

$$
\begin{bmatrix} a_n \\ a_t \end{bmatrix} =
\begin{bmatrix}
A & (\Phi^n)_{,q} H^{-1}(\Phi^t)_{,q}^T \\
(\Phi^t)_{,q} H^{-1}(\Phi^n)_{,q}^T & (\Phi^t)_{,q} H^{-1}(\Phi^t)_{,q}^T
\end{bmatrix}
\begin{bmatrix} \lambda^n \\ \lambda^t \end{bmatrix} +
\begin{bmatrix} b \\ (\Phi^t)_{,q} H^{-1}(U - C) + \eta^t \end{bmatrix}
$$

$$
= A_r \begin{bmatrix} \lambda^n \\ \lambda^t \end{bmatrix} + b_r
$$

[16] shows the following theorem:

> **Theorem :**
> In the case of sticking, if $rang \left(\Phi^n\right)_{,q}^T = min(N,Q)$, we can ensure the existence of a solution.
> If $rang \left(\Phi^n\right)_{,q}^T < min(N,Q)$ and if $b_r \in Im\left(\Phi^n\right)_{,q}^T$, we can still ensure the existence of a solution. Furthermore, taking into account the definition of b_r, this condition is always verified.

No uniqueness demonstration of the solution can be done in the general case, but it is important to note that no example of sticking contact, with several solutions, has ever been published in the literature.

Sliding
We must deal with a complementarity problem of the form:

$$
a_n \geq 0 \quad \lambda^n \geq 0 \quad (a_n)^T(\lambda^n) = 0
$$

(88)

With $a_n = (A + (\Phi^n)_{,q} H^{-1}\tilde{\mu}(\Phi^t)_{,q}^T)\lambda_n + b = A_g \lambda_n + b$

We could be tempted to apply here the preceding theorem results. The problem is that the modified Delassus operator A_g is no more semi-definite positive in general. We can thus state no results of existence nor uniqueness. Worse than that: contrary to sticking, there exists in the literature examples of sliding contacts which do not have a solution (inconsistency), or have several solutions (indeterminacy) [46][45][47][19]. We return the study of these two cases to paragraph (3.4.3.).

Under certain conditions, we can however show the existence and uniqueness of the solution in the case of sliding, but for that, the friction coefficient must not be «too large». To illustrate our matter, we will state a theorem which is valid in case of sliding, but also in the case of sticking.

Sticking-Sliding

> **Theorem [16]:** If it is supposed that rang $\left(\Phi^n\right)^T_{,q} = min(N,Q)$, then
>
> (1) There is a positive scalar $\overline{\mu}$ such that if $\mu \in [0,\overline{\mu}]$ for any contact (C_i) with $i \in I_s$, the 3D frictional contact problem admits a solution.
>
> (2) There is a positive scalar $\overline{\mu}$ such that if $\mu \in [0,\overline{\mu}]$ for any contact (C_i) with $i \in (1..Q)$, the 3D frictional contact problem admits a single solution.

When all the friction coefficients are bounded, we can thus ensure the existence and the uniqueness of the solution. All the difficulty lies in the evaluation of the term $\overline{\mu}$. Trinkle proposes in the same article, a calculation of this parameter. But according to this author, this estimate is too pessimistic to be truly usable.

We will not summarize the existence and uniqueness results which are given by [37], because these results are only valid in the case of a polyhedral approximation of the Coulomb cone. However, one finds in [37] a quadratic modelling of the cone Coulomb. In this article, the author states existence results for sliding contacts. It is interesting to notice that Anitescu also imposes bounded friction coefficients, but this restriction is expressed less rigorously (μ sufficient small) than Trinkle's.

More recently, Pang and Stewart [36] have proposed an existence demonstration for the frictional contact problem. Their demonstration is based on homotopy arguments and thus makes it possible to focus on the results existence without being concerned with their uniqueness. Unfortunately these results are stated within the framework of a velocity formulation (time-stepping formulation). Because of our preceding choices, this demonstration could not be useful to us. However, we will reconsider the problems of results existence for time-stepping formulations when we discuss the problems of inconsistency and indeterminacy.

Let us now report the convergence results which are at our disposal.

3.3.2 Convergence

We have chosen to solve an optimization problem under constraints with an Gauss-Seidel relaxation algorithm. We will see in which conditions we are assured that this algorithm converges. As before, let us make the distinction between contact with and without friction.

Contact without Friction

We must deal with a problem of the form:

$$\min_{(\lambda^n} J(\lambda^n) = \min_{(\lambda^n} \left(\frac{1}{2}(\lambda^n)^T A\lambda^n + (\lambda^n)^T b \right) \quad \lambda^n \geq 0 \qquad (QP2) \qquad (89)$$

The following result is stated:

> **Theorem [29]:** *If the functional J is elliptic, then the Gauss-Seidel relaxation method converges for problem (QP2).*

In our study, the functional J is quadratic, but we know that if the Delassus operator A is definite positive, then the functional J is elliptic and the preceding theorem applies. That is to say that convergence can only be proven for hypostatic systems.

Contact with Friction

No results of convergence are available.

We can draw from this paragraph the following conclusions:
- There does not exist in the literature theorem which ensures, in general, the existence and the uniqueness of a solution to the frictional contact problem. The theorems which we have stated are too restrictive to be exploitable.
- The results of convergence are quite thin. But, it is important to note that the choices we made (quadratic cone of Coulomb, Gauss-Seidel relaxation method) are not in question. A formulation (LCP, algorithm of Lemke) does not lead to better results.

We have to note here, that, it is not because we do not know how to demonstrate a result, that this result is not true in general. As a matter of fact, the Gauss-Seidel method has converged perfectly and has given a unique solution for all the examples we have treated. Even without convergence, existence or uniqueness results, our point of view consists in applying our approach directly, but with a certain number of precautions. Thus at each calculation step we check the results validity.

In practice, it is very easy to control the existence of the solutions and the convergence of the method. In fact, if one of these results is not checked, the algorithm stops and, the user is automatically informed of the situation. We will see however, paragraph (3.4.3.), how to detect the cases where no solution exists (inconsistency) and we will propose a manner of treating them. The problem of the uniqueness of the solution is much more alarming because, when the problem has several solutions, the algorithm generally converges towards one of them (local minimum) and the simulation continues without the user suspecting anything. The results obtained by this way are not reliable and can lead to erroneous conclusions. It is thus very important to be able to detect the multiple solution cases (indeterminacy); this study is also returned in paragraph (3.4.3.).

We now will specify more precisely our algorithm structure.

3.4 Our Approach

This paragraph is composed of three parts. First, we will give a pseudo code of the Gauss-Seidel method written in acceleration. We will see, then, how we have modified our code in order to treat bilateral constraints and to allow driven joints. We will conlude by the study of two indeterminacy and inconsistency examples.

We always consider a system (Σ) with (N) principal joints, (P) closed loop joints and (Q) contacts $(C_i)_{i=1..Q}$ with $i \in I_{nn}$.

3.4.1 Structure of the Gauss-Seidel Procedure in Acceleration

Given the pseudo code of the Gauss-Seidel method in acceleration, we must solve a problem of the form:

$$\begin{cases} \lambda^{n_i} = c_{n_i}{}^{n_i} a_F + b_{n_i} & i = 1..Q \\ \lambda^{x_i} = c_{x_i}{}^{x_i} a_F + b_{x_i} \\ \lambda^{y_i} = c_{y_i}{}^{y_i} a_F + b_{y_i} \end{cases} \Bigg\} \quad i \in I_r \qquad (90)$$

Subject to the conditions:

$$\begin{cases} \text{Signorini}({}^{n_i} a_F, \lambda^{n_i}) & i = 1..Q \\ \text{Coulomb}({}^{x_i} a_F, {}^{y_i} a_F, \lambda^{x_i}, \lambda^{y_i}) & i \in I_r \end{cases} \qquad (91)$$

For each contact (C_i), we define the parameters:
- status[i]=1 if the contact is active, zero if not,
- rolling[i]=1 if the contact is a sticking contact, zero if it is a sliding contact.

```
{ Gauss-Seidel procedure in acceleration }
{ Status classifying }
tri_contact(status,rolling)
{ Estimation of the invariant vectors and matrix }
calcul_aL
calcul_Ar
{ Initialization of the contact impulses }
for i=1 to Q
    λni=0
    λxi=0
    λyi=0
endfor
```

```
{ Relaxation method }
while a==0
    k=k+1
    { Unilateral constraints }
    for i=1 to Q
        { Active contact }
        if statut[i]==1
            calcul_bᵢ
            calcul_cᵢ
            { Normal contact }
            if bₙᵢ>0
                { Sticking }
                if roulement[i]==1
                    λⁿⁱ =bₙᵢ
                    { Sticking continue }
                    if (bₓᵢ²+bᵧᵢ²)<μ²(λⁿⁱ)²
                        λˣⁱ =bₓᵢ
                        λʸⁱ =bᵧᵢ
                    { Stick-slip transition }
                    else
                        λˣⁱ =μλⁿⁱbₓᵢ /(bₓᵢ²+bᵧᵢ²)^(1/2)
                        λʸⁱ =μλⁿⁱbᵧᵢ /(bₓᵢ²+bᵧᵢ²)^(1/2)
                    endif
                { Sliding }
                else
                    μ̃ ₓᵢ =-μvₓᵢ /(vₓᵢ²+vᵧᵢ²)^(1/2)
                    μ̃ ᵧᵢ =-μvᵧᵢ /(vₓᵢ²+vᵧᵢ²)^(1/2)
                    λⁿⁱ =f(bₙᵢ,μ̃ ₓᵢ ,μ̃ ᵧᵢ ,cₙᵢ)
                    λˣⁱ =μ̃ ₓᵢ λⁿⁱ
                    λʸⁱ =μ̃ ᵧᵢ λⁿⁱ
                endif
            { lift-off }
            else
                λⁿⁱ=0
                λˣⁱ=0
                λʸⁱ=0
            endif
        endif
    endfor
    { Bilateral constraints }
    (see paragraph 3.4.2.)
    { Driven joints }
    (see paragraph 3.4.2.)
    { Convergence test }
    if norm(Θ(k,k+1,λⁿⁱ,λˣⁱ,λʸⁱ))<ε
        a=1
    endif
endwhile
```

Before going further, we are going to specify an important point. The Gauss-Seidel method is an iterative method, it is thus appropriate to adopt a convergence criterion which ensures a sufficient precision to the results. As it is a delicate business, one could be tempted to impose a draconian criterion which strongl increases the iteration count. This strategy is, according to us, too expensive. To solve this problem, the surest manner to proceed is to adopt a convergence criterion which guarantees the status convergence of the contacts. Once this criterion is validated, we solve a differential-algebraic system, the form of which is dictated by the contact status. We thus obtain an acceleration field, with a maximum precision, that allows us to continue integration in full safety.

Note : Concerning the contact efforts, we do not use the values of the Lagrange multipliers which are obtained by the resolution of the differential-algebraic system. We proceed in this manner for two reasons. First is that, in the case of a hyperstatic system, the resolution of the differential-algebraic system would lead to contact efforts incompatible with the status obtained at the end of the iterative process. The second reason is that in the case of our explicit D&P algorithm, the efforts are not necessary to integration; their precision imports little. We are thus satisfied to provide to the user the contact efforts values obtained at the end of the Gauss-Seidel procedure.

Let us see now, how to introduce bilateral constraints and how to define joint drivers.

3.4.2 Modifications

Bilateral Constraints

Let us suppose that the system (Σ) has (P) closed loop joints (pin or slider). The dynamics equations are written:

$$H\ddot{q} = (U - C) + (\Phi^L)_{,q}\lambda^L + (\Phi^n)_{,q}\lambda^n + (\Phi^t)_{,q}\lambda^t \tag{92}$$

Where the matrix $(\Phi^L)_{,q}$ represents the matrix of kinematics constraints associated to the closed loop joints.

As in the case of unilateral constraints, we can define a system of the form of (78):

$$\begin{bmatrix} {}^L a_F \\ {}^n a_F \\ {}^t a_F \end{bmatrix} = \begin{bmatrix} {}^L a_L \\ {}^n a_L \\ {}^t a_L \end{bmatrix} + A_r \begin{bmatrix} \lambda^L \\ \lambda^n \\ \lambda^t \end{bmatrix} \tag{93}$$

Where ($^L a$) represents the relative acceleration associated with the permanent constraints.

Let us apply the Gauss-Seidel method to solve the system (93). The calculation of the (5×P) lagrangian multipliers, associated to the closed loop joints, can be written in the form of (5×P) simple problems:

$$\lambda^{L_i} = \frac{1}{A_{r(i,i)}}{}^{L_i}a_F + \frac{1}{A_{r(i,i)}}\left(-{}^{L_i}a_L + \sum_{\substack{j=1..P+3Q \\ j\neq i}} A_{r(i,j)}\lambda^j\right) \tag{94}$$

$$= c_{L_i}{}^{L_i}a_F + b_{L_i} \qquad\qquad \text{for } i = 1..5\times P$$

This (i-th) equation is subjected to the constraint ${}^{L_i}a_F = 0$, which implies $\lambda^{L_i} = b_{L_i}$.

Thus, to treat the bilateral constraints, it is only necessary to add to the preceding pseudo code the few following lines:

```
{ bilateral constraints }
for i=1 to 5P
     calcul_bLi
     λLi=bLi
endfor
```

Joint drivers

Let us consider that the user wants to control the (i)-th joint (L_i) of the system (Σ). He must specify the position, velocity and acceleration of this joint. The unknown of the problem is then the U_i effort which we must apply on the joint (L_i) to obtain the desired movement. The (i)-th line of the dynamic equations is written, after inversion of the matrix H:

$$\ddot{q}_i = H_{(i)}^{-1}(U - C) + H_{(i)}^{-1}\left((\Phi^L)_{,q}\lambda^L + (\Phi^n)_{,q}\lambda^n + (\Phi^t)_{,q}\lambda^t\right) \tag{95}$$

Where $H_{(i)}^{-1}$ represents the (i)-th line of the matrix H^{-1}.

A simple calculation gives us:

$$U_i = \frac{1}{H_{(i,i)}^{-1}}\left(\ddot{q}_i - H_{(i)}^{-1}(\tilde{U} - C) - H_{(i)}^{-1}\left((\Phi^L)_{,q}\lambda^L + (\Phi^n)_{,q}\lambda^n + (\Phi^t)_{,q}\lambda^t\right)\right) \tag{96}$$

Where the vector $\tilde{U}$ is equal to the vector U whose (i)-th component is set to zero.

The equation (96) enables us to calculate U_i at each iteration of our decision procedure. After this calculation, it is advisable to update the variable a_L which depends on U. We define, for the joint (L_i), the pilot variable: pilote[i]=1 if the joint is controlled, zero if not.

We then add the following lines to the preceding pseudo code:

```
{ Driven joint }
for i=1 to N
     if pilote[i]==1
```

115

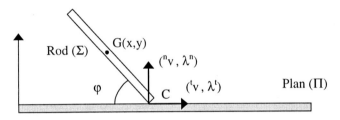

```
        calcul_U_i
      endif
    endfor
    calcul_a_L
```

We can see with which facility these modifications have been carried out. We now will study the cases of indeterminacy and inconsistency.

3.4.3 Particular Cases

We saw, paragraph (3.3.1.), that in the sliding case, we are unable to ensure the existence and the uniqueness of the results. To illustrate this, we will consider a simple example. It is a very classic example, suggested by Painlevé [46], which can be found in [19], [45], [47], [14], and whose detailed analysis is proposed in [48].
Let us consider (Σ), a homogeneous rod of length (2l), mass m and inertia $J=1/3ml^2$. This rod is in contact with the plan (Π) at the point C. We note ($^n v, ^t v$) the relative contact velocities represented figure 16. The position and the orientation of the rod are completely defined by the vector $q=[x,y,\varphi]^T$.

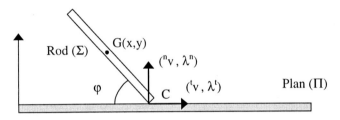

Fig.16. The rod example

The dynamics equations are written:

$$\begin{cases} m\ddot{x} = \lambda^t \\[2mm] m\ddot{y} = \lambda^n - mg \\[2mm] \dfrac{1}{3} ml^2 \ddot{\varphi} = -\lambda^t l \sin\varphi - \lambda^n l \cos\varphi \end{cases} \qquad (97)$$

Where (λ^n, λ^t) are the contact forces.
The constraints in acceleration are given by:

$$\begin{cases} {}^n a = \ddot{y} - l\ddot{\varphi}\cos\varphi + l\dot{\varphi}^2 \sin\varphi \\[2mm] {}^t a = \ddot{x} - l\ddot{\varphi}\sin\varphi - l\dot{\varphi}^2 \cos\varphi \end{cases} \qquad (98)$$

Let us suppose that the sliding velocity is strictly positive, then:

$$\lambda^t = -\mu\lambda^n \tag{99}$$

The equations (97), (98) and (99), allow us to express the normal contact force as a function of the relative normal acceleration:

$$\lambda^n = c_{ng}{}^n a + b_{ng} \tag{100}$$

With
$$\begin{cases} c_{ng} = \dfrac{m}{(1+3\cos\varphi(\cos\varphi-\mu\sin\varphi))} \\ b_{ng} = (-l\dot\varphi^2\sin\varphi+g)c_{ng} \end{cases}$$

Note : *We adopt here the index (ng) in order to specify that in this «normal equation», we consider also tangential parameters. Since we have only one contact, c_{ng} represents the inverse of the modified Delassus operator A_g defined by:*

$$A_g = (\Phi^n)_{,q} H^{-1}\left((\Phi^n)_{,q}^T + \tilde\mu(\Phi^t)_{,q}^T\right) = A + (\Phi^n)_{,q} H^{-1}\left(\tilde\mu(\Phi^t)_{,q}^T\right)$$

We will study the behavior of this system according to the coefficient of friction μ.

- If $\mu < \dfrac{1+3\cos^2\varphi}{3\cos\varphi\sin\varphi}$, the c_{ng} term is strictly positive.

  The intersection of the Signorini graph with the line of equation (100) can be pictured as in figure 14. We can thus ensure in this case the existence and the uniqueness of the solution.
  We will see that when the c_{ng} term is strictly negative, complexity increases.

- It is supposed subsequently that $\mu > \dfrac{1+3\cos^2\varphi}{3\cos\varphi\sin\varphi}$.

Note: We can notice that we are within the framework of the theorem of existence and uniqueness of Trinkle, paragraph (3.3.1.), with:

$$\bar\mu = \dfrac{1+3\cos^2\varphi}{3\cos\varphi\sin\varphi}.$$

Indeterminacy

Let us suppose that b_{ng} is strictly positive. The problem can be represented by the following figure:

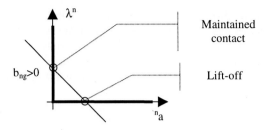

Fig.17. Indeterminacy case

We can see clearly that the line (100) cuts the Signorini graph in two points. Under these conditions, the problem has two solutions. We are thus in a characterized case of indeterminacy.

Inconsistency

If b_{ng} is strictly negative, we are then in the following situation:

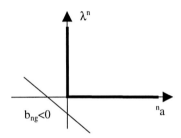

Fig.18. Inconsistency case

We do not have any intersection between the line (100) and the Signorini graph. The problem thus does not have any solution.

These problems raise two questions:
- First, can we detect such situations simply? It is clear that yes, since we have just to control the signs of the variables c_{ng} and b_{ng} after the convergence test of the decision procedure.
- The second question which arises, is what are we going to do once the problem is identified? It is necessary to be well aware that the problems of indeterminacy and inconsistency occur when we are in extreme cases, where the hypothesis (friction of Coulomb and rigid bodies) is no more valid. To come out of this bad step, it is thus necessary to release one of these two conditions. The simplest solution consists in requiring the user to decrease the friction coefficient (we can even say which one and suggest a value). If the user refuses this option, it is necessary to allow system

deformation. This solution is more or less simple to carry out, because it requires some modifications of the mechanism structure.

In practice, we can reach cases of indeterminacy or inconsistency in the course of simulation. It is thus advisable to be vigilant, and each time we detect a problem, we stop the simulation by informing the user of the situation and of the measures to be taken.

However, we can find in the literature certain methods which make it possible to deal with these problems. Among them we will recall here the Principle of Constraints of Kilmister&Reeve, which is stated in the following way:

> ***Proposal [45]:*** *The constraints must be checked by forces each time it is possible, if not we will have recourse to impulses.*

Is this code of conduct sufficient to solve our problems? The application of this principle to the case of inconsistency is direct. In this situation, the forces diverge and do not allow any more to verify the contact constraints. In accordance with the preceding proposal, we thus apply an impulse to the system. The jump which results from this operation is often called «the catastrophe of Lecornu », in honour of the author who first proposed to solve the paradox of Painlevé by means of impulses.

It appears clearly that this principle cannot solve a problem of indeterminacy, because it simply indicates to choose a force solution without however specifying which of these solutions must be chosen.

In order to deal with indeterminacy Painlevé has stated the following principle:

> ***Proposal [48]:*** *If two rigid bodies in contact without friction separate, then under the same conditions, but with friction, they will separate too.*

These principles are in fact recipes which allow to continue simulation at all cost, without any mechanical considerations. So as a conclusion, it appears reasonable to stop the simulation when we face such problems. The user will be informed, by a message of the situation and of the measures to be taken (to decrease the coefficient of friction or to introduce deformation). In this objective, we place tests at the end of the Gauss-Seidel procedures in velocity and in acceleration. These tests consist in controlling the sign of c_{ng} and b_{ng} for each sliding contact. We thus consider that we have an effective protection against the problems of inconsistency and indeterminacy.

Before ending this discussion, let us just say a word about the "time-stepping" method. When using such a method, a problem of indeterminacy or inconsistancy will always find a solution. The case of inconsistancy is particularly interesting because as a "time-stepping" method is working with impulses, it can thus verify the principle of Kilmister&Reeve without any external intervention. So the Painlevé paradox can be solved directly with this approach. But, it is really important to note that the solution then depends on the step of time. You will thus find different solutions as a function of the step of time. This phenomenon has been highlighted by Moreau in [14]. So even if a "time-stepping" method always finds a solution in the case of indeterminacy, as this solution is not unique, we are not interested by this property.

Let us now study the problem of impacts, this will be the subject of the next section.

4   Treatment of the Shocks

The object of this chapter is the treatment of the shocks. We again use the notations of the preceding chapter, and we place ourselves at the time t_n* where a shock occurs. At this instant, we suppose that the contact (C_i) is established with:

$$\begin{cases} \Phi^{n_i}(t_n*) = 0 \\ {}^{n_i}v(t_n*) < 0 \end{cases} \tag{101}$$

We admit here that the shock is an instantaneous phenomenon, which leaves the geometry invariant, but which involves a velocity discontinuity. This hypothesis of invariant geometry is justified in [49]. It is admitted that the velocity has bounded variations, we can thus define a velocity jump. Our objective is to calculate the velocity after the shock, therefore, we are interested in the jump of this variable:

$$\Delta^{n_i}v = {}^{n_i}v^+ - {}^{n_i}v^- \tag{102}$$

Where $\left({}^{n_i}v^-, {}^{n_i}v^+\right)$ represent respectively the normal relative velocities before and after the shock.

This chapter is composed of three paragraphs. First, we will establish the dynamic equations for an inelastic shock and we will again formulate an optimization problem under constraints. We will then study the definition of an impact law, elastic or partially plastic, and we will justify the choice of the coefficient of restitution which we have adopted. In a third paragraph, we will introduce a certain number of concepts which appear important to us. We will consider the problem of impulse propagation. We will then propose a procedure which makes it possible to interpret the impulsive forces in forces and to conclude we will be interested in the problem of the contact capture.

4.1   Formulation of the proble

We will see here how we can formulate an optimization problem under constraints. We will first establish the dynamic equations for an inelastic shock.

4.1.1   Formulation of the Dynamic Equations

We place ourselves on the time interval $[t_{n-1}, t_n]$. Let us consider the most general case of a system (Σ) with (P) closed loop joints and (Q) active frictional contacts. Under these conditions, the dynamic equations can be written:

$$H\ddot{q} - U + C - (\Phi^L)_{,q}^T \lambda^L - (\Phi^n)_{,q}^T \lambda^n - (\Phi^t)_{,q}^T \lambda^t = 0 \tag{103}$$

At time $t_n{}^*$ the system undergoes a shock. Let us see how the equations (103) behave. We will be interested in the limit of the following equation when t_{n-1} and t_n tend towards $t_n{}^*$:

$$\int_{t_{n-1}}^{t_n} (H\ddot{q})dt + \int_{t_{n-1}}^{t_n} (-U+C)dt - \int_{t_{n-1}}^{t_n} \left((\Phi^L)_{,q}^T \lambda^L\right)dt$$
$$- \int_{t_{n-1}}^{t_n} \left((\Phi^n)_{,q}^T \lambda^n\right)dt - \int_{t_{n-1}}^{t_n} \left((\Phi^t)_{,q}^T \lambda^t\right)dt = 0 \tag{104}$$

Successively let us consider each term of this equation. This procedure is borrowed from [18].

term 1: The operator H is a continuous function of time. The Average Value theorem thus enables us to ensure the existence of an instant t_k such as:

$$\int_{t_{n-1}}^{t_n} (H\ddot{q})dt = H(q(t_k))\int_{t_{n-1}}^{t_n} (\ddot{q})dt = H_{t_k}\left[\dot{q}\right]_{t_{n-}}^{t_n} \quad \text{with} \quad t_k \in \left[t_{n-1}, t_n\right] \tag{105}$$

If we consider the limit of this equation, we obtain:

$$\lim_{(t_{n-1}, t_n) \to t_n{}^*}\left(\int_{t_{n-1}}^{t_n} (H\ddot{q})dt\right) = H_{t_n{}^*}(\dot{q}^+ - \dot{q}^-)_{t_n{}^*} = H_{t_n{}^*}(\Delta\dot{q})_{t_n{}^*} \tag{106}$$

term 2: U and C are functions of the position and velocity (which has bounded variations), consequently we have:

$$\lim_{(t_{n-1}, t_n) \to t_n{}^*}\left(\int_{t_{n-1}}^{t_n} (-U+C)dt\right) = 0 \tag{107}$$

terms 3, 4, 5: The kinematic constraint matrices are continuous functions of time. We give here the principle of calculation for the closed loop joints. The Average Value theorem gives us:

$$\int_{t_{n-1}}^{t_n} \left((\Phi^L)_{,q}^T \lambda^L\right)dt = \left((\Phi^L)_{,q}^T\right)_{t_k} \int_{t_{n-1}}^{t_n} \left(\lambda^L\right)dt \quad \text{with} \quad t_k \in \left[t_{n-1}, t_n\right] \tag{108}$$

While passing to the limit, we have:

$$\lim_{(t_{n-1}, t_n) \to t_n{}^*}\left(\int_{t_{n-1}}^{t_n} \left((\Phi^L)_{,q}^T \lambda^L\right)dt\right) = \left((\Phi^L)_{,q}^T\right)_{t_n{}^*} \lim_{(t_{n-1}, t_n) \to t_n{}^*} \int_{t_{n-1}}^{t_n} \left(\lambda^L\right)dt$$
$$= \left((\Phi^L)_{,q}^T\right)_{t_n{}^*} S^L \tag{109}$$

Where the vector S^L represents the impulsive efforts associated to the λ^L closed loop efforts. In the same manner, we define the impulsive forces (S^n, S^t) associated with the contact forces (λ^n, λ^t).

According to these results, the dynamics equation (103) are written after passage to the limit:

$$H\Delta\dot{q} - (\Phi^L)^T_{,q}S^L - (\Phi^n)^T_{,q}S^n - (\Phi^t)^T_{,q}S^t = 0 \qquad (110)$$

Where we voluntarily omitted the index (t_n*), for it is understood that all these quantities are calculated at time (t_n*).

As you can see the inertia forces and the external efforts have been eliminated from equation (110). The principal unknowns of our problem are now, the velocity after the shock and the closed loop and contact impulses.

Let us see now how to formulate a contact law as a function of these variables.

4.1.2 Contact Law

We suppose here that all the contacts $(C_i)_{i=1..Q}$ are active at time t_n*.

Inelastic Normal Contact Law

The normal contact forces are subjected to the complementarity conditions in acceleration:

$$\begin{cases} \text{if} \quad i \in I_{nn} \quad {}^{n_i}a \geq 0 \quad \lambda^{n_i} \geq 0 \quad ({}^{n_i}a)(\lambda^{n_i}) = 0 \\ \text{else} \quad \lambda^{n_i} = 0 \end{cases} \quad \text{for} \quad i = 1,...,Q \qquad (111)$$

The condition $\lambda^{n_i} \geq 0$ implies:

$$S^{n_i} = \lim_{(t_{n-1},t_n) \to t_n*} \int_{t_{n-1}}^{t_n} \left(\lambda^{n_i}\right) dt \geq 0 \quad \text{for} \quad i = 1,...,Q \qquad (112)$$

Note : The reciprocal is not true in general.

According to this relation, we formulate complementarity conditions as a function of the velocity and of the impulses:

Proposal [13]: In the case of a shock, the following complementarity conditions must be checked:

$$\begin{cases} \text{if} \quad i \in I_n \quad {}^{n_i}v^+ \geq 0 \quad S^{n_i} \geq 0 \quad ({}^{n_i}v^+)(S^{n_i}) = 0 \\ \text{else} \quad \lambda^{n_i} = 0 \end{cases} \quad \text{for} \quad i = 1..Q \qquad (113)$$

In this form, this relation constitutes an inelastic normal contact law.
Note : *For only one contact, this law is equivalent to the following condition:*
$(^{n_i}v^+) = 0$.

Tangential Contact Law

The things are appreciably more complex here. At the time of impact, we must face several situations, in which the impulses will not always check the same inequalities as the forces. To illustrate our matter, we will suppose that the shock is not instantaneous, but that it is held on a very small interval of time $[t_n*-\delta, t_n*+\delta]$. We will see up to what point we can connect the forces, at time $(t_n*+\delta)$, with the impulses calculated on the interval $[t_n*-\delta, t_n*+\delta]$. Five cases arise:

The contact slips at the beginning and at the end of the time interval (no transition)
It is clear that one has the following relation:

$$\lambda^{t_j} = \tilde{\mu}\lambda^{n_j} \quad \Rightarrow \quad S^{t_j} = \tilde{\mu}S^{n_j} \tag{114}$$

The contact sticks at the beginning and at the end of the time interval (no transition)
It is shown easily that:

$$\left|\lambda^{t_j}\right| \leq \mu\lambda^{n_j} \quad \Rightarrow \quad \left|S^{t_j}\right| \leq \mu S^{n_j} \tag{115}$$

The contact slips initially and then sticks (slip-stick transition)
Here still, it is easy to show that:

$$\left|\lambda^{t_j}\right| \leq \mu\lambda^{n_j} \quad \Rightarrow \quad \left|S^{t_j}\right| \leq \mu S^{n_j} \tag{116}$$

The contact sticks initially and then slips (stick-slip transition)
The following relation is not always checked:

$$\lambda^{t_j} = \tilde{\mu}\lambda^{n_j} \quad \Rightarrow \quad S^{t_j} = \tilde{\mu}S^{n_j} \tag{117}$$

The contact slips initially and then reverse sliding occurs (reverse sliding transition)
Once again, the following relation is not true:

$$\lambda^{t_j} = \tilde{\mu}\lambda^{n_j} \quad \Rightarrow \quad S^{t_j} = \tilde{\mu}S^{n_j}$$

The following figure illustrates these two last problematic cases:

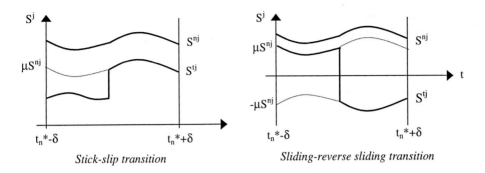

Fig.19. Variation of the contact impulses during a shock

The formulation of the Coulomb law as a function of impulses is thus not direct. Though it is, we prolong this law in impulse, it is an hypothesis which is important to note, here is its statement:

Proposal [13]: In the case of a shock, we adopt a tangential contact law of the form:

$$
\begin{cases}
\text{if} \quad i \in I_r \quad \left\| S^{t_i} \right\| = \sqrt{(S^{x_i})^2 + (S^{y_i})^2} \leq \mu \lambda^{n_i} \\[4mm]
\text{if} \quad i \in I_s \quad S^{t_i} = \begin{bmatrix} S^{x_i} \\ S^{y_i} \end{bmatrix} = \begin{bmatrix} -\dfrac{(^{x_i}v^+)}{\left\| ^{t_i}v^+ \right\|} \mu S^{n_i} \\[4mm] -\dfrac{(^{y_i}v^+)}{\left\| ^{t_i}v^+ \right\|} \mu S^{n_i} \end{bmatrix} = -\dfrac{^{t_i}v^+}{\left\| ^{t_i}v^+ \right\|} \mu S^{n_i} = \tilde{\mu} S^{n_i}
\end{cases}
\tag{118}
$$

The graph of this relation is given, in two dimensions, on the following figure:

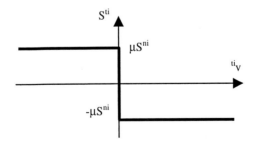

Fig.20. Tangential shock law, Coulomb law as a function of impulses

Let us see now how to use these laws to solve our problem.

4.1.3 Formulation and Resolution of an Optimization Problem under Constraints

Setting of the proble

The problem is very similar to the one of the preceding chapter and the procedure we are going to use, in order to solve it, is exactly that of Moreau (paragraph 3.2.2.). We will write an optimization problem under constraints (equalities and inequalities) of the form:

$$
\begin{bmatrix} {}^{L}v^{+} \\ {}^{n}v^{+} \\ {}^{t}v^{+} \end{bmatrix} = \begin{bmatrix} {}^{L}v^{-} \\ {}^{n}v^{-} \\ {}^{t}v^{-} \end{bmatrix} + A_r \begin{bmatrix} S^{L} \\ S^{n} \\ S^{t} \end{bmatrix} \tag{119}
$$

Where $({}^{L}v^{-}, {}^{L}v^{+})$ are the relative velocities associated to the closed loop joints before and after the shock.

$({}^{n}v^{+}, {}^{t}v^{+})$ represent respectively normal and tangential relative velocities after the shock.

$({}^{n}v^{-}, {}^{t}v^{-})$ are normal and tangential relative velocities before the shock.

Ar is the Delassus operator.

It is clear that we hav $({}^{L}v^{-}, {}^{L}v^{+}) = 0$, so the system (119) can thus be written:

$$
\begin{bmatrix} 0 \\ {}^{n}v^{+} \\ {}^{t}v^{+} \end{bmatrix} = \begin{bmatrix} 0 \\ {}^{n}v^{-} \\ {}^{t}v^{-} \end{bmatrix} + A_r \begin{bmatrix} S^{L} \\ S^{n} \\ S^{t} \end{bmatrix} \tag{120}
$$

This system is subjected to the conditions imposed by the contact law:

$$
\begin{cases} \text{Signorini}({}^{n}v^{+}, S^{n}) \\ \text{Coulomb}({}^{t}v^{+}, S^{t}) \end{cases} \tag{121}
$$

Resolution of the proble

The problem is formulated here as a function of the velocities, we thus adopt the Gauss-Seidel method of which the pseudo code was given at the paragraph (3.2.2.). We make to this procedure the necessary modifications to take into account bilateral constraints and driven joints (see paragraph 3.4.2.).

The results of existence and unicity of the solution, and convergence of the method, are exactly the same than in the paragraph (3.3.).

Up to now, we have made the hypothesis of an inelastic shock, let us see now how to model elastic or partially plastic shocks.

4.2   Taking into Account of a Normal Restitution Coefficient

In order to model an elastic shock or a partially plastic shock, we will introduce a restitution coefficient. We make here the hypothesis that the materials in contact don't undergo deformations in the tangent plane of contact. We thus do not consider tangential restitution coefficient, although experimental [50] and numerical studies (modelling by finite elements method) [51] allow to highlight this phenomenon. However, it seems that, in a first approximation, we can consider only a normal restitution coefficient, even if, with this hypothesis, we won't be able to simulate magic ball bounces [13]. We return the reader, who may be interested by this subject, to the work of Pfeiffer [19] for the modelling of a tangential restitution coefficient.

This paragraph is divided into three parts. First, we will point out the principal definitions of the restitution coefficient which can be found in the literature. Among these approaches, we will select the Poisson restitution coefficients. We will formulate a normal contact law using this coefficient. To finish, we will see how our algorithm have been modified to take into account the Poisson restitution coefficient.

4.2.1   Definitions

We consider here only one contact, we will see further how to prolong these laws in the case of multiple contacts. There are three different definitions of the restitution coefficient:

The Newton coefficient (kinematic definition) (see for example [52]).
This coefficient is defined very simply by the following relation:

$$e_N = -\frac{(^n v^+)}{(^n v^-)} \tag{122}$$

The coefficient of Poisson (dynamic definition) (see for example [53]).
The shock, although instantaneous, is divided into two phases, a phase o compression and a phase of restitution. We respectively associate with each of these phases the indices (c) and (r). The Poisson restitution coefficient is defined by the impulse at the end of the phase of restitution divided by the impulse at the end of the phase of compression. Which can be written:

$$e_P = \frac{S_r^n}{S_c^n} \tag{123}$$

The coefficient of Stronge (energy definition) [54][55].
This author proposes a definition of the restitution coefficient which is based on energy considerations. For that, once again the shock is divided into two phases of compression and of restitution and we have:

$$e_S = -\frac{\int_{t_c}^{t_r} \lambda^n ({}^n v^+) dt}{\int_{t_n^*}^{t_c} \lambda^n ({}^n v_c) dt} = -\frac{\int_{\lambda_c^n}^{\lambda_r^n} ({}^n v^+) d\lambda^n}{\int_{\lambda_n^*}^{\lambda_c^n} ({}^n v_c) d\lambda^n} \qquad (124)$$

Where (tc, tr) represent the instants of end of compression and restitution respectively. These instants are completely abstract since we consider that the shock is an instantaneous phenomenon. It simply makes it possible to express the work of the normal impulse. Pérès [20] gives a definition of the coefficient of restitution which is completely similar.

All these coefficients are by definition ranging between zero and one. Zero value corresponds to an inelastic shock and the value one means an elastic shock.

The Stronge restitution coefficient is the most rigorous and it is from a mechanical point of view the most satisfactory. But, on the other hand, its use leads to a system of non-linear equations very difficult to solve, this is why we have not chosen this approach.

Before eliminating this solution completely, it is advisable to be ensured of the validity of the other approaches. The Newton and Poisson coefficients lead to the resolution of a linear system, but they present certain defects about which we now will speak.

Note: let us see at which conditions the Newton and Poisson restitution coefficients are equivalent. We have equivalence [47] when:

- *μ=0 - No friction.*
- *e=0 - Inelastic shock.*
- *The final contact status is the sliding.*
- *$v_{to}=0$ - The initial tangential velocity is null, we say that we have a direct shock.*
- *The gravity centers of the two impacting bodies and the point of contact are in the same line. It is said that we have a generalized central impact.*

Apart from these five cases, we will obtain different results. Rigorously, it should be noted that these results are only true for one contact, in the event of multiple contacts, we obtain different results (see for example [19]).

It is now well known that the Newton restitution coefficient can lead to an increase of the total energy in the case of a non-central impact with friction (see for example [47] where the case of a rod impacting a plane is analysed using Routh method). As a consequence, we will never use the Newton restitution coefficient.

What about the Poisson restitution coefficient? It can be shown [56], in the case of a non-central impact with friction, that the work done by the normal contact impulse is negative even for restitution coefficient equal to one. It should be zero if the shock model was perfect. Here is thus the only weakness of the coefficient of restitution of Poisson, it dissipates a little too much energy. Is this a sufficient reason not to use it

From a practical point of view, the behaviour of the Poisson restitution coefficient appears completely satisfactory to us. We thus adopt this coefficient, like [53][37][19], because it seems the best compromise between the simplicity of calculations and the quality of the results. We can find in [53] experimental validations of this approach on three simple examples, which confirm that our choice is a good one.

It is important to note that there does not exist in the literature any demonstration which proves that the association "coefficient of restitution of Poisson, Coulomb law" leads systematically to a dissipative phenomenon. Some authors have tried to demonstrate this principle but without success, either because their demonstration comprised an error [19], or because their hypotheses were too strong (equal coefficients) [37]. Nevertheless, we can note that there does not exist in the literature examples which contradict this principle and it is quite hazardous to advance the opposite like [34].

There exists obviously other approaches of which we voluntarily have not spoken until now. We report here two of them.

The first consists in using the Newton restitution coefficient, but because of the energy gain problem, it is clear that it cannot be done without taking certain freedoms with respect to the law of Coulomb. Brach [57][52] and Smith [58] have chosen this approach. Brach proposes to control the total energy variation after each shock. This author derives some bounds for the friction coefficient, which is thus modified so that we always have a loss of energy. We can note that the calculus of the total energ variation requires an additional work. The interested reader will find in [56] a criticism of this model. In a way even more abrupt, Smith proposes to replace the Coulomb law by the following expression:

$$S^t = -\mu \left| S^n \right| \frac{\left\| {}^t v^- \right\| {}^t v^- + \left\| {}^t v^+ \right\| {}^t v^+}{\left\| {}^t v^- \right\|^2 + \left\| {}^t v^+ \right\|^2} \tag{125}$$

You will find a detailed presentation of these methods in [49] and [59]. You will also find in [59] a description of «new contact laws» whose style is very close to Smith laws.

A second approach consists in discretizing the solids in contact locally in order to model their deformations. This method is described in [56]. The principle of the method consists in connecting the displacement of the contact nodes to the normal and tangential contact forces, using Green functions of influence. This is a regularized approach, the shocks are modelled without velocity discontinuity. The problem is again formulated as an optimization problem under constraints, whose principal unknown are the relative accelerations and the contact efforts.

Each of these formulations has its own advantages and drawbacks, nevertheless none seems to us as compact and rigorous as the approach we have adopted.

We will see now, how to formulate a normal contact law using the Poisson restitution coefficient.

4.2.2 Normal Contact Law

Definition

According to the Poisson restitution coefficient definition, the shock will be separated in two phases, the compression and the restitution phases. The compression phase corresponds to an inelastic shock. The proposal, stated paragraph (4.1.2.), thus provided all the necessary elements to its treatment. We will reuse exactly the same contact law. As previously, the indices (c) and (r) mean that a quantity is calculated at the end of the compression phase or at the end of the restitution phase.

> ***Proposal [13]:*** *During the compression phase, the following complementarity conditions must be checked:*

$$\begin{cases} \text{if} \quad i \in I_n \quad {}^{ni}v_c \geq 0 \quad S_c^{ni} \geq 0 \quad ({}^{ni}v_c)(S_c^{ni}) = 0 \\ \text{else} \quad \lambda^{ni} = 0 \end{cases} \qquad \text{for} \quad i = 1..Q \qquad (126)$$

For the restitution phase, we state the following principle:

> ***Proposal [13]:*** *During the restitution phase, the following complementarity conditions must be checked:*

$$\begin{cases} \text{if} \quad i \in I_n \quad {}^{ni}v^+ \geq 0 \quad (S_r^{ni} - e_{P_i}S_c^{ni}) \geq 0 \quad ({}^{ni}v^+)(S_r^{ni} - e_{P_i}S_c^{ni}) = 0 \\ \text{else} \quad \lambda^{ni} = 0 \qquad\qquad\qquad \text{for} \quad i = 1..Q \end{cases} \qquad (127)$$

The following figure illustrates these two relations:

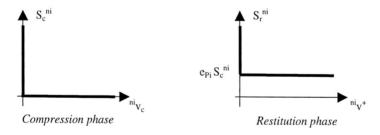

Compression phase   *Restitution phase*

Fig.21. Normal shock la

We will say, by a language abuse, that we have obtained the «modified Signorini conditions».

Let us see now how our algorithm was modified in order to take into account the Poisson restitution coefficient.

4.2.3 Implementation

The compression phase is very simple to model, since it corresponds to an inelastic shock, i.e. a shock with a Newton restitution coefficient equal to zero. The system (120) can thus be written:

$$\begin{bmatrix} 0 \\ {}^n v_c \\ {}^t v_c \end{bmatrix} = \begin{bmatrix} 0 \\ {}^n v^- \\ {}^t v^- \end{bmatrix} + A_r \begin{bmatrix} S_c^L \\ S_c^n \\ S_c^t \end{bmatrix} \qquad (128)$$

This system is subject to the conditions imposed by the contact law:

$$\begin{cases} \text{Signorini}({}^n v_c, S_c^n) \\ \text{Coulomb}({}^t v_c, S_c^t) \end{cases} \qquad (129)$$

For the restitution phase, the system (120) is written:

$$\begin{bmatrix} 0 \\ {}^n v^+ \\ {}^t v^+ \end{bmatrix} = \begin{bmatrix} 0 \\ {}^n v_c \\ {}^t v_c \end{bmatrix} + A_r \begin{bmatrix} S_r^L \\ S_r^n \\ S_r^t \end{bmatrix} \qquad (130)$$

With the conditions:

$$\begin{cases} \text{Signorini\_modified}({}^n v^+, S_r^n) \\ \text{Coulomb}({}^t v^+, S_r^t) \end{cases} \qquad (131)$$

Note : The Coulomb law is prolonged in impulse (see paragraph 4.1.2.). As we have supposed it, we do not introduce a tangential restitution coefficient. However, it is interesting to note that the introduction of this parameter will lead to modifications of the Coulomb law during the restitution phase [13].
These two problems are solved using the Gauss-Seidel projection method, formulated at the velocity level, described in the paragraphs (3.3.2.) and (4.1.3.). Once again, when the Gauss-Seidel method has converged, we assemble a differential-algebraic system, which form is dictated by the new contact status. We thus calculate the velocity jump with a maximum precision.

Our algorithm modifications are thus extremely simple and can be summarized by the fact that we solve two successive contacts problems (compression and then restitution). At the end of this process, our algorithm gives the velocity after the shock, as well as the impulses and the contact status. We thus obtain new initial conditions which enable us to continue integration. The problem of the shocks is thus completely solved.

We now will insist on some problems which appear important to us.

4.3   Important Concepts

We will tackle successively here the problems of impulse propagations, of the interpretation of impulses in forces and finally of capture, which is a big problem to deal with when an acceleration formulation is chosen.

4.3.1 Propagation of the Impulses

This problem is seldom mentioned in the literature, however it is characteristic of the contact model we use. We will highlight it on the famous example of «the Newton cradle» (figure 22 below). Let us consider a set of three spheres (1), (2) and (3). The spheres (2) and (3) are initially in contact and at rest, the sphere (1) is unstuck and dropped without initial speed.

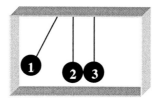

Fig.22. The Newton cradle in initial position

The restitution coefficients between the spheres are fixed to one and we do not consider friction. When the sphere (1) comes into contact with the sphere (2), a shock occurs. We then calculate the velocity after the shock by using the procedure described in the paragraph (4.2.3.). We obtain the following results:

$$\left(\dot{q}_1^+ = -\frac{1}{3} \quad \dot{q}_2^+ = \frac{2}{3} \quad \dot{q}_3^+ = \frac{2}{3} \right) \tag{132}$$

Which corresponds to the following situation:

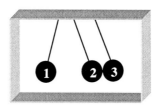

Fig.23. The Newton cradle after the first shock using
unilateral contact theor

It is clear that it is not the result which we expected. In fact, we note that the contact impulse is not propagated in the system. [53] presents a detailed analysis of this example and proposes to solve this problem by considering sequences of shocks. Therefore, they introduce the concept of external and internal contacts. An external contact is a new contact, which creates a shock. An internal contact is an active contact at the time of the shock. The principle of the method consists in numbering the internal contacts as a function of the number of bodies which separate them from an external contact. To illustrate this principle, we give the following figure:

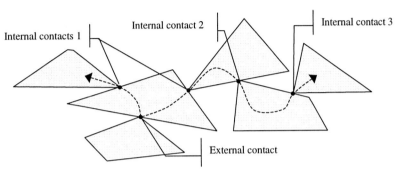

Fig.24. Sequence of internal contacts

Once this operation carried out, we deal first with a problem where only the external contact is active. Then, we solve a second problem in which all the contacts are released, except the internal contacts with the number (1). We continue this operation until all the internal contacts are reviewed. We thus artificially propagate the impulses through system.

If we apply this principle to our problem, we obtain two successive shocks (external contact, then internal contact (1)). What gives us the following solutions:

$$\left(\dot{q}_1^+ = 0 \quad \dot{q}_2^+ = 1 \quad \dot{q}_3^+ = 0\right) \quad \text{then} \quad \left(\dot{q}_1^+ = 0 \quad \dot{q}_2^+ = 0 \quad \dot{q}_3^+ = 1\right) \tag{133}$$

We thus obtain a solution which is fully satisfactory. Nevertheless, this method has three essential drawbacks :

- It is clear that this method quickly becomes very heavy when the number of internal contacts (I) becomes large. We are then obliged to deal with (I) successive problems.
- Simple examples (such as 3-ball Newton's cradle) prove that it may not converge to any outcome, or that it may provide the user with two solutions [49].
- This method is not applicable to the systems which contain closed chains, formed indifferently by bilateral connections or internal contacts. To explain this, we represent the following figure:

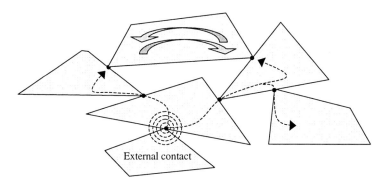

Fig.25. Propagation of the impulses in a closed loop syste

In the light of these remarks, it appears clearly that the method of [53] is not usable in the general case. Thus, we do not have any tool at our disposal to face the problem of the impulse propagations. It is a major defect of the analytical contact model, it is important to be informed of it. There has been recently an increase of interest on that subject and new contact laws are studied in order to overcome this problem. Here again, the only way to avoid this difficulty is to introduce a certain amount of deformation. Nevertheless, we refuse this solution and we take the party to accept this defect.

4.3.2 Interpretation of the Impulses in Forces

At the time of a shock, our algorithm calculates a set of contact impulses which does not have obvious mechanical significance. However, it can be interesting to have at our disposal an estimate of the contact forces, to carry out various types of analyses (FEM, fatigue analysis). It would be thus desirable to find a method which makes it possible to translate the impulses into efforts. To our knowledge, this problem has not yet been treated in the literature. We thus have developed our own approach. The principle of the method consists in interpreting a normal impulse of an "external" contact (see preceding paragraph for the definition of an external contact) in a normal force. For that, we will consider a problem with only one contact without friction. Let us see how to proceed.

We consider the system (Σ) made up of (N) principal joints, (P) closed loop joints and (Q) active contacts. We place ourselves at the time $t_n{}^*$, just after the shock resolution by the Gauss-Seidel procedure. At this time, the relative velocities $({}^n v^+, {}^t v^+)$ and the contact impulses (S^n, S^t) are known parameters.

Among the (Q) active contacts, we select the (i)-th external contact (C_i) such that :

$${}^{n_i} v^- = \max_{j=1,\ldots,Q} ({}^{n_j} v^-).$$ Just remember that, at the time of a shock, we have to solve a

system of the form:

$$\begin{bmatrix} 0 \\ ^n v^+ \\ ^t v^+ \end{bmatrix} = \begin{bmatrix} 0 \\ ^n v^- \\ ^t v^- \end{bmatrix} + A_r \begin{bmatrix} S^L \\ S^n \\ S^t \end{bmatrix} \qquad (134)$$

We will be interested in the $(5 \times P + i)$-th line of this equations system, which we write:

$$A_{r(i,i)}^{-1} (^{n_i} v^+ - ^{n_i} v^-) =$$

$$A_{r(i,i)}^{-1} \left(\sum_{j=1..5P} A_{r(i,j)} S^{Lj} + \sum_{\substack{j=5P+1..5P+Q \\ j \neq 5P+i}} A_{r(i,j)} S^{nj} + \sum_{j=5P+Q+1..5P+2Q} A_{r(i,j)} S^{tj} \right) + S^{n_i} \qquad (135)$$

Or:

$$A_{r(i,i)}^{-1} (^{n_i} v^+ - ^{n_i} v^-) = G(S^{n_i}) \qquad (136)$$

We can consider that this equation corresponds to the problem of a particle which impacts a plane. From this observation, we will make a parallel between this equation and the equivalent equation which is given when we use a penalization method. With this approach, we set:

$$m\ddot{y} + T\dot{y} + Ky^{3/2} = 0 \quad \text{with} \quad \begin{cases} m = A_{r(i,i)}^{-1} \\ \dot{y} = -^{n_i} v \end{cases} \qquad (137)$$

The (y) variable represents the penetration between the bodies (positive in the event of contact). So that the equation (47) is complete, we have to define the parameters (K,T). The term of stiffness K is classically calculated very simply using the Hertz theory (see definition in introduction).

Note: The calculation of the term of stiffness K implies that one at least of the bodies in contact has a circular form. We will take into account this remark for the choice of contact (C_i). If no contact of this type exists, we ask the user to enter a radius of curvature.

The choice of the term of damping T is more delicate, because its form determines the existence of an analytical solution. Marhefka [60] proposes to use a term of damping of the form:

$$T = \frac{3}{2} \Omega K y^{3/2} \qquad (138)$$

With this definition, the equation has an analytical solution of the form:

$$y = 5/2\sqrt{-\frac{5m}{9K\Omega^2}}\,5/2\sqrt{3\Omega(\dot{y}-\dot{y}_0)+2\ln\left|\frac{2+3\Omega\dot{y}_0}{2+3\Omega\dot{y}}\right|} \tag{139}$$

Where $\dot{y}_0 = -^{n_i}v^-$

With the help of this expression, we can calculate the term Ω. Indeed, it is enough to notice that at the end of the shock the penetration (y) is null. The parameter Ω must thus satisfy the relation:

$$3\Omega(^{n_i}v^- -\,^{n_i}v^+) + 2\ln\left|\frac{2-3\Omega(^{n_i}v^-)}{2-3\Omega(^{n_i}v^+)}\right| = 0 \tag{140}$$

This relation makes it possible to connect the term of damping T an the normal relative velocity after the shock.
All the problem (137) parameters are now determined. By analogy between the equations (136) and (137), we have:

$$G(\lambda^{n_i}) = \max_{y\geq 0}(T\dot{y} + Ky^{3/2}) = \max_{y\geq 0}\left(Ky^{3/2}(\frac{3}{2}\Omega\dot{y}+1)\right) \tag{141}$$

We thus deduce from this equation an equivalent normal contact force. A proportionality factor Λ is defined by the relation:

$$\Lambda = \frac{\lambda^{n_i}}{S^{n_i}} \tag{142}$$

To translate all the other impulses into efforts, we generalize this relation in the following way:

$$\begin{bmatrix}\lambda^L\\\lambda^n\\\lambda^t\end{bmatrix} = \Lambda\begin{bmatrix}S^L\\S^n\\S^t\end{bmatrix} \tag{143}$$

The calculation of Λ is carried out each time a shock occurs.
We compared the results obtained by this procedure with those given by MECHANICA and DADS on some simple examples. In all the cases, the results obtained are of the same order of magnitude. But, it is important to note that this procedure will fail if (E) external contacts are activated at exactly the same time with very different materials (consider for example a steel ball and a rubber ball dropped on a plane from the same height). In this case, if the (E) external contacts are independent the procedure should be run for the (E) independent problems. In the

case of dependent contacts, we have no solution. But we have to say here that, in the case of an acceleration formulation with a high accuracy event detection, external contacts will very seldom occur at the same time.

This approach is our first trial to answer this difficult question. But despite all our efforts and to be perfectly honest, we have to notice that contact force values during a shock must be considered as a rough idea and not exact values. This remark is not only valid for SDS (Solid Dynamics), but also for ADAMS (MDI), DADS (LMS) and Mechanica Motion (PTC). Indeed, even for a simple ball dropped on a plane, you have some differences on the maximum normal contact force greater than 30% (for the same movement: same velocity before and after the shock) depending on the software you use. All what we can say today is that our procedure gives normal contact forces very close to Mechanica Motion (PTC), not worse, not better.

4.3.3 Problem of Capture

We have made the choice to work in acceleration for reasons of treatment speed. This method imposes that one systematically seeks all the events which occur during the simulation. However, in certain cases, this strategy can be dangerous. Let us consider the example of a sphere which bounces on a plane with a restitution coefficient ($e_P<1$). After a certain number of shocks, the amplitude of the rebounds becomes very small, it is then impossible to detect each event. It is thus advisable to define a criterion which enables us to decide when the rebounds must cease. For that, we place ourselves at the end of the shock procedure, and we systematically test the relative normal velocities of the released contacts. If these velocities are significant, we accept that the corresponding contacts are released, otherwise their restitution coefficients are set to zero and the restitution phase is done again. Here is the pseudo code of our capture procedure :

```
{ Procedure of shock }
{ Storage of the restitution coefficient}
for i=1 to Q
    e_temp_Pi= e_Pi
endfor
{ Phase of compression }
    Gauss-Seidel procedure in velocity
    Differential-algebraic system
    capture_number=0
    while capture_number<=1
        label 1
        { Phase of restitution }
        Gauss-Seidel procedure in velocity
        Differential-algebraic system
        { Capture procedure}
        capture_number=capture_number+1
        capture=0
        for i=1 to Q
        { Modification of the restitution coefficient }
        if (status⁻[i]==1) & (status⁺[i]==0) & (niv⁺<εi) & (ePi>0)
            ePi=0
```

```
            capture=capture+1
        endif
    endfor
    if capture==0
        capture_number=2
    endif
endwhile
```
{ Update of the restitution coefficient}
```
for i=1 to Q
    ePi=e_tempPi
endfor
```
{ End shock }

Where ε is a vector specified by the user.

Status⁻ and status⁺ are respectively the contact status before and after the shock.

You can notice that the variable capture_number can not exceed one, otherwise the shock ends. That is to say that we just have a restitution phase, some capture tests and eventually another restitution phase, but no more. In fact we should have done many successive procedures "capture tests & restitution phase" until no capture problem is detected. But this process can be time consuming. Furthermore, we have to say that this procedure is really difficult to implement because of the round-off errors. We are working here on very small quantities, close to the integration precision, so we must be careful. Today, this procedure gives very good results and it seems that it is sufficient to stabilize the system in the case of capture problems.

5 Industrial Example - C60 Circuit Breaker

The C60 is a low voltage circuit breaker (domestic circuit breaker) which has been developed by the Schneider Electric company in Grenoble, France. This example is especially interesting because it is really relevant of the problems we have to treat. The structure of this mechanism is as follows:

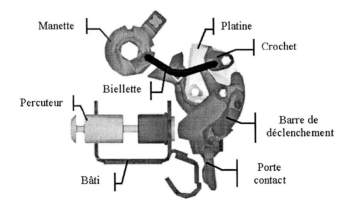

Fig.26. The C60 circuit breaker

137

The C60 circuit breaker looks like a simple mechanism with only eight bodies and eight joints (one closed loop joint). But the complexity of this mechanism relies in the number of contacts which occur during the movement. Indeed, we have to define twelve contacts to model correctly the behaviour of this mechanism. All these contacts are 2D contacts, and they are defined as point-line, point-segment or point-curve contacts, circle-line, circle-segment or circle-curve contacts.

We have to note that this mechanism is in equilibrium only because we have a contact between the "crochet" and the "barre de déclenchement" (see figure below). Without friction the equilibrium position of this mechanism changes.

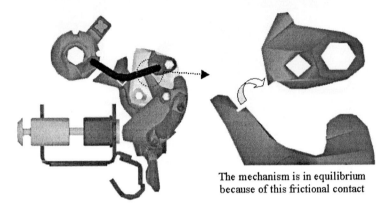

The mechanism is in equilibrium
because of this frictional contact

Fig.27. Equilibrium of the C60 circuit breaker

So this mechanism is both interesting because we have a lot of contacts but also because it is very sensitive to friction.

The simulation model has been built using STL geometry imported from Pro-Engineer (PTC) in SDS (Solid Dynamics). It takes approximately ten hours to import parts, define mass properties, define joints, contacts and apply loads. For the contact parameters, we have used an approximation of the restitution coefficients measured by one of our specialists (on a very simple system ball-plane), and the friction coefficients have been estimated using values found in the literature.

In order to validate our approach we have made some measurements on a prototype when the currant intensity grows abnormally. When such an event occurs, the "percuteur" is submitted to an electrodynamic force which can be measured or calculated (using E-mag software). We then obtain a force which is defined as a function of the current intensity and of the "percuteur" displacement. We are going to apply this force on the "percuteur" in order to simulate the movement of the C60, and we will then compare the simulation results with the experimental results.

We have made two kinds of experimental measures which we are now going to describe.

5.1 First Experiment

A X-ray analysis has been performed in order to record the distance OD between the "bati" and the "porte-contact" where the electrical contact is realized:

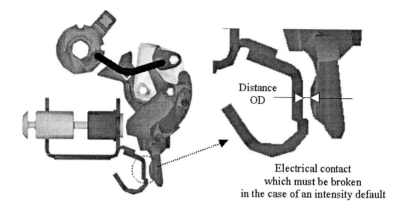

Fig.28. Electrical contact of the C60 circuit breaker

Here are the results we have obtained:

	Simulation measures	Experimental measures
Time Impact (t_I)	0.944 ms	0.912 ms
I Impact	4012 A	3833 A
OD=0.9 m	t_I + 67 µs	t_I + 73 µs
OD=2.9 m	t_I + 223 µs	t_I + 220 µs
OD=3.6 m	t_I + 279 µs	t_I + 288 µs

Where the time of impact is the time when the "percuteur" bumps into the "porte-contact".

This experiment validates both the electrodynamic force model we have used, and also, of course our frictional contact model.

5.2 Second Experiment

In order to get a better idea of the behaviour of the C60, we have also realized a movie with a high speed camera (4400 images per second, 225 µs between each image). We have thus obtained the distance OD as a function of time. Here are the results we have obtained :

Porte-contact Displacement (m)

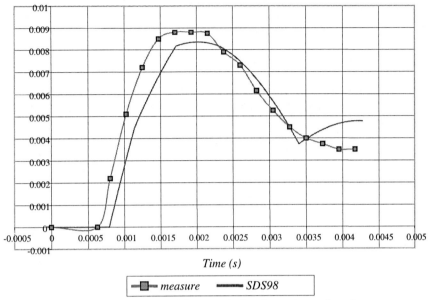

Time (s)

measure ▬▬▬ SDS98

Fig.29. Comparison of the numerical and experimental results

We obtain a very good correlation between the simulation and the experimental results.

Today, we have tested our approach on two other industrial examples and we have validated the results of the simulation using experimental measures. One of these mechanism was especially complicated, with 20 bodies and up to 35 frictional contacts. Each time, we have been able to predict the opening velocity of the electrical contacts as a function of time with an error inferior to 15%.

Of course, we have also treated many academic examples (most of them where found in [19]), such as the rocking problem [15], the woodpecker toy [19] or other industrial examples like emergency stop button or medium and high voltage circuit breakers. Each time it was possible, we have made some comparizon with the Moreau algorithm.

6   Conclusion

This document describes a frictional contact model based on unilateral constraint theory. To our mind, this approach has some very interesting properties.

Our algorithm has been implemented in the SDS software (Solid Dynamics), and a commercial version is available since may 1999. This version only treats 2D frictional contacts, but a version with 3D contacts should be available during the spring 2000.

The main characteristics of the method we have implemented are as follows:

- We use a high order integration scheme (RK5 Dormand&Prince) with a variable step size. This high accuracy integration scheme guarantees the validity of the results. Furthermore, we have implemented some very useful tools like dense output for event finding, automatic stiffness detection and also a PI control of the step size [8].
- We have been able to couple this high order integration scheme with a rigorous frictional contact model. The structure of our program is interesting, because the frictional contact treatment is independent of the integration scheme. We can thus change the integration scheme or plug this contact module on another program very easily.
- We have chosen to write the frictional contact problem as two coupled optimization under constraints, which are solved simultaneously using a Gauss-Seidel method. This method is extremely robust and allows to model the quadratic Coulomb cone without any approximation (polyhedral Coulomb cone). We have modified this method in order to use an acceleration formulation, to take into account bilateral joints and driven joints, and also to introduce a Poisson restitution coefficient.
- We have considered the important cases of inconsistency and indeterminacy. We have implemented some tests to detect these situations, the validity of the results is thus guaranteed.
- The numerical results we have obtained with our approach are perfectl satisfactory, they are both stable and precise. We have no more oscillations (unlike penalization methods) and the contact forces or the accelerations are only perturbed by the events which occurred during the simulation. Furthermore, we have been able to validate our approach on industrial examples.
- All the events are perfectly defined and reported to the user. Each event can be located accurately, so the results analysis is really easier. You will no more miss an event because of the step size (as in the classical result reports where all the results are pictured with a constant step size).

Of course our approach has some drawbacks:

- We can only treat a limited contact number. In fact, this is not really the number of contacts which is limited, it is the number of events which occurred during a step of time. If this number is to big, our strategy, which consists in finding systematically all the events, won't be effective.
- As we model frictional contacts by the addition of constraints, the system may become hyperstatic in the course of the simulation. In order to deal with this problem, the hyperstatism index is reported to the user at each step of time.

These two points have to be well understood by the user.

Acknowledgement

This research was partly supported by the French Council for Technical Research (ANRT) under the grant-in-aid n°95-681.

References

[1] DADS, *"User's Manual"*, Computer Aided Design Software Incorporated, Oakdale, Iowa.

[2] K.E. Brenan, S.L. Campbell, L.R. Petzold, *"Numerical Solution of Initial-Value Problems in Differential-Algebraic Equations"*, Elsevier Science Publishing Co., 1989.

[3] J.J. Moreau, *"Some numerical methods in Multibody Dynamics : Application to Granular Materials"*, European Journal of Mechanics, A/Solids, vol.13, n°4, p.93-114, 1994.

[4] M.W. Walker, D.E. Orin, *"Efficient Dynamic Simulation of Robotic Mechanisms"*, Journal of Dynamic Systems, Measurement and Control, p. 205-211, 1982.

[5] E.J. Haug, "Computer-Aided Kinematics and Dynamics of Mechanical Systems, Volume I, Basic Methods", Allyn and Bacon, 1989.

[6] J. Garcia De Jalon, E. Bayo, *"Kinematic and Dynamic Simulation of Multibody Systems"*, Springer-Verlag, 1993.

[7] M. Crouzeix, A.L. Mignot, *"Analyse Numérique des Equations différentielles"*, Collection Mathématiques Appliquées pour la Maîtrise, Masson, 1992.

[8] E. Hairer, G. Wanner, *"Solving Ordinary Differential Equations I&II, Nonstiff Problems, Stiff and Differential-Algebraic Problems, 2 nd edition"*, Springer-Verlag, Springer Series in Computational Mathematics, 1996.

[9] P. Lotstedt, L. Petzold, *"Numerical Solution of Nonlinear Differential Equations with Algebraic Constraints I : Convergence Results for Backward Differentiation Formulas"*, Mathematics of Computation, Vol. 46, N°174, p. 491-516, 1986.

[10] M. Abadie, *"Simulation Dynamique de Mécanismes, Prise en Compte du Contact Frottant"*, PhD Thesis, Montpellier University, France, 1998.

[11] J. Baumgarte, *"Stabilization of Constraints and Integrals of Motion in Dynamical Systems"*, Comp. Math. Appl. Mech. Eng., Vol. 1, p. 1-16, 1972.

[12] M. Jean, *"Frictional Contact in Collections of Rigid or Deformable Bodies : Numerical Simulation of Geomaterial Motions"*, Mechanics of Geomaterial Interfaces, Elsevier Science Publisher, 1994.

[13] C. Glocker, *"Dynamik von Starrkörpersystemen mit Reibung und Stö βen"*, Thèse de Doctorat, VDI Verlag, München, 1995.

[14] J.J. Moreau, *"Dynamique des Systèmes à Liaisons Unilatérales avec Frottement Sec Eventuel, Essais Numériques"*, Université de Montpellier II, Laboratoire de Mécanique et de Génie Civil, Note Technique 85-1, 1986.

[15] M. Jean, J.J. Moreau, *"Unilaterality and Dry Friction in the Dynamics of Rigid Body Collections"*, Proceedings Contact Mechanics International Symposium, Edited by A. Curnier, Presses Polytechniques et Universitaires Romandes, Lausanne, p. 31-48, 1992.

[16] J.C. Trinkle, J.S. Pang, S. Sudarsky, G. Lo, *"On dynamic Multi-Rigid-Body Contact Problems with Coulomb Friction"*, Zeitschrift Angewandte Mathematik und Mechanik, vol.77, p. 267-279, 1997.

[17] D. Baraff, *"Fast Contact Force Computation for Nonpenetrating Rigid Bodies"*, SIGGRAPH 94, Computer Graphics Proceedings, Annual Conference Series, p. 23-34, 1994.

[18] E.J. Haug, S.C. Wu, S.M. Yang, *"Dynamics of Mechanical Systems With Coulomb Friction, Stiction, Impact and Constraint Addition-Deletion, Part I : Theory"*, Mechanism and Machine Theory, Vol. 21, N°5, p. 401-406, 1986.

[19] F. Pfeiffer, C. Glocker, *"Multibody Dynamics With Unilateral Contacts"*, John Wiley & Sons, 1996.

[20] J. Pérès, *"Mécanique Générale"*, Masson et Cie, 1953.

[21] E.J. Haug, S.C. Wu, S.M. Yang, *"Dynamics of Mechanical Systems With Coulomb Friction, Stiction, Impact and Constraint Addition-Deletion, Part II : Planar Systems"*, Mechanism and Machine Theory, Vol. 21, N°5, p. 407-416, 1986.

[22] E.J. Haug, S.C. Wu, S.M. Yang, *"Dynamics of Mechanical Systems With Coulomb Friction, Stiction, Impact and Constraint Addition-Deletion, Part III : Spatial System"* Mechanism and Machine Theory, Vol. 21, N°5, p. 417-425, 1986.

[23] P. Joli, M. Pascal, J.R. Gibert, *"Système avec Changement de Topologie dus à des Liaisons de Contact"*, Actes du 11ème Congrés Français de Mécanique, Vol. 5, p. 429-432, 1993.

[24] T. Klisch, *"Contact Mechanics in Multibody Systems"* , Mechanism and MachineTheory, Vol. 34, p. 665-675, 1999.

[25] J.J. Moreau, *"Les Liaisons Unilatérales et le Principe de Gauss"*, Compte Rendu de l'Académie des Sciences, Tome 256, p. 871-874, 1963.

[26] J.J. Moreau, *"Quadratic Programming in Mechanics : Dynamics of One-sided Constraints"*, J. SIAM Control, Vol. 4, N°1, 1966.

[27] P. Lotstedt, *"Mechanical Systems of rigid Bodies subject to Unilateral Constraints"*, SIAM J. Appl. Math., Vol. 42, N°2, 1982.

[28] P. Lotstedt, *"Numerical Simulation of Time-dependent Contact and Friction Problem in Rigid Body Mechanics"*, SIAM J. Sci. Stat. Comput., Vol. 5, N°1, 1984.

[29] P.G. Ciarlet, *"Introduction à l'Analyse Matricielle et à l'Optimisation"*, Collection Mathématiques Appliquées pour la Maîtrise, Masson, 1990.

[30] J.S. Pang, J.C. Trinkle, *"Complementarity Formulation and Existence of Solutions of Dynamic Multi-Rigid-Body Contact Problems with Coulomb Friction"*, Mathematical Programming, vol. 73, p. 199-226, 1996.

[31] J.S. Pang, J.C. Trinkle, *"Dynamic Multi-Rigid-Body Systems with Concurrent Distributed Contacts"*, Journal of Applied Mechanics, à paraître, 1996.

[32] C. Glocker, *"Formulation of Spatial Contact Situations in Rigid Multibody Systems"*, To appear in: Special Issue of CMAME on Computational Modelling of Contact and Friction, Ed. J.A.C. Martins and A. Klarbring.

[33] M. Ferris, *"Complementarity Problems – Applications, Modelling and Solution"*, Informs National Meeting, Seattle, 1998.

[34] D.E. Stewart, J.C. Trinkle, *"Dynamics, Friction, and Complementarity Problems"*, International Conference on Complementarity Problems, John Hopkins University, Baltimore, 1995.

[35] D.E. Stewart, J.C. Trinkle, *"An Implicit Time-Stepping Scheme for Rigid Body Dynamics with Inelastic Collisions and Coulomb Friction"*, International Journal of numerical Methods in Engineering, 1995.

[36] J.S. Pang, D.E. Stewart, *"A Unified Approach to Discrete Frictional Contact Problems"*, International Journal of Engineering Science, Vol. 37, p.1747-1768, 1999.

[37] M. Anitescu, F.A. Potra, *"Formulating Dynamic Multi-Rigid-Body Contact Problems with Friction as Solvable Linear Complementarity Problems"*, Reports on Computational Mathematics, N° 93/1996, Dept. of Mathematics, University of Iowa, 1996.

[38] M. Anitescu, F.A. Potra, D.E. Stewart, *"Time-stepping for Three-dimensional Rigid Body Dynamics"*, To appear in Computational Methods in Applied Mechanics and Engineering.

[39] M Wösle, F. Pfeiffer, *"Dynamics of Multibody Systems Containing Dependent Unilateral Constraints with Friction"*, Journal of Vibration and Control, Vol. 2, p. 161-192, 1996.

[40] M Wösle, F. Pfeiffer, *"Dynamics of Multibody Systems with Unilateral Constraints"*, Proceedings of the Annual Scientific Conference of Gesellschaft für Mathematik und Mechanik, Charles University, Prague, 1996.

[41] A. Klarbring, *"Mathematical Programming and Augmented Lagrangian Methods for Frictional Contact Problems"*, Proceedings of the International Symposium Contact Mechanics, Ed. A. Currier, Lausanne, 1992.

[42] J.J. Moreau, *"Unilateral Contact and Dry Friction in Finite Freedom Dynamics"*, Non-smooth Mechanics and Applications, CISM Courses and Lectures, N°302, Springer-Verlag, 1987.

[43] J.J. Moreau, *"Numerical Aspects of the Sweeping Process"*, To appear in Computational Methods in Applied Mechanics and Engineering.

[44] D. Baraff, *"Coping with Friction for Non-penetrating Rigid Body Simulation"*, SIGGRAPH 91, Computer Graphics Proceedings, Vol. 25, N°4, p. 31-40, 1991.

[45] D. Baraff, *"Issues in Computing Contact Forces for Non-Penetrating Rigid Bodies"*, Algorithmica, Springer-Verlag, p. 292-352, 1993.

[46] P. Painlevé, *"Leçons sur le Frottement"*, Librairie Scientifique A. Hermann, Paris, 1895.

[47] Y. Wang, M.T. Mason, *"Two dimensional Rigid-Body Collisions with Friction"*, Journal of apllied Mechanics, vol. 59, p. 635-643, 1992.

[48] F. Génot, B. Brogliato, *"New Results on Painlevé Paradoxe"*, European Journal of Mechanics A/Solids, vol.18, p. 653-677, 1999.

[49] B. Brogliato, *"Nonsmooth Impact Mechanics. Models, Dynamics and Control"*, Lecture Notes in Control and Information Sciences, Springer-Verlag, Vol. 220, 1996.

[50] N. Maw, J.R. Barber, J.N. Fawcett, *"The Role of Elastic Tangential Compliance in Oblique Impact"*, Journal of Lubrification Theory, Vol. 103, p. 74-80, 1981.

[51] C.E. Smith, P.P. Liu, *"Coefficients of Restitution"*, Journal of Applied Mechanics, Vol. 59, p. 963-969, 1992.

[52] R.M. Brach, *"Rigid Body Collisions"*, Journal of Applied Mechanics, Vol. 56, p. 133-138, 1989.

[53] I. Han, B.J. Gilmore, *"Multi-Body Impact Motion With Friction. Analysis, Simulation, and Experimental Validation"*, Journal of Mechanical Design, Vol. 155, p. 331-340, 1993.

[54] W.J. Stronge, *"Rigid Body Collisions with Friction"*, Proc. R. Soc. Lond., Vol. A431, p. 169-181, 1990.

[55] W.J. Stronge, *"Unraveling Paradoxical Theories for Rigid body Collisions"*, Journal of Applied Mechanics, Vol. 58, p. 1049-1055, 1991.

[56] Y.T. Wang, V. Kumar, *"Simulation of Mechanical Systems With Multiple Frictional Contacts"*, Journal of Mechanical Design, Vol. 166, p. 571-580, 1994.

[57] R.M. Brach, *"Friction, Restitution, and Energy Loss in Planar Collisions"*, Journal of Applied Mechanics, Vol. 51, p. 164-170, 1984.

[58] C.E. Smith, *"Predicting Rebounds Using Rigid-Body Dynamics"*, Journal of Applied Mechanics, Vol. 58, p. 754-758, 1991.

[59] A. Chatterjee, *"Rigid body Collisions, Some General Considerations, New Collision Laws, and some Experimental Data"*, PhD Thesis, Cornell University, 1997.

[60] D.W. Marhefka, D.E. Orin, *"Simulation of Contact Using a Nonlinear Damping Model"*, Proceedings of the 1996 IEEE, International Conference on Robotics and Automation, Minneapolis, Minnesota, p. 1662-1668, 1996.

Stability of Periodic Motions
with Impacts

Alexander P. Ivanov

Moscow State Technical University, 107005, 2-nd Baumanskaya st., 5

Abstract. Periodic motions in multibody systems with colliding elements are considered. Such motions arise in various technical devices as an essential condition of their work or, on the contrary, as undesirable and dangerous effect. Anyway, the stability problem is of great practical importance. In the present chapter various approaches to stability analysis are discussed. The simplest class of impact motions form those with non-degenerate collisions. The usual linearization technique can be adopted for the analysis of such motion. Conditions of asymptotic stability and structural stability are given in terms of eigenvalues of monodromy matrix. There exist several types of singularities, which cannot be treated by this method, such as grazing incidence, multiple collision, and motions including intermittent contact phases. General features of these singularities are described and some examples of stability analysis presented.

1 Introduction

In multibody dynamics, rigid bodies can move under applied forces separately or contacting one another. Moreover, if two bodies move simultaneously towards the same region, they collide. From a physical point of view, the collision is associated with the appearance of large shortcoming forces which prevent mutual penetration of the bodies. In mathematical formulation, such collision is treated as instantaneous action of an impulsive force. After it, the bodies continue separate motion. We discuss formal description of impact motions in Sect. 2.

In some technical devices, such as percussion machines, typewriter carriages, vibratory feeders, etc. repeated collisions are used for practical needs. On the contrary, the collisions of pipes in heat exchangers with walls or with one another is very undesirable. This is the reason why the problem of stability of periodic impact motions are of big practical value.

An intensive study of impact oscillations began in 1940. A number of works were issued devoted to periodic motions of one-degree of freedom forced impact oscillator. In spite of apparent simplicity of such system, it possesses very complicated dynamics and may have along with periodic motions of arbitrary large period non-periodic, or chaotic trajectories (cf. [1]). The monograph [2] is devoted to the impact oscillations in systems with two degrees of freedom. More details and references may be found in [3]. We present general results on periodic motions with non-degenerate impacts in Sect. 3.

In the last three decades, the studies of degenerate cases started. One of such singularities is related to grazing incidence of two rigid bodies. From a physical point of view, such phenomena accompanies the transition from free oscillation to the impact ones. It leads in general to instability of periodic motions or their disappearance (cf. [4–17]. We discuss grazing singularity in Sect. 4.

The next type of singularity, considered in Sect. 5, is connected with multiple collisions. Such collisions are observed in multibody systems with several impact pairs. For instance, a string with concentrated masses which is forced to oscillate near a wall hits the wall with several masses simultaneously [18, 19]. As was shown in [20], standard analytical techniques are inapplicable to the analysis of such motion, and new approaches are to be developed.

2   System Formulation

In this section, we show how to describe the motion of given multibody system with impacts.

2.1   Unilateral Constraints

We consider a multibody system with n degrees of freedom and generalized co-ordinates $\mathbf{q} = (q_1, q_2, \ldots, q_n)$, i.e. some independent parameters which determine uniquely the system configuration. It is supposed usually that the values q_i are arbitrary. However, one should account that different bodies can not penetrate one into another. This condition implies certain restrictions on the values q_i, known as one-sided or unilateral constraints. They are expressed by the inequalities

$$f_j(t, \mathbf{q}) \geq 0, \quad j = 1, 2, \ldots k. \tag{2.1}$$

Examples.   1. In the system of two balls the distance between their centers can not be less than the sum of radii:

$$\sqrt{(x_1 - x_2)^2 + (y_1 - y_2)^2 + (z_1 - z_2)^2} - R_1 - R_2 \geq 0, \tag{2.2}$$

where (x_i, y_i, z_i) are cartesian co-ordinates of the center of i-th ball $(i = 1, 2)$.

2. If the system consists of $l > 2$ balls, then it is subject to $k = l(l-1)/2$ unilateral constraints similar to (2.2), with one constraint for each pair of balls. If the balls can occupy any position in 3D or 2D, all these constraints are mutually independent. However, if we have a collinear system of l balls, each of them may contact its neighbours (or neighbour) only. Hence, in this case there is only $l - 1$ independent constraints.

3. A heavy particle, moving inside two-dimentional convex region D, is known as Birkhoff billiard (Fig. 1). In polar co-ordinates ρ, ϕ the unilateral constraint has form

$$f(\phi) - \rho \geq 0, \tag{2.3}$$

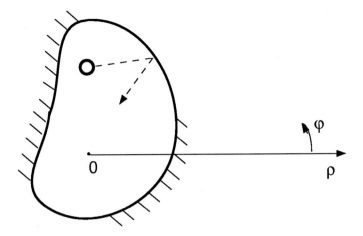

Fig. 1. Birkhoff billiard.

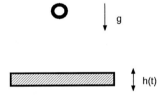

Fig. 2. A bouncing ball.

where equality corresponds to the boundary of D.

4. Let a ball move upon a horizontal table, oscillating along the vertical (Fig. 2). Then

$$z - R - A \sin \omega t \geq 0, \tag{2.4}$$

where A and ω are the amplitude and frequence of table oscillations.

5. Consider two bodies connected with a tether (Fig. 3). Then the distance between the points $M_1(x_1, y_1, z_1)$ and $M_2(x_2, y_2, z_2)$ where the tether is attached can not exceed its length L:

$$L - \sqrt{(x_1 - x_2)^2 + (y_1 - y_2)^2 + (z_1 - z_2)^2} \geq 0.$$

2.2 Equations of Smooth Motion

Suppose that all unilateral constraints are relaxed, i.e. all relations (2.1) are satisfied as inequalities. Then the motion is "free" in the sense that it is

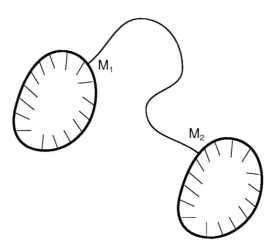

Fig. 3. Tethered bodies.

unaffected by the presence of the constraints. The equations of such smooth motion may be presented in Lagrangian form

$$\frac{d}{dt}\left(\frac{\partial T}{\partial \dot{q}_i}\right) - \frac{\partial T}{\partial q_i} = Q_i, \quad i = 1, 2, \ldots n, \tag{2.5}$$

where T is kinetic energy, Q_i – generalized forces.

Another type of smooth motion corresponds to such cases where one (or more) relations (2.1) turn to be equalities while other relations are satisfied as inequalities. Let, for examlpe, $f_1 \equiv 0, f_2 > 0, \ldots f_k > 0$. Then system (2.5) includes additional reaction force R :

$$\frac{d}{dt}\left(\frac{\partial T}{\partial \dot{q}_i}\right) - \frac{\partial T}{\partial q_i} = Q_i + R_i, \quad i = 1, 2, \ldots n. \tag{2.6}$$

In particular, if the constraint is ideal (there is no friction between contacting bodies), then $R_i = \lambda \partial f_1/\partial q_i$ with Lagrangian multiplier $\lambda \geq 0$. This multiplier can be determined from the system (2.6) in account of the identity $f_1 \equiv 0$.

2.3 Impacts

When two bodies collide at a moment $t = t'$, they come into contact with non-zero approach velocity, i.e.

$$f_1 = 0, \quad \frac{df_1}{dt} < 0. \tag{2.7}$$

Obviously, the preservation of conditions (2.1) for $t > t'$ is possible provided the generalized velocity $\dot{q}(t)$ is discontinuous at $t = t'$. The general equations of impact have the following form:

$$\mathbf{q}^+ = \mathbf{q}^-, \quad \dot{\mathbf{q}}^+ = \dot{\mathbf{q}}^- + \mathbf{I}(\mathbf{q}^-, \dot{\mathbf{q}}^-), \qquad (2.8)$$

where superscripts "minus" and "plus" denote one-side limits at $t = t'$.

It is important to note that the impact rule (2.8) cannot be derived from equations of smooth motion (2.5) or (2.6). The impact phenomenon depends significantly on physical properties of colliding solids, while these dynamical equations are independent of them. Consider, for example, a ball which drops to a concrete base. As was shown by Galileo, laws of falling down for different bodies are similar, and we can calculate velocity before collision. On the contrary, the result of such a collision cannot be predicted without specifying material properties of the ball since cannon-ball, tennis ball, and glass decanter will response differently.

Generally speaking, proper determination of the impulse I in (2.8) is a self-standing important scientific problem. Moreover, it is far more complicated than solution of equations (2.5) since impact phenomena involve stress waves, expanding in colliding solids and reflecting from their boundaries. To describe these waves one is to use partial differential equations. One can find furter details in the monographs [21, 22] where collision on free solids is discussed. Note that the impact rule (2.8) depends not upon the physical properties of colliding bodies only, but on their connections with other bodies (if any) as well (cf. [23, 24]).

In practice, one often uses simplified impact rules. The most popular one is a rule based on Newtonian coefficient of restitution $e \in [0, 1]$. Being an ideal constraint expressed by the inequality $q_1 \geq 0$, the impulse has components $I_1 = -(1+e)\dot{q}_1^-, I_2 = \ldots = I_n = 0$. One should realize that such rule is rather inadequate in many cases and this can cause apart from other additional difficulties in qualitative analysis (cf. [25]).

2.4 Multiple Impacts

In the previous subsection, we discussed "simple" impacts, where only one of constraints (2.1) takes part. In general, there exist such situations where three or more bodies collide simultaneously, i.e. two or more constraints have non-zero reaction. There exist a traditional approach to the solution of multiple impact problem: the impulsive reaction of the constraints are treated independently (cf. [26]). A subtle analysis shows that such treatment is correct only in the case where certain orthogonality conditions are satisfied. Such conditions mean that the impulsive reaction of each consrtaint depends on the corresponding approach velocity only: see Sect. 5. They are sufficient for the correctness of multiple impact. Besides, there exist other exceptional cases of such correctness, which are realised for specially constructed both system configuration and impact rule (cf. [20, 23]).

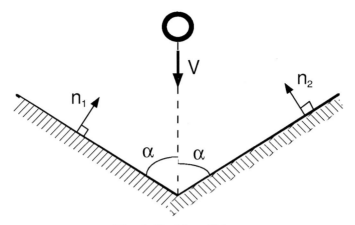

Fig. 4. Double collision.

Example. Consider a ball of unit mass which strikes two walls simultaneously (Fig. 4). Both walls are smooth, and impulsive reactions are collinear to the normal vectors $\mathbf{n_1}, \mathbf{n_2}$. We can try to calculate the impulses $\mathbf{I_1}, \mathbf{I_2}$ similar to the case of simple collision by means of the coefficients of restitution e_1, e_2 :

$$\mathbf{I_1} = \mathbf{I_1^o} = (1 + e_1) V \mathbf{n_1} \sin \alpha, \quad \mathbf{I_2} = \mathbf{I_2^o} = (1 + e_2) V \mathbf{n_2} \sin \alpha, \qquad (2.9)$$

where α is a half of the angle between the planes. The total impulse is usually understood as the sum of the quantities in (2.9):

$$\mathbf{I} = \mathbf{I_1^o} + \mathbf{I_2^o}. \qquad (2.10)$$

In particular, if $e_1 = e_2$ then formula (2.10) implies that the ball rebounds in the direction which is opposite to the initial one. Suppose now that the initial position of the ball is slightly moved to the left. In such a case, the ball hits initially the left wall and rebounds from it with the velocity $\mathbf{V^*}$, where

$$\mathbf{V^*} = \mathbf{V} + \mathbf{n_1}(1 + e_1) V \sin \alpha \qquad (2.11)$$

After that, the ball hits the right wall provided its velocity is forwarded towards it, i.e.

$$(\mathbf{V^*}, \mathbf{n_2}) = -[1 + (1 + e_1) \cos 2\alpha] V \sin \alpha < 0. \qquad (2.12)$$

The inequality (2.12) is true if $\alpha < \pi/3$, otherwise it might be broken. In particular, if $e_1 = 1$ (the collision is absolutely elastic) then for any $\alpha \geq \pi/3$ the ball will not collide with the right wall. In the case (2.12) is satisfied, the second impulse has the form

$$\mathbf{I_2} = -(1 + e_2)(\mathbf{V^*}, \mathbf{n_2})\mathbf{n_2}.$$

Now formula (2.10) leads to the following result:

$$\mathbf{I} = \mathbf{I_1^o} + \mathbf{I_2^o} + (1 + e_1)(1 + e_2)\mathbf{n_2}\sin\alpha\cos 2\alpha \tag{2.13}$$

The comparison of (2.10) and (2.13) shows that small changes in initial position of the ball imply significant changes in the resulting impulse unless $\alpha = \pi/4$, i.e. the walls are orthogonal. Therefore, correct definition of multiple collision in this example is possible only in that case.

In the remainder of this paper, we shall suppose that equations (2.5), (2.6), and (2.8) are given.

3 Stability of Regular Periodic Impact Motions

In this section, we discuss a wide class of periodic motions with impacts. In fact, most of practical examples of impact systems are relative to the class. However, certain restriction should be made to ensure the applicability of analytical techiques, presented below. We shall suggest in Sect. 3.3 that

(i) the solution include only a finite number of impacts (2.8) per period;
(ii) each impacts is non-degenerate, i.e. if $f_j(t') = 0$, then $\dot{f}_j(t'-0) \neq 0$ and $\dot{f}_j(t'+0) \neq 0$;
(iii) none of the impact is multiple, i.e. no more that one of inequalities (2.1) turn to equality at the same time;
(iv) there is no sliding intervals where one (or more) of the inequalities (2.1) turns to equality.

In the Sects 3.1-3.2 no restriction is made, and they have general nature.

3.1 Concepts of Stability of Motion with Impacts

The classical definitions of stability of certain kind (Lyapunov, asymptotic, orbital, structural) were formulated initially for smooth motions (cf. [27]). Let the system be described by the equations

$$\dot{\mathbf{x}} = \mathbf{F}(t,\mathbf{x}), \quad \mathbf{x} \in R^k \tag{3.1}$$

with a continuously differentiable function $\mathbf{F}$. Let $\mathbf{x}^*(t)$ be a solution of (3.1) existing for all $t \geq t_0$.

Definition 1. A solution $\mathbf{x}^*(t)$ is called stable (in the Lyapunov sence) if for any value $\varepsilon > 0$ there exist a $\delta > 0$ such that for any solution $\mathbf{x}(t)$ of (3.1), the inequality $\|\mathbf{x}(t_0) - \mathbf{x}^*(t_0)\| < \delta$ implies that

$$\|\mathbf{x}(t) - \mathbf{x}^*(t)\| < \varepsilon \tag{3.2}$$

for any $t > t_0$. Otherwise, $\mathbf{x}^*(t)$ is called unstable.

Definition 2. A solution $\mathbf{x}^*(t)$ is called asymptotically stable if it is stable and for any solution $\mathbf{x}(t)$ to system (3.1) the inequality $\|\mathbf{x}(t_0) - \mathbf{x}^*(t_0)\| < \delta$ implies that $\|\mathbf{x}(t) - \mathbf{x}^*(t)\| \to 0$ for $t \to \infty$.

Definition 3. A solution $\mathbf{x}^*(t)$ to the autonomous system (3.1) (i.e. the function $\mathbf{F}$ does not depend on t) is called orbital stable if for any value $\varepsilon > 0$ there exist such $\delta > 0$ that for any solution $\mathbf{x}(t)$ of system (3.1) the inequality $\|\mathbf{x}(t_0) - \mathbf{x}^*(t_0)\| < \delta$ implies that

$$\inf_{t' > t_o} \|\mathbf{x}(t) - \mathbf{x}^*(t')\| < \varepsilon. \tag{3.3}$$

The last definition means that the disturbed trajectory $\mathbf{x}(\mathbf{t})$ is close to the undisturbed one if time is disregarded.

A straightforward attempt to apply the Def.1 to impact motion would fail since the solutions $\mathbf{x}^*(t)$ and $\mathbf{x}(t)$ experience impacts at different instants t' and $t' + \Delta t$. During short imterval $t \in (t', t' + \Delta t)$ the difference between these two motions is of order I, which indicates instability in the sense of Def.1. Obviously, such conclusion does not agree with our intuitive idea of stability.

Example. Consider the example 4 of previous section once more. There exist simple periodic motions of the ball where it bounces from the table at the same oscillation phase t' and with the same velocity v [28]. Under certain conditions, such motions are stable in the sense that the ball, under small initial perturbations, will bounce from the table at instants $t' + \Delta t_j$ with velocities $v + \Delta v_j$ $(j = 1, 2, \ldots)$ where the values Δt_j and Δv_j vanish with the perturbations.

To adopt the Definitions 1, 2 to impact motions, it was proposed in [29] to require the correctness of the ineq.(3.2) for all time except ε – vicinities of those instants when impacts occur. In other words, if $\mathbf{x}^*(t)$ experiences an impact at $t = t'$, then on the interval $t \in (t' - \varepsilon, t' + \varepsilon)$ the ineq. (3.2) is not to be satisfied.

Another approach to the concept of stability [30] is based on Def.3. The uderlying idea is the independence of orbital stability of non-simultaneity of impacts for two motions $\mathbf{x}^*(t)$ and $\mathbf{x}(t)$ which is caused by the autonomy. General case where $\mathbf{F}$ depends on t can be treated similarly after inclusion of independent variable t to the set of variables $\mathbf{x}$ by means of formula $x_{k+1} = t$. We obtain the following definition of stability which is equivalent to the orbital stability in the extended phase space.

Definition 4. A solution $\mathbf{q}^*(t)$ to the impact system (2.6)–(2.8) will be called stable if the orbit $\Upsilon = (t, \mathbf{q}^*(t), \dot{\mathbf{q}}^*(t))$ is a stable set in the extended phase space. In other words, for any value $\varepsilon > 0$ there exist such $\delta > 0$ that for any solution $\mathbf{q}(t)$ of this system inequality $d(\mathbf{x}(t_0), \Upsilon) < \delta$ implies that

$$d((t, \mathbf{x}(t)) - \Upsilon) < \varepsilon$$

for any $t > t_0$. Here d denotes euclidean metrics in the extended phase space.

In a similar way the definition of asymptotic stability can be adopted.

Another concept is the structural stability which is related not to a particular solution to system (3.1), but to the phase portrait as a whole (cf. [1]). Suppose that the right-hand side of this system depends on some parameter μ. Structural stability in a region $D \in R^k$ means that small changes of μ do not cause qualitative changes in the phase portrait within D. The lack of structural stability for certain values μ is called bifurcation.

For instance, if the system possesses for $\mu \in (\mu_1, \mu_2)$ a periodic solution which is asymptotically stable, then it is structurally stable in a vicinity of the closed orbit which corresponds to this solution.

The structural stability in impact systems can be introduced similarly to the smooth case.

3.2 Poincaré Maps

A convenient tool for the analysis of periodic motions is the Poincaré map. We start with basics of theory in smooth (no impact) systems, while impact systems will be discussed later. To construct the map, one has to choose a surface of section π in the extended phase space $R \times R^k$ and consider the points of successive intersections of a given trajectory with this surface. Let $\chi, \bar{\chi}$ be such two subsequent points. Then the map $P : \pi \mapsto \pi$ is defined as $P(\chi) = \bar{\chi}$.

Given a τ-periodic motion $x^*(t)$, the classical approach is based on the stroboscopic section

$$\pi_\tau : \quad t = t_0 (\mathrm{mod}\, \tau). \tag{3.4}$$

This means that the image $P(x)$ of any $x \in R^k$ is the point that the trajectory of system (3.1), starting at x at $t = 0$, attains at $t = \tau$. In particular,

$$P(x^*(0)) = x^*(\tau) = x^*(0), \tag{3.5}$$

i.e. the periodic orbit $x^*(t)$ yields a fixed point of the map P.

Such an approach is relevant to the stability analysis since the stability of the periodic orbit is equivalent (provided the trajectory depends on initial point x in a continuous way, which is always true in smooth systems but may be not true in impact systems: cf. Sect. 2.4 and Sect. 5) to the stability of fixed point which corresponds to it. In fact, there is no need to determine the map P in explicit form: most often its Jacobian DP governs the local behaviour and stability [1]. To calculate this Jacobian, the system (3.1) is to be linearized around the orbit $x^*(t)$:

$$\dot{\mathbf{y}} = \mathbf{A}(t)\mathbf{y} + o(y), \quad \mathbf{y} = \mathbf{x} - \mathbf{x}^*, \quad \mathbf{A}(t) = \left\| \frac{\partial \mathbf{F}(t, \mathbf{x}^*)}{\partial \mathbf{x}} \right\| \qquad (3.6)$$

After omitting nonlinear terms, the system (3.6) can be extended to the matrix equation

$$\dot{\mathbf{Y}}(0, t) = \mathbf{A}(t)\mathbf{Y}(0, t), \quad \mathbf{Y}(0, 0) = \mathbf{E}, \qquad (3.7)$$

where $\mathbf{E}$ is the unit matrix of order k. The matrix $\mathbf{Y}$ in formula (3.7) is called *the general solution matrix*, and $\mathbf{Z} = \mathbf{Y}(\tau)$ is called *the monodromy matrix*. The difference between the periodic solution and a nearby one can be evaluated by means of the following statement.

Theorem 3.1. For any $t > 0$ the following formula is valid

$$\mathbf{y}(t) = \mathbf{Y}(0, t)\mathbf{y}(0) + o(\|\mathbf{y}(0)\|). \qquad (3.8)$$

Theorem 3.2. 1. If all the eigenvalues of the monodromy matrix $\mathbf{Z}$ lie inside the unit circle in the complex plane, then the periodic solution $\mathbf{x}^*$ is asymptotically stable.

2. If at least one of these eigenvalues lies outside the unit circle, the periodic solution is unstable.

These two results belong to Poincaré. In the following subsection, they will be used for the analysis of impact motions.

Another possible type of section is

$$\pi_x : \quad f(\mathbf{x}) = 0. \qquad (3.9)$$

It is applied usualy in autonomous systems where periodic solutions may have continuum of periods. Besides, it is often used in system subjected to the unilateral constraint $f(\mathbf{x}) \geq 0$, (cf. [7–9]). It should be noted that this seemingly natural section need to be used with care since it does not satisfy, in general, to the definition of Poincaré map. This means that some trajectories do not intersect the surface (3.9) or are tangent to it. This may cause artificial bifurcations as a periodic orbit may change the number of intersections with the section (3.9) without actual changes in qualitative behaviour.

Example. The motion of a ball, bouncing elastically (i.e. without energy losses) from an immovable table, is described by the equation (cf. (2.4), where $A = 0$)

$$\ddot{z} = -g, \quad z - R \geq 0.$$

If we choose the section of type (3.9): $z = R$ then $\dot{z}_a = -\dot{z}_r$, where the superscripts a and r indicate approach and rebound velocities. After the collision, new rebound velocity will be the same as the previous one. Therefore, the associated map is indentical, and its Jacobian (in fact, derivative) is unity.

We next construct for the system a section of type (3.4) in a vicinity of periodic trajectory $z^*(t)$ with period τ. We choose the initial moment t_0 so that $\dot{z}^*(t_0) = 0$, therefore, $z^*(t_0) = R + g\tau^2/8$. A nearby trajectory $z(t)$ (such that it includes just one collision within interval $t \in (t_0, t)$) includes an impact at a moment t' being determined from the following quadratic equation:

$$z_0 + \dot{z}_0(t^* - t_0) - \frac{1}{2}g(t^* - t_0)^2 = R,$$

hence,

$$t^* = t_0 + \left(\dot{z}_0 + \sqrt{\dot{z}_0^2 + 2g(z_0 - R)} \right)/g. \qquad (3.10)$$

The approach velocity is

$$\dot{z}(t^* - 0) = \dot{z}_0 - g(t^* - t_0) = -\sqrt{\dot{z}_0^2 + 2g(z_0 - R)},$$

while rebound velocity has opposite sign:

$$\dot{z}(t^* + 0) = -\dot{z}(t^* - 0). \qquad (3.11)$$

After the collision, the disturbed trajectory is described so:

$$z(t) = \dot{z}(t^* + 0)(t - t^*) - \frac{1}{2}g(t - t^*)^2, \quad \dot{z}(t) = \dot{z}(t^* + 0) - g(t - t^*) \qquad (3.12)$$

Subsituting (3.10), (3.11) into (3.12) and supposing there $t = t_0 + \tau$, we obtain explicit form of the map P:

$$P(z_0, \dot{z}_0) = \left(R + \bar{S}(\tau - (\dot{z}_0 + \bar{S})/g) - g(\tau - (\dot{z}_0 + \bar{S})/g)^2/2, \dot{z}_0 + 2\bar{S} - g\tau \right), \qquad (3.13)$$

where $\bar{S} = \sqrt{\dot{z}_0^2 + 2g(z_0 - R)}$. To calculate the Jacobian DP, we set $z_0 = z_0^* + \xi$, $\dot{z}_0 = \psi$, where ξ, ψ are small increments, and neglect in (3.13) the terms of second and higher orders:

$$P(z_0^* + \xi, \dot{z}_0 + \psi) = (z_0^* + \xi, 4\xi/\tau + \psi). \qquad (3.14)$$

Therefore,

$$DP = \begin{pmatrix} 1 & 0 \\ 4/\tau & 1 \end{pmatrix}$$

Note that the first map has the single multiplier which is equal to unity, while (3.14) has two unit multipliers. In both cases the Theorem 3.2 does not solve question of stability. In fact, in this simple example the Definitions $1 - 3$ can be verified directly: the periodic motion turned to be unstable, but orbitally stable. We will consider more complicated example of a ball, bouncing from an oscillating table, in the Sect. 3.3.3.

Any solution to system (3.5) corresponds to a periodic solution to (3.1). Such solutions usually cannot be constructed analytically, and numerical or approximate methods are to be implemented. We restrict ourselves to the following result and refer to [31] for more details.

Theorem 3.3 (Banach). Let (M, d) be a metric space, $P : M \to M$ be such a map that for some $q \in (0, 1)$ and any $x', x'' \in M$

$$d(P(x'), P(x'')) \le qd(x', x'') \tag{3.15}$$

(in this call P is called contraction). Then the map P has unique fixed point x^*, and for any $x_0 \in M$ the sequence $x_1 = P(x_0), x_2 = P(x_1), \ldots$ converges to x^*.

3.3 Non-degenerate Periodic Motions

3.3.1 Definition
Let $\mathbf{q}^*(t)$ be a periodic solution to impact system (2.5), (2.6), (2.8). We call it regular, or non-degenerate, provided all the following conditions are met:

(i) the solution include only a finite number of impacts (2.8) per period;
(ii) each impact is non-degenerate, i.e. if $f_j(t') = 0$, then $\dot{f}_j(t' - 0) \ne 0$ and $\dot{f}_j(t' + 0) \ne 0$;
(iii) none of the impact is multiple, i.e. no more that one of inequalities (2.1) turn to equality at the same time;
(iv) there is no sliding intervals where one (or more) of the inequalities (2.1) turns to equality.

3.3.2 Linearization
Let $\mathbf{q}^*(t)$ be a non-degenerate periodic solution to system (2.6), (2.8) with s impacts per period which occur at instants $0 < t'_1 < t'_2 < \ldots < t'_s < \tau$. We denote for simplicity $\mathbf{x} = (\mathbf{q}, \dot{\mathbf{q}})$ and rewrite (2.6) in the form (3.1) with $k = 2n$. Our aim is to extend the above presented linearization technique to such discontinuous orbit.

First of all, note that the Theorem 3.1 can be applied on any interval of impactless motion. Therefore, formula (3.8) is valid for $t \in (0, t'_1)$. An obstacle for further prolongation of this formula is the asynchronism of impacts: while the periodic trajectory experiences it at $t = t'_1$, the disturbed trajectory $\mathbf{x}(t)$ comes to impact at $t = t'_1 + \Delta t_1$. If the impact is caused by an attempt to violate the unilateral constraint $f_j \ge 0$, then the value Δt_1 can be determined from the equation

$$f_j\left(\mathbf{x}^*(t'_1 + \Delta t_1) + \mathbf{y}(t'_1 + \Delta t_1)\right) = 0. \tag{3.16}$$

In account with non-degeneracy condition (ii), $\dot{f}_j(\mathbf{x}^*(t'_1 - 0)) = -V_1 < 0$. Hence, the left-hand side in (3.16) can be extracted in Taylor formula as follows:

$$f_j(\mathbf{x}(t'_1 + \Delta t_1)) + (\mathbf{y}(t'_1 + \Delta t_1), \text{grad } f_{j_1}(\mathbf{x}(t'_1 + \Delta t_1))) + o(\|\mathbf{y}\|). \tag{3.17}$$

Here the first term can be considered as a function of t, since $f_j(\mathbf{x}(t_1')) = 0$, $\dot{f}_j(\mathbf{x}^*(t_1' - 0)) = -V_1$, we get

$$f_j(\mathbf{x}(t_1' + \Delta t_1)) = -V_1 \Delta t_1 + o(\Delta t_1).$$

The second term in (3.17) can be simplified by omitting of Δt_1 in $\mathbf{x}(t_1' + \Delta t_1)$, the error is $o(\|\mathbf{y}\|)$, since f is continuously differentiable. Thus, the equations (3.16) gets form

$$-V\Delta t_1 + (\mathbf{y}^-, \operatorname{grad} f_{j_1}(\mathbf{x}(t_1'))) + o(\Delta t_1) + o(\|\mathbf{y}\|) = 0, \quad \mathbf{y}^\pm = \mathbf{y}(t_1' \pm 0). \quad (3.18)$$

The relation (3.18) implies that

$$\Delta t_1 = (\mathbf{y}^-, \operatorname{grad} f_j(\mathbf{x}^*(t_1'))) / V_1 + o(\|\mathbf{y}(0)\|). \qquad (3.19)$$

We see that the interval between impacts in two close trajectories has duration of the first order with respect to initial disturbances. However, this interval is very important in the calculations of the monodromy matrix since within it the two trajectories differ by non-small value. Indeed, we can assume without loss of generality that $\Delta t_1 > 0$. For $t \in (t_1', t_1' + \Delta t_1)$ we have

$$\dot{\mathbf{x}} = \mathbf{F}_1^- + \ldots, \quad \dot{\mathbf{x}}^* = \mathbf{F}_1^+ + \ldots, \quad \mathbf{F}_1^\pm = \mathbf{F}(t_1', \mathbf{x}^*(t_1' \pm 0)),$$

where $\mathbf{F}$ is the vector field in (3.1) and omitted terms vanish as $\Delta t_1 \to 1$. Therefore,

$$\begin{aligned}
\mathbf{y}(t_1' + \Delta t_1 + 0) &= \mathbf{x}(t_1' + \Delta t_1 + 0) - \mathbf{x}^*(t_1' + \Delta t_1) \\
&= \mathbf{x}(t_1') + \Delta t_1 \dot{\mathbf{x}}(t') + \mathbf{I}\left(\mathbf{x}(t_1') + \Delta t_1 \dot{\mathbf{x}}(t_1')\right) \\
&\quad -\mathbf{x}^*(t_1' - 0) - \mathbf{I}\left(\mathbf{x}^*(t_1'),\right) - \Delta t_1 \dot{\mathbf{x}}^*(t_1') + o(\mathbf{y}^-).
\end{aligned} \qquad (3.20)$$

The formula (3.20) implies that the vector of disturbances $\mathbf{y}$ experiences a jump close to the impact instant t_1'. In matrix form, this jump can be described by the relation

$$\begin{aligned}
\mathbf{y}^+ &= \mathbf{J}\mathbf{y}^- + o(\mathbf{y}^-), \\
\mathbf{J} &= \mathbf{E} + \mathbf{I_x} + V_1^{-1}(\mathbf{F_1^-} - \mathbf{F_1^+}) \times \operatorname{grad} f_j + V_1^{-1}\mathbf{I_x}(\mathbf{F_1^-} \times \operatorname{grad} f_j),
\end{aligned} \qquad (3.21)$$

where $\mathbf{E}$ is the unit matrix of order k, $\mathbf{I_x} = \|\partial \mathbf{I}/\partial \mathbf{x}\|$, and $\times$ denotes the external vector product:

$$\mathbf{a} \times \mathbf{b} = \|a_i b_j\| \quad (i, j = 1, 2, \ldots, k).$$

We would like to stress that the jump formula (3.21) is more complex than impact law (2.8) because it involves not only system geometry, pre-impact velocity, and impact rule, but vector field $\mathbf{F}$ as well. This is the reason why we refer to the matrix $\mathbf{J}$ as to the *jump matrix* but not *restitution* one.

We arrive at the following conclusion.

Theorem 3.4. For motion with non-degenerate impacts, the formula (3.8) is valid where $\mathbf{Y}$ is piecewise-continuous solution to (3.7). The breakpoints of $\mathbf{J}$ coincide with impact instants for the motion in discussion, by that

$$\mathbf{Y}^+ = \mathbf{J}\mathbf{Y}^-, \quad \mathbf{Y}^\pm = \mathbf{Y}(t' \pm 0), \tag{3.22}$$

where matrix $\mathbf{J}$ is defined in (3.21).

Examples.

1. Consider one-degree-of freedom system with unilateral constraint $q \geq 0$ and Newton impact rule $\dot{q}^+ = -e\dot{q}^-$. In this case

$$\mathbf{x} = (q, \dot{q})^T, \quad f(\mathbf{x}) = x_1, \quad \mathbf{F} = (x_2, F_2)^T, \quad \mathbf{I} = (0, -(1+e)x_2)^T,$$

$$\operatorname{grad} f = (1,0)^T, \quad \mathbf{I_x} = \begin{pmatrix} 0 & 0 \\ 0 & -1-e \end{pmatrix}.$$

Calculations by formula (3.21) give

$$\mathbf{J} = \begin{pmatrix} -e & 0 \\ (F_2^+ + eF_2^-)/x_2^- & -e \end{pmatrix} \tag{3.23}$$

2. In more general case of one-degree-of-freedom systems with impacts $\dot{q}^+ = \dot{q}^- + I(\dot{q}^-)$ we have similar calculations as before with

$$\mathbf{I_x} = \begin{pmatrix} 0 & 0 \\ 0 & I' \end{pmatrix}, \quad I' = \frac{dI}{d\dot{q}^-},$$

and the formula (3.23) transfers to

$$\mathbf{J} = \begin{pmatrix} 1 + I/x_2^- & 0 \\ (F_2^+ - (1+I')F_2^-)/x_2^- & 1+I' \end{pmatrix} \tag{3.24}$$

3. In a system with n degrees of freedom and an ideal (frictionless) unilateral constraint $q_1 \geq 0$, the impacts are described by relations (cf. [32])

$$\dot{q}_1^+ = \dot{q}_1^- + I(\dot{q}_1^-), \quad p_j^+ = p_j^-, \quad p_j = \frac{\partial T}{\partial \dot{q}_j} \quad (j = 2,\ldots,n),$$

where p_j - s are the generalized momentum entries. Denote $\mathbf{x} = (q_1, \dot{q}_1, q_2, \ldots, q_n, p_2, \ldots, p_n)^T$. By analogy with the first example, we have

$$\operatorname{grad} f = (1,0,\ldots,0)^T, \quad \mathbf{I_x} = \begin{pmatrix} 0 & 0 & \cdots & 0 \\ 0 & I' & \cdots & 0 \\ \vdots & \vdots & \ddots & \vdots \\ 0 & 0 & \cdots & 0 \end{pmatrix}.$$

Therefore,

$$
\mathbf{J} = \begin{pmatrix}
1 + I/x_2^- & 0 & 0 & \cdots & 0 \\
\left(F_2^+ - (1+I')F_2^-\right)/x_2^- & 1 + I' & 0 & \cdots & 0 \\
(F_3^+ - F_3^-)/x_2^- & 0 & 1 & \cdots & 0 \\
\vdots & \vdots & \vdots & \ddots & \vdots \\
(F_{2n}^+ - F_{2n}^-)/x_2^- & 0 & 0 & \cdots & 1
\end{pmatrix} \qquad (3.25)
$$

3.3.3 Stability Conditions

Consider once more non-degenerate periodic impact motion $\mathbf{q}^*(t)$. We use Theorem 3.4 to construct the general solution matrix within the interval $t \in (0, \tau)$. As a result, we derive the monodromy matrix in the form

$$
\mathbf{Z} = \mathbf{Y}(t_s', \tau)\mathbf{J}_s\mathbf{Y}(t_{s-1}', t_s')\mathbf{J}_{s-1}\cdots\mathbf{J}_1\mathbf{Y}(0, t_1') \qquad (3.26)
$$

Now the Theorem 3.2 allows to formulate conditions of asymptotic stability and instability in terms of the eigenvalues of matrix (3.18).

Example. The motion of a ball, bouncing from an oscillating table, is described by the equation (see (2.4))

$$
\ddot{z} = -g, \quad z - R - h(t) \geq 0, \quad h(t) = A\sin\omega t \geq 0.
$$

Let $x_1 = z - R - h(t)$, $x_2 = \dot{x}_1$, then the system transfers to the following form:

$$
\dot{x}_1 = x_2, \quad \dot{x}_2 = -g - \ddot{h}(t), \quad x_1 \geq 0. \qquad (3.27)
$$

Consider the simplest periodic motion with one impact per period $\tau = 2\pi l/\omega$, where l is an integer. The matrix $\mathbf{A}(t)$ in system (3.6) and the solution to that system are given by:

$$
\mathbf{A} = \begin{pmatrix} 0 & 1 \\ 0 & 0 \end{pmatrix}, \quad \mathbf{Y}(t_1, t_2) = \exp\mathbf{A}(t_2 - t_1) = \begin{pmatrix} 1 & t \\ 0 & 1 \end{pmatrix} \qquad (3.28)
$$

The impact matrix $\mathbf{J}$ is expressed by (3.24) with $F_2^\pm = -g - \ddot{h}(t')$. We set the impact instant $t = t'$ as the initial one. Then the matrix (3.26) is

$$
\mathbf{Z} = \mathbf{Y}(t', t' + \tau)\mathbf{J} = \begin{pmatrix} 1 + (I + \tau u I')/x_2^- & \tau(1 + I') \\ u I'/x_2^- & 1 + I' \end{pmatrix}, \quad u = g + \ddot{h}(t').
$$

The characteristic equation has the form

$$
\rho^2 - a_1\rho + a_2 = 0, \qquad (3.29)
$$

$$
a_1 = \operatorname{Tr}\mathbf{Z} = 2 + I' + (I + \tau u I')/x_2^-, \quad a_2 = \det\mathbf{Z} = (1 + I')(1 + I/x_2^-).
$$

The conditions of asymptotic stability $|\rho_{1,2}| < 1$ are satisfied in two cases: either both roots are real and belong to the interval $(-1, 1)$, or they are conjugate complex numbers and their product (i.e. a_2) is less than unity. Simple calculations show that both cases can be described by the following double inequality:

$$|a_1| < 1 + a_2 < 2. \tag{3.30}$$

In particular, for Newton impact rule we obtain

$$a_1 = -2e - (1+e)\tau u/x_2^-, \quad a_2 = e^2, \tag{3.31}$$

and the conditions (3.30) transfer to the single inequality

$$|2e + (1+e)\tau u/x_2^-| < 1 + e^2. \tag{3.32}$$

Furthemore, the conditions of existence of the periodic motion under study have the form

$$x_1 > 0 \text{ for any } t \in (t', t'+\tau), \quad \dot{z}(t'+0) = -\dot{z}(t'-0) - \tau g/2. \tag{3.33}$$

Here the inequality means that between the collisions, the ball moves above the table and the equality indicates that the duration of flight equals τ. As $\dot{z}(t'-0) = x_2^- + \dot{h}(t')$, $\dot{z}(t'+0) = -ex_2^- + \dot{h}(t')$, we obtain

$$\dot{x}_2^- = -\tau g/(1+e), \quad \dot{h}(t') = 0.5\tau g(1-e)/(1+e). \tag{3.34}$$

The second equation (3.34) can be used for determination of impact instant t'.

Substituting (3.34) to the stability condition (3.32), we arrive finally at the following result:

$$-2g(1+e^2)/(1+e)^2 < \ddot{h}(t') < \tag{3.35}$$

3.3.4 Bifurcations

We have seen in Sect. 3.2 how the periodic orbits correspond to fixed points of Poincaré maps. An important property of non- degenerate impact motions is the differentiability of the Poincaré map in a vicinity of the fixed point (see Theorem 3.4). Moreover, as was shown in [33], the higher order derivatives exist too, provided the functions $\mathbf{F}, \mathbf{I}$ are smooth enough. Therefore, typical bifurcations of regular periodic impact orbits are similar to that of of fixed points of smooth maps: saddle-node, period doubling, pitchfork, exchange of stability, etc (cf. [1]). All bifurcations are related to the existence of Floquet multipliers (e.g. eigenvalues of monodromy matrix) which lie exactly at the unit circle. On the contrary, if all the multipliers are inside or outside the unit circle (so called hyperbolic case), the system is locally structurally stable.

Example. Consider again the ball bouncing from the oscillating table and let us discuss the behaviour on the boundary of the stability region (3.35). We shall study two basic cases.

1. Let $\ddot{h} = 0, d^3h/dt^3 \neq 0$ for $t = t'$. Introducing (3.34) into (3.31) one obtains: $a_1 = 1 + e^2, a_2 = e^2$. Thus, the characteristic equation has a root, equal to unity. Suppose that the parameter e gets a small alteration Δe, then the right-hand side of second equality (3.34) gets an alteration of opposite sign. Since

$$\dot{h}(t' + \Delta t) = \dot{h}(t') + \frac{1}{2}\frac{d^3h(t')}{dt^3}(\Delta t)^2 + o(\Delta t)^2,$$

the equations (3.34) have two solutions in case where $d^3h(t')/dt^3 \Delta e < 0$ and no solution in the opposite case. Such behaviour is typical for the saddle-node bifurcation.

2. In the case where $\ddot{h}(t') = -2g(1 + e^2)/(1 + e)^2$, the coefficients of the characteristic equation are $a_1 = -1 - e^2, a_2 = e^2$. Thus, the characteristic equation has a root equal to -1, which indicates a flip bifurcation. The detailed analysis of such a case was carried out in [34]. It was shown that if

$$w^* = h^{(3)}(t')\left(\frac{2}{3}\kappa - \tau w\right) - \frac{1}{6}\tau h^{(4)}(t') > 0,$$

where

$$w = -\frac{1}{\ddot{h}(t')}\left(h^{(3)}(t') - \frac{4}{\tau\kappa}\left(g + \ddot{h}(t')\right)\right), \quad \kappa = \frac{1-e}{1+e},$$

the bifurcation is subcritical. In the opposite case $w^* < 0$ the bifurcation is supercritical, i.e. for $\Delta e < 0$ we have one stable solution of period τ while for $\Delta e > 0$ appear two stable solutions of period 2τ and the former solution loses stability.

4 Grazing Incidence

4.1 Appearance of Grazing

To understand better the nature of grazing phenomena, consider at first two examples.

1. Forced oscillations of system with clearance are represented schematically in Fig. 5. The mass M can move in the horizontal direction under external force $P(t) = A\sin\omega t$, it is connected from the left side with a wall by means of a spring and a dashpot. If the amplitude A is small, the mass experiences free oscillations without collisions upon the rigid stop S on the right. On the contrary, for large values of A, the free motion will be

162

interrupted for sure by collisions. Obviously, there exist a critical value A_0, such that the collisions start at it. In fact, for $A = A_0$ the mass periodically touches the stop with zero approach velocity. By that $I = 0$, and no collision occurs. Such soft touch will be called grazing later on.

2. A pendulum in a regular flow exhibits non-linear self-oscillations [35] (Fig. 6). The amplitude of such oscillations increases as the speed u of the flow grows, and at certain critical value u_0 the pendulum will experience grazing incidences with both walls.

We can conclude that the grazing appears as transitional behaviour which is intermediate between free motion and impact one. In the case of constraint $f(\mathbf{q}) \geq 0$, the grazing is described by the relations

$$f(\mathbf{q}(t')) = 0, \quad \dot{f}(\mathbf{q}(t')) = 0, \quad \ddot{f}(\mathbf{q}(t')) > 0. \tag{4.1}$$

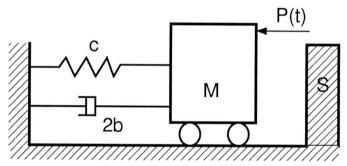

Fig. 5. Linear vibro-impactor.

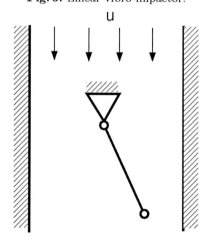

Fig. 6. Pendulum in a flow.

4.2 An Autonomous One-Degree of Freedom System

4.2.1 One Unilateral Constraint

The simplest system where the grazing might occur is one with single degree of freedom where neither active forces, nor the unilateral constraint depend on time explicitly. Such system may be presented in the form

$$\ddot{q} = F_2(q, \dot{q}), \quad q \geq 0. \tag{4.2}$$

The phase space of system (4.2) is the half-plane $(q, \dot{q})$ with $q > 0$. The trajectories can not have self-intersections; they are continuous for $q > 0$ and have jumps at vertical line $q = 0$. As far as the impulse function $I(\dot{q}^-)$ is monotone, any periodic orbit includes either no impacts at all, or one such impact per period. The grazing orbit is an intermediate case where the orbit goes through the origin (Fig. 7).

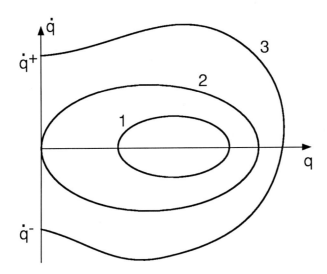

Fig. 7. Periodic orbits: 1) Non-impact, 2) Grazing, 3) Impact.

A specific property of any automous system is that at least one of its multipliers equals to unity. If all the other multipliers lie inside the unit circle, the orbit is orbitally asymptotically stable, and if there exist a multiplier outside the circle, it is unstable. In the case under discussion we have $\rho_1 = 1$, and $\rho_2 = \det \mathbf{Z}$. If the orbit is impactless, the value $\det \mathbf{Z}$ can be calculated with the help of the Liouville formula:

$$\det \mathbf{Z} = \exp \oint_0^\tau \frac{\partial F_2}{\partial \dot{q}}(t)\, dt, \tag{4.3}$$

where the integral is taken along the periodic orbit. We can see from the formula (4.3) that $\rho_2 > 0$, therefore, stability takes place provided $\rho_2 \in (0, 1)$.

Consider now discontinuous orbit. According to Theorem 3.4 and formula (3.24),

$$\det \mathbf{Z} = (1 + I')(1 + I/\dot{q}^-) \exp \oint_0^\tau \frac{\partial F_2}{\partial \dot{q}}(t)\, dt. \tag{4.4}$$

For usual impact rules the values $(1+I')$ and $(1+I/\dot{q}^-)$ belong to the interval $(0, 1)$, e.g. for Newton impact rule they both are equal to $-e$.

At last, for the grazing orbit (line 3 in Fig. 7) the disturbed trajectory lies either inside given orbit, or outside it. In the first case, the motion is impactless, and formula (4.3) is to be used. In the second case, the formula (4.4) should be applied with $\dot{q}^- \to 0$. Since $I(0) = 0$, we have

$$\lim_{\dot{q}^- \to 0} \frac{I(\dot{q}^-)}{\dot{q}^-} = I'(0).$$

Therefore, from the impacting side of the grazing trajectory the second Floquet multiplier has form

$$\rho_2 = (1 + I'(0))^2 \exp \oint_0^\tau \frac{\partial F_2}{\partial \dot{q}}(t)\, dt. \tag{4.5}$$

Note that from physical restrictions $I'(0) \in [-2, -1]$, and the right-hand side in (4.5) is less than that in (4.3).

We are in a position now to formulate the stability conditions for grazing orbit. Denote U the curvilinear integral being situated at the right-hand side of formula (4.5) (the integral is calculated along this orbit).

Proposition 4.1. The grazing orbit is orbitally asymptotically stable if $U < 0$, unstable if $\exp\{U\} > (1 + I'(0))^{-2}$, and half-stable if $1 < \exp\{U\} < (1 + I'(0))^{-2}$. (Half-stability means that such trajectories which start within the grazing orbit, wander from it for $t \to \infty$, while outside trajectories tend to the grazing orbit.)

Example. Consider oscillator with nonlinear damping

$$\ddot{q} + z[\dot{q}^2 + (q - 1)^2 - 1]\dot{q} + q - 1 = 0, \quad q \geq 0, \tag{4.6}$$

where $z = \pm 1$. The impacts are described by Newton rule where $e \in (0, 1]$. The equation (4.6) has a periodic solution

$$q = 1 - \cos t, \quad \dot{q} = \sin t, \tag{4.7}$$

since the term in the square brackets vanishes. The orbit (4.7) comes to the origin at $t = 0, 2\pi, 4\pi, \ldots$. In given example,

$$F = -z[\dot{q}^2 + (q-1)^2 - 1]\dot{q} - q + 1, \quad \partial F / \partial \dot{q} = -z[\dot{q}^2 + (q-1)^2 - 1] - 2z\dot{q}^2. \quad (4.8)$$

Substituting (4.7) into (4.8), we obtain

$$U = -2z \int_0^{2\pi} \sin^2 t \, dt = -2\pi z \quad (4.9)$$

According to the Proposition 4.1, the orbit is stable if $z = 1$, unstable if $z = -1, e^2 \exp(2\pi) > 1$, and half-stable if $z = -1, e^2 \exp(2\pi) < 1$.

We discuss now the grazing bifurcation. Suppose that system (4.2) depends on certain parameter μ :

$$\ddot{q} = F_2(q, \dot{q}, \mu), \quad q \geq 0. \quad (4.10)$$

For $\mu < 0$ there exist impactless periodic orbit (line 1 in Fig. 7), and for $\mu = 0$ it transfers to grazing orbit (line 2). The problem is: if there exist motion with impact for small μ? To answer this question, one is to make rather cumbersome calculations (cf. [34]). We give here the final result only.

Proposition 4.2. If $U < 0$, then there exists asymptotically stable periodic impact motion for small $\mu > 0$. If $\exp(U) > (1 + I'(0))^{-2}$, then there exist unstable periodic impact motion for small $\mu > 0$. If $1 < \exp(U) < (1 + I'(0))^{-2}$, then there exist asimptotically stable periodic impact motion for small $\mu < 0$ and no periodic motion exist for $\mu > 0$.

Example. Include the parameter μ in formula (4.6):

$$\ddot{q} + z[\dot{q}^2 + (q - 1)^2 - (1 + \mu)^2]\dot{q} + q - 1 = 0, \quad q \geq 0, \quad (4.11)$$

The equation (4.11) has a periodic solution

$$q = 1 - (1 + \mu) \cos t, \quad \dot{q} = (1 + \mu) \sin t,$$

which is impactless for $\mu \in (-2, 0)$. Similar to (4.9) we obtain the value U. The three variants of Proposition 4.2 correspond to the cases $z = 1$, $z = -1, e^2 \exp\{2\pi\} > 1$, and $z = -1, e^2 \exp\{2\pi\} < 1$.

4.2.2 Double Unilateral Constraint

Consider now a modification of the system (4.2) with double unilateral constraint:

$$\ddot{q} = F_2(q, \dot{q}), \quad -a \leq q \leq a. \quad (4.12)$$

The phase space of system (4.12) is the vertical unbounded stripe (Fig. 8). An example of such system is presented in Fig. 6. Along with the impactless periodic motions, which are represented by closed curves (line 1 on Fig. 8), such system may possess the periodic motions with impact upon one of the constraints (line 2) or both of them (line 3). Besides, in general such orbits might exist which graze one of the constraints (line 4) or both of them (line 5). The orbits of type 4 were considered previously, and the propositions 4.1 and 4.2 may be used for their analysis. The new case is the orbit of type 5 which is typical to systems with symmetry similar to that shown on Fig. 6.

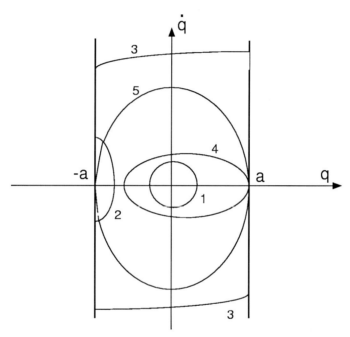

Fig. 8. Periodic orbits in double-constrained system: 1) Non-impact, 2) Impact, 3) Double-impact, 4) Grazing, 5) Double-grazing.

Note that all the nearby orbits to the grazing orbit 5 belong either to type 1, or to type 3, since different trajectories can not have common points. In the first case, the multiplier ρ_2 is expressed by the formula (4.3), while in the second case we have the motion with two impacts per period. In view of the Theorem 3.4 and formula (3.17), we can generalize the equlity (4.5) in a such way:

$$\rho_2 = (1 + I_1'(0))^2 (1 + I_2'(0))^2 \exp \oint_0^\tau \frac{\partial F_2}{\partial \dot{q}}(t)\, dt, \qquad (4.13)$$

where the subscripts 1 and 2 indicate the impact laws for the constraints. The Proposition 4.1 should be modified in the following way:

Proposition 4.1'. The double-grazing orbit (line 5 on Fig. 8) is orbitally asymptotically stable if $U < 0$, unstable if $\exp(U) > (1+I_1'(0))^{-2}(1+I_2'(0))^{-2}$, and half-stable if $1 < \exp(U) < (1+I_1'(0))^{-2}(1+I_2'(0))^{-2}$.

The bifurcations of double-grazing orbits may in general case lead to the appearance of orbits of types 2 (or similar orbit with impacts on the right only) or 3. We restrict ourselves by symmetrical systems, supposing that

$$F_2(-q, -\dot{q}, \mu) \equiv F_2(q, \dot{q}, \mu), \quad I_1(\dot{q}) = -I_2(-\dot{q}),$$

i.e. for any solution $q(t)$ to system (4.12) $-q(t)$ is a solution too. Due to such symmetry, the orbits of type 2 can not exist in a vicinity of the double-grazing orbit (otherwise two conjugate orbits of such type would have common points). Therefore, only two types of periodic orbits may appear: 1 or 3. Proposition 4.2 is to be changed as follows.

Proposition 4.2'. If $U < 0$, then in a vicinity of double-grazing orbit there exist asimptoticaly stable double-impact orbit for small $\mu > 0$. If $\exp(U) > (1+I_1'(0))^{-4}$, then there exist unstable periodic impact motion for small $\mu > 0$. If $1 < \exp(U) < (1+I_1'(0))^{-4}$, then there exist asimptoticaly stable periodic impact motion for small $\mu < 0$ and no periodic motion exist for $\mu > 0$.

Example. Consider the system presented on Fig. 6. It is symmetrical provided the suspension point lies in the axe of symmetry and the impact laws are similar for both walls. It is understood that impactless self-oscillations which exist for $u < u_0$, are stable. Therefore, the impact oscillations exist for $u > u_0$ and are stable at least if the difference $u - u_0$ is small. (To study further bifurcation one should apply techniques presented in Sect. 3.)

4.3 One-Degree-of-Freedom System with Periodic Forcing

Consider now non-autonomous system

$$\ddot{q} = F_2(q, \dot{q}, t), \quad q \geq 0, \tag{4.14}$$

where the right-hand side is a 2π - periodic function with respect to time t. The behaviour of such system even in absence of impacts might be very complicated and is not fully understood yet (cf. [1]). However, stability and local bifurcations of periodic orbits can be studied by elaborated techniques, which can be applied to non-degenerate impact motions too (see Sect. 3). As concern to grazing orbits, the nearby dynamics is very complicated as a rule.

We refer to papers [4–17, 36–43] and others which are devoted to grazing in specific systems and to attempts to obtain general properties of grazing orbits.

From a mathematical point of view, the source of difficulties in stability and bifurcation analysis is that in formulas (3.16)–(3.18) the approach velocity x_2^- is situated in denominator. Therefore, if such velocity is arbitrary small, some of elements of impact matrix J are infinitely large. In fact, this behaviour does not ensure instability, it just indicates that the linerization technique can not be used. Indeed, the Propositions 4.1, 4.1' show that in some cases the impacts lead to the stabilization of periodic orbits. The following proposition was stated in [9].

Proposition 4.3. Let $\mathbf{x}^*(t) = (\mathbf{q}^*(t), \dot{\mathbf{q}}^*(t))$ be a periodic solution to system (4.14) which includes one grazing per period. Let Z be the monodromy matrix which is calculated disregarding the possibility of impacts near grazing point, and ρ_1, ρ_2 be its eigenvalues. If both numbers $\rho_{1,2}$ are real and belong to the interval $(0, 1)$, the grazing orbit is asymptotically stable. Otherwise, it is unstable.

Note that the stability here is ensured by the absence of repeated collisions with small approach velocity: any disturbed orbit experiences such collisions no more than two times.

To clarify the multiformness of grazing bifurcation consider the system in Fig. 5. Such system (with $M = 1$) is a particular example of linear vibro-impactor

$$\ddot{q} + 2b\dot{q} + cq = P(t), \quad q \geq 0. \tag{4.15}$$

We denote $p(t)$ a periodic solution of system (4.15) (the inequality $p(t) < 0$ is possible here) and denote $\mathbf{Y}(t)$ the fundamental solution matrix of the homogeneous system (cf.(3.7)):

$$\mathbf{Y}(t) = \begin{pmatrix} \varphi_1(t) & \varphi_2(t) \\ \dot{\varphi}_1(t) & \dot{\varphi}_2(t) \end{pmatrix}, \tag{4.16}$$

$$\varphi_1(t) = (\cos \delta t + \frac{b}{\delta} \sin \delta t) \exp(-bt), \quad \varphi_2(t) = \frac{1}{\delta} \sin \delta t \exp(-bt), \quad \delta = \sqrt{c - b^2}.$$

We look for simplest periodic impact solutions to the system (4.15) with one impact per period $\tau = 2\pi N$. Such solution is described within the interval of impactless motion $(t', t' + \tau)$ by the following relations:

$$q(t) = p(t) + (q(t') - p(t'))\varphi_1(t - t') + (\dot{q}(t') - \dot{p}(t'))\varphi_2(t - t'),$$
$$\dot{q}(t) = \dot{p}(t) + (q(t') - p(t'))\dot{\varphi}_1(t - t') + (\dot{q}(t') - \dot{p}(t'))\dot{\varphi}_2(t - t'), \tag{4.17}$$

We have the following periodicity conditions (Newton impact rule is used):

$$q(t' + \tau) = q(t') = 0, \quad \dot{q}(t' + 0) = eV, \quad V = -\dot{q}(t' + \tau - 0). \tag{4.18}$$

Introducing formulas (4.17) into conditions (4.18) leads to the following result:

$$p(t')(1 - \varphi_1(\tau)) + (eV - \dot{p}(t'))\varphi_2(\tau) = 0,$$

$$p(t')\dot{\varphi}_1(\tau) + (eV - \dot{p}(t'))(1 - \dot{\varphi}_2(\tau)) = (1 + e)V.$$

$$(4.19)$$

The system (4.19) is linear with respect to $p(t'), \dot{p}(t')$, and its solution can be expressed in the appropriate matrix form:

$$\begin{pmatrix} p(t') \\ \dot{p}(t') \end{pmatrix} = \begin{pmatrix} 0 \\ eV \end{pmatrix} - (1 + e)V(\mathbf{E} - \mathbf{Y}(\tau))^{-1} \begin{pmatrix} 0 \\ 1 \end{pmatrix}. \qquad (4.20)$$

The (4.20) can be solved geometrically. For that purpose we represent on the phase plane the closed orbit $(p(t'), \dot{p}(t'))$ as well as straight half-line which is the graph of the right-hand side. (The last graph is only half-line because the approach velocity V is positive with necessity.) Any solution of period τ corresponds to a common point of these two lines (we can determine the impact moment t' from the position of the point of intersection on the closed orbit and the value V from its position on the half-line). Note that such correspondence is not reversible because any point of intersection determines the periodic orbit $q(t)$, but the property $q(t) \geq 0$ is not guaranteed and is to be checked additionally.

A governing role in the position of the half-line with respect to the vertical axes is played by the element y_{12} of matrix $\mathbf{Y}(\tau)$, i.e. $y_{12} = \varphi_2(\tau)$. If $y_{12} > 0$, the half-line issues for $V = 0$ from the origin and goes to the left half-plane. In fact, this direction is determined uniquely by the element in the first row and the second column of the matrix $(\mathbf{E} - \mathbf{Y}(\tau))^{-1}$. This element has the same sign as y_{12}, and the total in the right-hand side in (4.20) has the opposite sign. Similarly, if $y_{12} < 0$, the half- line issues from the origin to the right half-plane. At last, in case where $y_{12} = 0$ the half-line coincides with the vertical axis.

If the half-line belongs to the right half-plane, it can intersect the impact-less orbit $(p(t), \dot{p}(t))$ at two points (if the orbit is non-convex, the number of possible intersections increases). That indicates the coexistence of impact motions of given period with the impactless motion. On the contrary, if the half-line belongs to the left half-plane, it can not intersect the impactless orbit $(p(t), \dot{p}(t))$, and impact motions of the corresponding period do not exist. As can be seen from (4.16), the sign of $y_{12} = \varphi_2(\tau)$ depends first of all on the value δ. If the damping is large, so that $c < b^2$, δ will be imaginary complex number. In such a case we have $\varphi_2(t) > 0$ for any t. Thus, the half-line corresponding to any period $\tau = 2\pi N$ belongs to the left half-plane (see draft on Fig. 9a). If the damping is not too large and $c > b^2$, then δ is a real number, and $\varphi_2(t)$ may have both signs. In fact, if $\sin \delta \tau > 0$, the half-line is situated in the left half-plane, and if $\sin \delta \tau < 0$, it is situated in the right half-plane (a possible portrait is depicted on Fig. 9b).

Suppose now that the excitation force $P(t)$ depends on certain parameter μ, then the periodic solution depends on μ too. Say,

$$P(t) = \mu \sin t + B,$$

then

$$p(t) = Bc^{-1} + \alpha \sin t + \beta \cos t, \qquad (4.21)$$

$$\alpha = \mu(c-1)[4b^2 + (1-c)^2]^{-1}, \quad \beta = -2\mu b[4b^2 + (1-c)^2]^{-1},$$

The orbit (4.22) is a circle with the center at the point $(Bc^{-1}, 0)$ and of radius $R = (\alpha^2 + \beta^2)^{1/2} = \mu[4b^2 + (1-c)^2]^{-1/2}$ (see Fig. 9). For small values of μ this circle belongs to the right half-plane, i.e. the orbit is impactless. If $\mu = \mu_0 = Bc^{-1}[4b^2 + (1-c)^2]^{1/2}$ the orbit includes the origin, and if $\mu > \mu_0$, it encircles the origin (Fig. 9).

We can see what happens as the parameter μ is increased from zero. There are two different cases: large and small damping. If damping is large, no periodic impact motion exist for $\mu < \mu_0$. However, for $\mu > \mu_0$ a large number of such motions (maybe, infinite number) appears at once (Fig. 9a). Such situation is typical for chaotic dynamics. In case where damping is not too large, the grazing is preceeded by a number of saddle-node bifurcations, at which the periodic impact motions of different periods appear by pairs. Namely, these are such motions for which $\sin \delta\tau < 0$. At grazing bifurcations half of these orbits disappear (those with vanishing approach velocity) while a number of impact motions with $\sin \delta\tau > 0$ appear (Fig. 9b). Clearly, the picture may change considerably as the value δ is altered.

We conclude from this simplest (but non-trivial) example that there exist an infinite number of possible scenarios of grazing bifurcation. Only small number of general conclusion can be made concerning the post-bifurcation dynamics. First of all, we can see that if $c < b^2$ or $c > b^2$, $\sin \delta\tau > 0$, then the motion of period 2π does not disappear for $\mu > \mu_0$. In general, it gets instability. In case where $\sin \delta\tau < 0$, given periodic orbit experiences the saddle-node bifurcation and disappears for $\mu > \mu_0$. At last, if $\sin \delta\tau = 0$, stable impact motion exist for $\mu > \mu_0$ (such situation takes place in the autonomous case, considered above).

The similar concusions can be made with respect to general nonlinear system (4.14) with impact rule (2.8). Note that a comprehensive consideration of low-speed collisions shows [42] that in account of non-zero impact duration and the presence of usual (non-impulsive) forces, the following estimation holds

$$I = O(|\dot{q}^-|^3). \qquad (4.22)$$

(To compare: in the case of Newton impact rule $I = -(1+e)\dot{q}^- = O(|\dot{q}^-|)$.) In view the formula (4.22), the trajectories on Fig. 9 are not straight half-lines, but certain curves which issue from the origin and tangent to the $\dot{q}$ - axis. The grazing bifurcation comes to a sequence of typical simple bifurcations, such as flips, saddle-nodes and so on. The following conclusion can be made [42]:

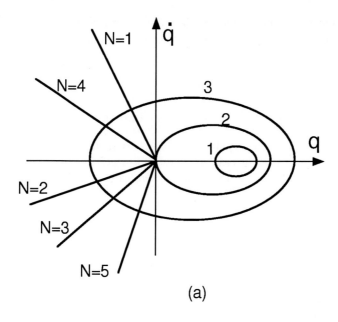

(a)

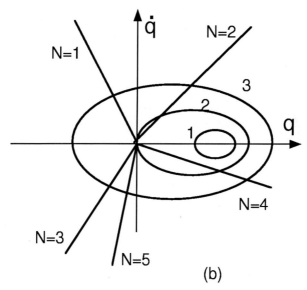

(b)

Fig. 9. Grazing bifurcation of vibro-impactor: a) Large damping, b) Small damping.
1) Impactless orbit, 2) Impactless motion impossible.

Proposition 4.4. The evolution of a stable periodic orbit which undergoes grazing bifurcation depends first of all on the element y_{12} of the monodromy matrix $Y(t', t' + \tau)$:

(i)   if $y_{12} > 0$, then the orbit survives but loses stability;
(ii) if $y_{12} < 0$, then the orbit disappears;
(iii) if $y_{12} = 0$, then the orbit survives and preserves stability.

Note that in papers [11, 40-41] global consequences of grazing were studied. It was shown that in case (i) where both multipliers ρ_1 and ρ_2 are real numbers, several kinds of chaotic attractors exist due to the maximal value of these multipliers.

4.4   Multiple Degrees of Freedom

Similar to one-degree-of-freedom case, linear analysis is insuffucient to investigate stability of motions with grazing, since the jump matrix (3.25) includes infinitely large elements. This does not, however, imply instability: as was shown in [16], periodic motions with grazing can be stable. They derived sufficient conditions of stability. Unfortunately, such conditions include an infinite number of requirements, and one would got a headache trying to apply them to a specific problem.

We turn now to bifurcations and start our discussion with smooth systems. It is known that typical bifurcations, such as folds, flips, and Hopf bifurcations, are essentially one- or two-dimentional (cf. [1]). This means that there exist a low-dimensional center manifold which contains all qualitative changes, while in the transversal direction the system behaviours in a regular way. Unfortunately, such simplification seems to be impossible in case of grazing bifurcations, since the matrix (3.25) contains too many singularities. This multidimentioness is a serious obstacle for classification of grazing bifurcation, and not much can be said about it.

Note that in account of non-zero impact duration the grazing bifurcation can be reduced to a sequence (infinite in general) of standard bifurcations (cf. previous subsection). In comparison with one-degree-of-freedom case, such sequence may contain additionally Hopf bifurcations. Such approach allows to obtain the following statement, similar to Proposition 4.4 (cf. [34]):

Proposition 4.5. The evolution of a stable periodic orbit which undergoes grazing bifurcation depends on the elements of the characteristic matrix $H(\rho) = Y(t', t' + \tau) - \rho Id$ as follows. Let $\chi_1 = \det H(1) M_{21}(1)$ with M_{21} being the cofactor.

(i)   if $\chi_1 < 0$, then the orbit survives but loses stability;
(ii) if $\chi_1 > 0$, then the orbit disappears;
(iii) if $\chi_1 = 0$, then the orbit survives and preserves stability.

Examples of multidimensional grazing bifurcations can be found in [15, 17, 44].

5 Multiple Collisions

5.1 Orthogonality Conditions

Another type of degeneracy is related to violation of condition (iii) presented in Sect. 3.3. As was mentioned in Sect. 2.4, the first thing to be done is to check the multiple collision correctly defined for the periodic motion under study. Such check-up is connected with verification of so called orthogonality conditions. To explain the meaning of such conditions, discuss at first the case where two ideal constraints $f_1(\mathbf{q}) \geq 0$ and $f_2(\mathbf{q}) \geq 0$ are involved (Fig. 10). The unperturbed trajectory hits the vertice of the angle between two surfaces $f_1(\mathbf{q}) = 0$ and $f_2(\mathbf{q}) = 0$ and undergoes an impulse $\mathbf{I}^* = \mathbf{I}(\dot{\mathbf{q}}^-)$. A perturbed trajectory can experience two impacts: at first upon one of the constraints, then upon another constraint. The first impact is described by relation

$$\dot{\mathbf{q}}^{+-} = \dot{\mathbf{q}}^- + \boldsymbol{\Delta}\dot{\mathbf{q}} + \mathbf{I}(\mathbf{q}^- + \boldsymbol{\Delta_1}\mathbf{q}, \dot{\mathbf{q}}^- + \boldsymbol{\Delta}\dot{\mathbf{q}}), \qquad (5.1)$$

where $\dot{\mathbf{q}}^{+-}$ is post-impact velocity and $\Delta_1\mathbf{q}$ is such perturbation that $f_1(\mathbf{q}+\boldsymbol{\Delta_1}\mathbf{q}) = 0$, $f_2(\mathbf{q} + \boldsymbol{\Delta_1}\mathbf{q}) > 0$. Similarly, the second impact matches the formula

$$\dot{\mathbf{q}}^{++} = \dot{\mathbf{q}}^{+-} + \mathbf{I}(\mathbf{q}^- + \boldsymbol{\Delta_2}\mathbf{q}, \dot{\mathbf{q}}^{+-}), \qquad (5.2)$$

where $\Delta_2\mathbf{q}$ is a perturbation such that $f_1(\mathbf{q} + \boldsymbol{\Delta_2}\mathbf{q}) > 0$, $f_2(\mathbf{q} + \boldsymbol{\Delta_1}\mathbf{q}) = 0$.

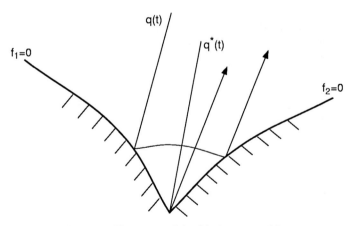

Fig. 10. Variation of double-impact orbit.

Now we tend the perturbations to zero, then

$$\dot{\mathbf{q}}^{++} = \dot{\mathbf{q}}^- + \mathbf{I}_1(\mathbf{q}^-, \dot{\mathbf{q}}^-) + \mathbf{I}_2(\mathbf{q}^-, \dot{\mathbf{q}}^- + \mathbf{I}_1(\mathbf{q}^-, \dot{\mathbf{q}}^-)), \qquad (5.3)$$

where subscripts indicate that only corresponding constraint is involved in the collision. To ensure continuous dependence of solution curve upon initial conditions, the equality

$$\mathbf{I}(\mathbf{q}^-,\dot{\mathbf{q}}^-) = \mathbf{I_1}(\mathbf{q}^-,\dot{\mathbf{q}}^-) + \mathbf{I_2}(\mathbf{q}^-,\dot{\mathbf{q}}^- + \mathbf{I_1}(\mathbf{q}^-,\dot{\mathbf{q}}^-)) \tag{5.4}$$

must hold; otherwise, arbitrary small initial perturbations will lead to considerable post-impact alterations. Similarly, in case there the trajectory $\mathbf{q}(t)$ hits at first the second constraint and then the first one, we obtain the following condition:

$$\mathbf{I}(\mathbf{q}^-,\dot{\mathbf{q}}^-) = \mathbf{I_2}(\mathbf{q}^-,\dot{\mathbf{q}}^-) + \mathbf{I_1}(\mathbf{q}^-,\dot{\mathbf{q}}^- + \mathbf{I_2}(\mathbf{q}^-,\dot{\mathbf{q}}^-)). \tag{5.5}$$

The relations (5.4), (5.5) express well-posedness of multiple collision problem. In the considered case of ideal constraints the reactions R_i in (2.6) are collinear to $\operatorname{grad} f_i (i = 1, 2)$. Thus, the impulses $\mathbf{I}_i$ are collinear to $C^{-1}\operatorname{grad} f_i$, where

$$C = \left\| \frac{\partial^2 T}{\partial \dot{q}_i \partial \dot{q}_j} \right\|_{i,j=1}^n \tag{5.6}$$

is the Hessian of the kinetic energy. Moreover, $\mathbf{I}_i$ depends on $\mathbf{q}$ and on approach velocity $\dot{f}_i^-$ only:

$$\mathbf{I_i}(\mathbf{q}^-,\dot{\mathbf{q}}^-) = \mathbf{I_i}(\mathbf{q}^-,\dot{f}_i^-), \quad (i = 1, 2)$$

We can see that to ensure conditions (5.4)–(5.5), the single equality must be valid:

$$\operatorname{grad} f_1 C^{-1}(\operatorname{grad} f_2)^T = 0, \tag{5.7}$$

which is known as the *orthogonality* in the kinetic Jacobi metrix.

The orthogonality condition (5.7) at given point $\mathbf{q}^-$ implies that

$$\mathbf{I}(\mathbf{q}^-,\dot{\mathbf{q}}^-) = \mathbf{I_1}(\mathbf{q}^-,(\dot{\mathbf{q}}^-,\operatorname{grad}\ f_1)) + \mathbf{I_2}(\mathbf{q}^-,(\dot{\mathbf{q}}^-,\operatorname{grad}\ f_2)). \tag{5.8}$$

To make this more clear, consider once more the example from Sect. 2.4. There

$$\begin{aligned}
\mathbf{I_1}(\mathbf{q}^-,\dot{\mathbf{q}}^-) &= \mathbf{I_1^o}, \\
\mathbf{I_2}(\mathbf{q}^-,\dot{\mathbf{q}}^- + \mathbf{I_1}(\mathbf{q}^-,\dot{\mathbf{q}}^-)) &= \mathbf{I_2^o} + (1+e_1)(1+e_2)\mathbf{n_2}\sin\alpha\cos 2\alpha, \\
\mathbf{I_2}(\mathbf{q}^-,\dot{\mathbf{q}}^-) &= \mathbf{I_2^o}, \\
\mathbf{I_1}(\mathbf{q}^-,\dot{\mathbf{q}}^- + \mathbf{I_2}(\mathbf{q}^-,\dot{\mathbf{q}}^-)) &= \mathbf{I_1^o} + (1+e_1)(1+e_2)\mathbf{n_1}\sin\alpha\cos 2\alpha,
\end{aligned} \tag{5.9}$$

where the values $\mathbf{I_1^o}, \mathbf{I_2^o}$ are defined in (2.9). Since the vectors $\mathbf{n_1}$ and $\mathbf{n_2}$ are not equal, the conditions (5.4) and (5.5) are satisfied in single case where $\alpha = \pi/4$ (then the total impulse is expressed by (2.10)) which means the orthogonality of two planes. In given case C is a scalar matrix, hence, (5.7) is equivalent to the orthogonality of the planes too.

Note that a perturbed motion might experience three or more subsequent simple collisions, and in such cases the multiple collision could not be defined correctly (not to say of very few exceptions, cf. [20, 23]).

Similar concept of orthogonality can be applied to collision which involves three or more constraints: in that case each pair of constraints involved must be orthogonal.

5.2 Variation of Periodic Orbit

The first step in the stability and bifurcation analysis is the study of variations $\mathbf{y}(t)$. The Theorems 3.1 and 3.4 show how to do it in the cases where the observed trajectory $\mathbf{x}^*(t)$ is either continuous or experiences non-degenerate discontinuities. Now we discuss the motion with multiple impacts. It is supposed that above considered orthogonality condition (5.7) is valid.

Consider a trajectory $\mathbf{q}^*(t)$ which hits at $t = t'$ the vortex between two surfaces $f_1(\mathbf{q}) = 0$ and $f_2(\mathbf{q}) = 0$. We try to perform analysis of close trajectories by analogy to Sect. 3.2. Let $x^* = (q^*, \dot{q}^*)$, $x = (q, \dot{q})$, $y = x^* - x$. A perturbed orbit $\mathbf{x}(t)$ may behave in three ways: it can collide with the surfaces $f_1 = 0$ and $f_2 = 0$ in turn (two different possibilities, see Fig. 6) or simultaneously.

We start with the case where $\mathbf{x}(t)$ hits at first the surface $f_1 = 0$ and then the surface $f_2 = 0$ (orbit $q(t)$ on Fig. 6). We denote Δt_1 time delay of the first collision, i.e. $f_1(\mathbf{x}(t' + \Delta t_1) = 0$. Much similar to $(3.16) - (3.21)$, we can conclude that after the first collision, the vector of disturbances $\mathbf{y}$ changes in the following way:

$$\mathbf{y}^{+-} = \mathbf{J_1}\mathbf{y}^- + o(y^-)$$

$$\mathbf{J_1} = \mathbf{E} + \mathbf{I_{1x}} + V_1^{-1}(\mathbf{F}^- - \mathbf{F}^{+-}) \times \operatorname{grad} f_1 + V_1^{-1}\mathbf{I_{1x}}(\mathbf{F}^- \times \operatorname{grad} f_1),$$

$$(5.10)$$

where $V_1 = -(\operatorname{grad} f_1, \mathbf{x}^*(t' - 0))$, $\mathbf{I_{1x}} = \|\partial \mathbf{I_1}(\mathbf{x}^*(t' - 0)/\partial \mathbf{x}\|$, $\mathbf{F}^{+-} = \mathbf{F}(t', \mathbf{x}^*(t' - 0)) + \mathbf{I_1}(\mathbf{x}^*(t' - 0))$. After that, the perturbed trajectory collides with the surface $f_2 = 0$. The final disturances $\mathbf{y}^+$ can be calculated similar to (5.10), with the intermediate vector $\mathbf{y}^{+-}$ being treated as initial one:

$$\mathbf{y}^+ = \mathbf{J_2}\mathbf{y}^{+-} + o(y^-) = \mathbf{J_2}\mathbf{J_1}\mathbf{y}^- + o(y^-)$$

$$\mathbf{J_2} = \mathbf{E} + \mathbf{I_{2x}} + V_2^{-1}(\mathbf{F}^{+-} - \mathbf{F}^+) \times \operatorname{grad} f_2 + V_2^{-1}\mathbf{I_{2x}}(\mathbf{F}^{+-} \times \operatorname{grad} f_2),$$

$$(5.11)$$

where, in account with relation (5.8), $V_2 = -(\operatorname{grad} f_2, \mathbf{x}^*(t' - 0))$, $\mathbf{I_{2x}} = \|\partial \mathbf{I_2}(\mathbf{x}^* (t' - 0)/\partial \mathbf{x}\|$ (i.e. these quantities are not affected by the first collision).

An opposite case is related to the first collision with $f_2 = 0$ and the second collision with $f_1 = 0$. The same calculations as above lead to the following result:

$$\mathbf{y}^+ = \mathbf{J}'_1 \mathbf{J}'_2 \mathbf{y}^- + o(y^-)$$

$$\mathbf{J}'_1 = \mathbf{E} + \mathbf{I}_{1x} + V_1^{-1}(\mathbf{F}^{-+} - \mathbf{F}^+) \times \operatorname{grad} f_1 + V_1^{-1}\mathbf{I}_{1x}(\mathbf{F}^{-+} \times \operatorname{grad} f_1),$$

$$\mathbf{J}'_2 = \mathbf{E} + \mathbf{I}_{2x} + V_2^{-1}(\mathbf{F}^- - \mathbf{F}^{-+}) \times \operatorname{grad} f_2 + V_2^{-1}\mathbf{I}_{2x}(\mathbf{F}^- \times \operatorname{grad} f_2),$$

$$(5.12)$$

where $\mathbf{F}^{-+} = \mathbf{F}(t', \mathbf{x}^*(t'-0)) + \mathbf{I}_2(\mathbf{x}^*(t'-0))$. To determine which of scenarios is realized, one should compare the values

$$\Delta t_1 = (\mathbf{y}^-, \operatorname{grad} f_1(\mathbf{x}^*(t')))/V_1 + o(y^-),$$

$$\Delta t_2 = (\mathbf{y}^-, \operatorname{grad} f_2(\mathbf{x}^*(t')))/V_2 + o(y^-) \qquad (5.13)$$

(see (3.19)): if $\Delta t_1 < \Delta t_2$, we have the first case, while the second case corresponds to the opposite inequality.

In addition to these two main cases, the equality $\Delta t_1 = \Delta t_2$ is possible too: it indicates that the perturbed trajectory hits the vertice as well as unperturbed one (but non-simultaneously). In such a case we can use directly formulas (3.20)-(3.21):

$$\mathbf{y}^+ = \mathbf{J}_3 \mathbf{y}^- + o(y^-), \quad \mathbf{J}_3 = \mathbf{E} + \mathbf{I}_{1x} + \mathbf{I}_{2x} + V_1^{-1}(\mathbf{F}^- - \mathbf{F}^+) \times \operatorname{grad} f_1$$

$$+ V_1^{-1}(\mathbf{I}_{1x} + \mathbf{I}_{2x})(\mathbf{F}^- \times \operatorname{grad} f_1),$$

$$(5.14)$$

(or, similarly, V_1, f_1 might be substituted with V_2, f_2). We come to the following result.

Proposition 5.1. Given an orbit including multiple collision, the variation $\mathbf{y}(t)$ depends on $\mathbf{y}(0)$ in a continuous, but non-differentiable way: a vicinity of the origin $y(0) = 0$ can be divided in two parts G_1 and G_2 such that Jacobi matrices $\|\partial \mathbf{y}(t)/\partial \mathbf{y}(0)\|$ exist in G_1, G_2 and on common boundary of these regions, but they can differ (see (5.11), (5.12), (5.14)).

Example. Consider a ball of unit mass which moves in the quadrant, constituted by two mutually orthogonal smooth planes under given force $\mathbf{F}$ and strikes both planes at $t = t'$ (Fig. 4). We introduce cartesian coordinates q_1, q_2 so that the imposed unilateral constraints are expressed by the inequalities $q_1 \geq 0$, $q_2 \geq 0$ and let $x_1 = q_1, x_2 = \dot{q}_1, x_3 = q_2, x_4 = \dot{q}_2$. The equations of motion are

$$\dot{x}_1 = x_2, \quad \dot{x}_3 = x_2,$$

$$\dot{x}_2 = F_2, \quad \dot{x}_4 = F_4, \qquad (5.15)$$

For simplicity, we accept Newton impact rules for each constraint, with the coefficients of restitution e_1 and e_2. According to (5.10)-(5.11) and due to the orthogonality we obtain (cp. (3.25))

$$
\mathbf{J_1} = \begin{pmatrix}
-e_1 & 0 & 0 & 0 \\
(F_2^{+-} + e_1 F_2^{-})/x_2^{-} & -e_1 & 0 & 0 \\
0 & 0 & 1 & 0 \\
(F_4^{+-} - F_4^{-})/x_2^{-} & 0 & 0 & 1
\end{pmatrix},
$$

$$
\mathbf{J_2} = \begin{pmatrix}
1 & 0 & 0 & 0 \\
0 & 1 & (F_2^{+} - F_2^{+-})/x_4^{-} & 0 \\
0 & 0 & -e_2 & 0 \\
0 & 0 & (F_4^{+} + e_2 F_4^{+-})/x_4^{-} & -e_2
\end{pmatrix}.
$$

(5.16)

Thus,

$$
\mathbf{J_2 J_1} = \begin{pmatrix}
-e_1 & 0 & 0 & 0 \\
(F_2^{+-} + e_1 F_2^{-})/x_2^{-} & -e_1 & (F_2^{+} - F_2^{+-})/x_4^{-} & 0 \\
0 & 0 & -e_2 & 0 \\
-e_2(F_4^{+-} - F_4^{-})/x_2^{-} & 0 & (F_4^{+} + e_2 F_4^{+-})/x_4^{-} & -e_2
\end{pmatrix}.
$$
(5.17)

Similarly,

$$
\mathbf{J_1' J_2'} = \begin{pmatrix}
-e_1 & 0 & 0 & 0 \\
(F_2^{+} + e_1 F_2^{-+})/x_2^{-} & -e_1 & -e_1(F_2^{-+} - F_2^{-})/x_4^{-} & 0 \\
0 & 0 & -e_2 & 0 \\
(F_4^{+} - F_4^{-+})/x_2^{-} & 0 & (F_4^{-+} + e_2 F_4^{-})/x_4^{-} & -e_2
\end{pmatrix}.
$$
(5.18)

The comparison of matrices (5.17) and (5.18) shows that they are different unless certain additional conditions are satisfied. Due to the orthogonality, the value $\dot{x}_2$ does not change in collision with $q_1 \geq 0$, and vice versa. Therefore, if F_2 does not depend on x_4 and F_4 does not depend on x_2, we have

$$
\mathbf{J_2 J_1} = \mathbf{J_1' J_2'} = \begin{pmatrix}
-e_1 & 0 & 0 & 0 \\
(F_2^{+} + e_1 F_2^{-})/x_2^{-} & -e_1 & 0 & 0 \\
0 & 0 & -e_2 & 0 \\
0 & 0 & (F_4^{+} + e_2 F_4^{-})/x_4^{-} & -e_2
\end{pmatrix}.
$$
(5.19)

Moreover, in this case the formula (5.14) leds to the same result for J_3. For instance, if $\mathbf{F} = \mathbf{F_0}(\mathbf{q}) + F_1(\mathbf{q})\dot{\mathbf{q}}$, where $\mathbf{F_0}(\mathbf{q})$ is a vector function and $F_1(\mathbf{q})$ is a scalar function which do not depend on $\dot{\mathbf{q}}$, the equation (5.19) is valid. Such situation occurs in case where the given force is a combination of potential forces and linear damping. On the contrary, in the presense of aerodynamic (quadratic) resistence we have $J_2 J_1 \neq J_1' J_2'$.

Up to now, we restricred our discussion by the simplest case of multiple impact where two unilateral constraints are involved. However, Proposition 5.1 can be extended easily to the case where $k \geq 3$ constraints take part in multiple collision. Note that the number of different scenarios of subsequent collisions for perturbed trajectory grows rapidly with k: there are $k!$ regions G_j, mentioned in Proposition 5.1, plus their boundaries. For instance, if we have three constraints $1, 2$ and 3, the following 13 scenarios are possible: $1 - 2 - 3$ (i.e. given sequence of simple impacts), $1 - 3 - 2$, $2 - 3 - 1$, $2 - 1 - 3$, $3 - 1 - 2$, $3 - 2 - 1$, $(1 + 2) - 3$ (i.e. double collision with 1 and 2 followed by simple collision with 3), $(1 + 3) - 2$, $(2 + 3) - 1$, $1 - (2 + 3)$, $2 - (1 + 3)$, $3 - (1 + 2)$, and $(1 + 2 + 3)$ (triple collision).

5.3 Stability Conditions

Consider a τ - periodic motion $\mathbf{q}^*(t)$ which satisfies non-degeneracy conditions (i), (ii), and (iv) (cf. Sect. 3), but includes one multiple collision per period, i.e. (iii) is violated. For simplicity, we suppose that this collision is double one. We can use the Theorem 3.4 and Proposition 5.1 to construct the local Poincaré map $P : \pi_\tau \to \pi_\tau$ in a vicinity U of fixed point $\mathbf{x}^*$, corresponding to the periodic motion (without loss of generality we suggest $\mathbf{x}^* = \mathbf{0}$). A specific feature of such a map is the lack of smoothness: there exist two regions G^+ and G^- with common boundary Γ such that $U = G^+ \cup G^- \cup \Gamma$ and

$$P(\mathbf{x}) = \begin{cases} P^+(\mathbf{x}), & \text{if } \mathbf{x} \in G^+, \\ P^-(\mathbf{x}), & \text{if } \mathbf{x} \in G^-, \end{cases} \tag{5.20}$$

where

$$P^\pm(\mathbf{x}) = \mathbf{A}_\pm \mathbf{x} + o(\mathbf{x}). \tag{5.21}$$

The matrices A^+ and A^- yield linear maps, which are similar on the tangent subspace to Γ, but differ in the transversal direction.

As the following simple examples show, a straightforward attempt to apply the theorem 3.2 to map (5.20) fails (cf. Ivanov, 1998):

Examples. 1. Consider piecewise-linear map in 2D:

$$\mathbf{A}(\mathbf{x}) = \begin{cases} \mathbf{A}_+\mathbf{x} & x_2 > 0 \\ \mathbf{A}_-\mathbf{x} & x_2 \leq 0 \end{cases} \tag{5.22}$$

$$\mathbf{A}_+ = \left\| \begin{matrix} \alpha & 1 \\ 0 & \alpha \end{matrix} \right\|, \quad \mathbf{A}_- = \left\| \begin{matrix} \alpha & 0 \\ 1 & \alpha \end{matrix} \right\|, \quad \alpha \in (-1, 0)$$

The eigenvalues of both matrices $\mathbf{A}_+$ and $\mathbf{A}_-$ belong to the unit circle, indicating stability of these two components (being considered separately). However, the product

$$\mathbf{A}_+\mathbf{A}_- = \left\| \begin{matrix} \alpha^2 + 1 & \alpha \\ \alpha & \alpha^2 \end{matrix} \right\|$$

possesses an eigenvector $\mathbf{l}^* = (1, \alpha - \alpha^3 + \ldots)^T$ with eigenvalue $\lambda^* = 1 + 2\alpha^2 - \alpha^4 + \ldots > 1$, with $\mathbf{l}^*$ belonging to the half-plane $x_2 < 0$, and $\mathbf{Al}^*$ belonging to $x_2 > 0$. Therefore, $\mathbf{A}$ is unstable.

2. Now we present an example of *continuous* unstable piecewise-linear mapping with stable components. We set in (5.21)

$$\mathbf{A}_+ = \begin{Vmatrix} 0 & -0.95 \\ 1 & -1.9 \end{Vmatrix}, \quad \mathbf{A}_- = \begin{Vmatrix} 0 & -0.6 \\ 1 & 1 \end{Vmatrix}$$

Each of the matrices $\mathbf{A}_+, \mathbf{A}_-$ is stable; moreover, on the boder line $x_2 = 0$ both submappings coinside. Nevertheless, the map $\mathbf{A}^5$ has an eigenvector $\mathbf{l}^* = (1; 0)$:

$$\mathbf{A}^5(\mathbf{l}^*) = \mathbf{A}_-^3 \mathbf{A}_+^2 (\mathbf{l}^*) = 1,0261^*$$

Since the eigenvalue is greater than unity, the map is unstable.

3. To build an opposite example, we suggest in (5.21)

$$\mathbf{A}_+ = \mathrm{diag}\{-2; -0, 1\}, \quad \mathbf{A}_- = \mathrm{diag}\{-0, 1; -2\}$$

In such a case both matrices $\mathbf{A}_+$ and $\mathbf{A}_-$ are clearly unstable. At the same time the "composed" map (5.21) is asymptotically stable. In fact, this map reverses signs of both co-ordinates of any vector. Therefore, elements of the sequence $\mathbf{l}, \mathbf{Al}, \mathbf{A}^2\mathbf{l}, \mathbf{A}^3\mathbf{l}, \ldots$ lie in turn in the half-planes $x_2 > 0$ and $x_2 < 0$ (excluding the case where l belongs to the line $x_2 = 0$: then all iterations belong to the same line while the first co-ordinste behaviors as geometric progression with factor $0,1$). Th stability follows from the following equality

$$\mathbf{A}_+ = \mathbf{A}_- = \mathrm{diag}\{0, 2; 0, 2\}$$

Now we formulate certain sufficient and necessary conditions of stability [30].

Proposition 5.2. If the matrices $\mathbf{A}_+$ and $\mathbf{A}_-$ in (5.21) have norms less than unuty (with respect to a norm $\|\mathbf{x}\|$ in R^2), then the fixed point of map (5.20) is asymptotically stable.

Proposition 5.3. If at least one of the matrices $\mathbf{A}_+$ and $\mathbf{A}_-$ in (5.21) has an eigenvector $\mathbf{l}^*$ inside the corresponding region G^+ (or G^-) with eigenvalue $\rho > 1$, then the fixed point of map (5.20) is unstable.

Note that the presence of eigenvalues $\rho < -1$ or complex eigenvalues with $|\rho| > 1$ does not ensure instability (see above Example 3). Therefore, the elaborated Floquet technique is not quite appropriate to stability analysis in the presence of multiple collisions. Nevertheless, there exist a special case where this method succeed (cf. [30]):

Proposition 5.4. Let the matrices $\mathbf{A}_\pm$ have in some basis the following block-triangle form

$$\mathbf{A}_\pm = \begin{pmatrix} C_{11} & C_{12}^\pm \\ 0 & C_{22} \end{pmatrix}, \tag{5.23}$$

with diagonal blocks C_{11}, C_{22}, similar for both components. Then the stability of map (5.21) is governed by the multipliers ρ_j precisely as in case of smooth maps (see Theorem 3.2).

Similar results (Propositions 5.2-5.4) are valid in case the map (5.20) has three or more components of smoothness.

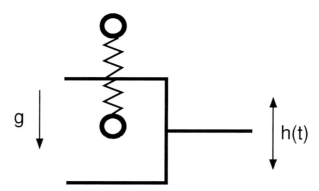

Fig. 11. Two bouncing balls.

Example. Consider a system of two identical elastically connected balls, bouncing from an oscillating "fork" (see Fig. 11). If the length of relaxed spring equals to the distance between the "pronges", there exist a number of translational periodic motions similar to periodic motions of one ball, bouncing from an oscillating table. In particular, there exist motions where the impacts occur at the same phase of oscillations (see example in Sect. 3.3.3). The matrices $\mathbf{A}_\pm$ were shown in [30] to have the form (5.23). Therefore, a full-scale stability analysis is possible. We omit rather cumbersome calculations (refer [30] for more details) and give final result only. The necessary and sufficient conditions of asymptotic stability consist of double inequality (3.35) and the following relation:

$$\left| \frac{(1+e)^2}{2\pi l} \left(1 + \frac{\ddot{h}(t')}{g} \right) \sin(2\pi l\omega) - 2e\cos(2\pi l\omega) \right| < 1 + e^2, \tag{5.24}$$

where ω is the natural frequency of the system in flight.

An alternative approach is connected with Lyapunov functions. Though construction of such function in a specific problem requires certain qualification and sometimes art, some problems of practical interest were studied by means of it. In particular, the proof of Proposition 5 is based upon that technique. The following system was considered in [45].

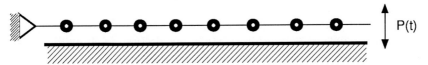

Fig. 12. A string with concentrated masses, hitting a wall.

Example. Forced oscillations of a string with concentrated masses near a wall (Fig. 12) can be described by the following equation:

$$\ddot{q} + \mathbf{A}q = \mathbf{Q}(t), \quad \mathbf{q}, \mathbf{Q} \in R^n, \quad q_k \geq 0(k = 1, \ldots n). \tag{5.25}$$

Here q_k is the distance from k-th mass to the wall, $\mathbf{A}$ is a positive stiffness matrix, and $\mathbf{Q}(t)$ is a τ - periodic excitation force. As was shown in [3, 18, 19], there exist periodic motions with simultaneous collisions in impact pairs.

Suppose $\mathbf{q}^*(t)$ is $N\tau$ - periodic so;ution of (5.25) with one multiple collision per period (all the masses hit the wall simultaneously). The jump matrices $\mathbf{J}_k$ describing each impact pair has according to (3.23) – (3.25) the following form:

$$\mathbf{J}_k = \begin{pmatrix} \mathbf{E}_n - (1 + e_k)\mathbf{D}_k & 0 \\ \Phi_k \mathbf{D}_k & \mathbf{E}_n - (1 + e_k)\mathbf{D}_k \end{pmatrix}, \quad \Phi_k = \frac{1 + e_k}{\dot{q}_k(t' - 0)} Q_k(t'), \tag{5.26}$$

where t' is the impact time, $\mathbf{D}_k$ is the matrix with all elements equal zero except unity at the intersection of k - th row with k - th column, and e_k is the coefficient of restitution in k - th impact pair. It is easy to show that the matrices $\mathbf{J}_i$ and $\mathbf{J}_j$ commute and the result of multiple collision does not depend on the order of simple impacts. Therefore, the multiple collision can be described by the following jump matrix:

$$\mathbf{J} = \begin{pmatrix} -\sum\limits_{k=1}^{m} e_k \mathbf{D}_k & 0 \\ \sum\limits_{k=1}^{m} \Phi_k \mathbf{D}_k & -\sum\limits_{k=1}^{m} e_k \mathbf{D}_k \end{pmatrix}. \tag{5.27}$$

The monodromy matrix has form

$$\mathbf{Y}(t', t' + N\tau) = \mathbf{J} \exp(N\tau \mathbf{A}). \tag{5.28}$$

Verification of stability conditions (Theorem 3.2) is connected with calculation of eigenvalues of the matrix (5.28). This problem is rather cumbersome for large n, and it is useful to obtain more simple sufficient conditions of stability. For such purpose we introduce a norm in R^{2n} by the formula

$$\|\mathbf{x}\|^2 = (\dot{\mathbf{q}}, \dot{\mathbf{q}}) + (\mathbf{A}\mathbf{q}, \mathbf{q}). \tag{5.29}$$

Obviously, the value $L = \|\mathbf{x}(t) - \mathbf{x}^*(t)\|$ is preserved during smooth motion (5.25) but it changes at impacts. Keeping proposition 5.2 in mind, we arrive at the inequality

$$(\mathbf{A}\mathbf{z}, \mathbf{z}) + (\mathbf{w} - \dot{\mathbf{z}}, \mathbf{w} - \dot{\mathbf{z}}) \le (\dot{\mathbf{q}}, \dot{\mathbf{q}}) + (\mathbf{A}\mathbf{q}, \mathbf{q}) \tag{5.30}$$

with $z_k = e_k q_k$, $w_k = \Phi_k q_k$ $(k = 1, \dots, m)$. If relation (5.30) holds $\forall \mathbf{x} \in R^{2m}$, then L is Lyapunov function, and given motion is stable. Verification of this property (Silvester criterion) is much easier than direct evaluation of eigenvalues (Schur inequalities).

Let for instance $m = 1$, $A = \omega^2$. The sufficient condition (5.30) looks so:

$$\left|\frac{\Phi}{\omega}\right| < 1 - e^2 \tag{5.31}$$

while criterion of Theorem 3.2 is equivalent to

$$\left|\frac{\Phi}{\omega} \sin N\omega\tau - 2e \cos N\omega\tau\right| < 1 + e^2. \tag{5.32}$$

Note that the condition (5.31) does not include the values N, τ. Moreover, if (5.32) is valid for any τ, then it coincides with (5.31). Therefore, the proposed analytical technique can be useful.

5.4 Bifurcations

Now we discuss possible response of a periodic motion with multiple impacts to small changes of equations of motion. We suggest that these changes are caused by alterations of certain parameter μ, and the Poincare map (5.20) can be rewritten in the form

$$P(\mathbf{x}, \mu) = \begin{cases} P^+(\mathbf{x}, \mu), & \text{if } \mathbf{x} \in G_\mu^+, \\ P^-(\mathbf{x}, \mu), & \text{if } \mathbf{x} \in G_\mu^-. \end{cases} \tag{5.33}$$

It is supposed that for $\mu = 0$ the map (5.33) has fixed point at the origin. Besides, the map (5.33) is continuous, i.e. $P^+ = P^-$ at any point on the common boundary of G^+ and G^-, however, $DP^+ \ne DP^-$.

Dynamics of map (5.33) has been studied in [4-6] irrespective to motions with multiple collisions. Nevertheless, obtained results are applicable to the problem under discussion, and it seems resonable to summarize them. First of all, choose local coordinates so that the common boundary of G^+ and G^- is given by $x_n = 0$. Then (5.33) can be rewritten in the form

$$P(\mathbf{x}, \mu) = \begin{cases} A^+\mathbf{x}, +\mu\mathbf{b} + \cdots, & \text{if } x_n \ge 0, \\ A^-\mathbf{x}, +\mu\mathbf{b} + \cdots, & \text{if } x_n < 0, \end{cases} \tag{5.34}$$

where nonlinear terms are omitted. The first $n-1$ columns of matrices A^+ and A^- coincide while the last n-th columns differ. The form (5.34) is relevant to the subsequent analysis, but the results will be formulated in an invariant form.

We will look for fixed points of (5.34) for $\mu \neq 0$. Suppose that there exist a point bfx^+ with $x_n^+ > 0$ such that $P(\mathbf{x}^+, \mu) = \mathbf{x}^+$. Then $\mathbf{x}^+$ corresponds to a solution of linear system

$$(\mathbf{E_n} - \mathbf{A^+})\mathbf{x} = \mu \mathbf{b}, \quad \mathbf{x}, \mathbf{b} \in R^n. \tag{5.35}$$

This system has unique solution provided the matrix $\mathbf{E_n} - \mathbf{A^+}$ is non-singular. The solution can be presented by Cramer formulas; in particular, $x_n^+ = \Delta_n/\Delta^+$ where $\Delta^+ = \det(\mathbf{E_n} - \mathbf{A^+})$ and Δ_n is determinant of the matrix, obtained from $\mathbf{E_n} - \mathbf{A^+}$ by substitution of the last column with $\mathbf{b}$. Clearly, $\mathbf{x}^+$ exists if and only if Δ^+ and Δ_n have the same sign. Similarly, we look for a fixed point bfx^- with $x_n^- < 0$ as a solution to system

$$(\mathbf{E_n} - \mathbf{A^-})\mathbf{x} = \mu \mathbf{b}, \quad \mathbf{x}, \mathbf{b} \in R^n. \tag{5.36}$$

We get $x_n^- = \Delta_n/\Delta^-$ where $\Delta^- = \det(\mathbf{E_n} - \mathbf{A^-})$ and Δ_n is the same number as befor (keep in mind that A^+ and A^- differ at the last column only). In case where the product $\Delta^+\Delta^-$ is positive, x_n^+ and x_n^- have the same sign, and precisely one of the points $\mathbf{x}^+$ and $\mathbf{x}^-$ is situated in the "correct" region. Another situation occure in the case $\Delta^+\Delta^- < 0$: depending on sign of μ, either none or both of the points belongs to the "correct" subspace. We have proven the following result.

Proposition 5.5. If $\Delta^+\Delta^- > 0$, there for small μ there exist unique fixed point $\mathbf{x}(\mu)$ of map (5.33) in a vicinity of the origin: $\mathbf{x}(0) = \mathbf{0}$. If $\Delta^+\Delta^- < 0$, there for small μ there exist either two fixed points $\mathbf{x}^\pm(\mu)$ or (for opposite sign of μ) no such points in a vicinity of the origin.

Note that conditions of this proposition can be stated in terms of the eigenvalues of matrices $\mathbf{A^+}$ and $\mathbf{A^-}$. For that purpose, consider a function $\chi(\delta) = \det(\mathbf{E} - \delta\mathbf{A}, \ \delta \in R$. Clearly, $\chi(0) = 1 > 0$, and the function $\chi(\delta)$ has zeros which are inverse to the eigenvalues of $\mathbf{A}$. Therefore, $\chi(1) = \det(\mathbf{E} - \mathbf{A} > 0$ if and only if $\mathbf{A}$ has an even number of real eigenvalues greater than unity. We conclude that the first (regular) case mentioned in Proposition 5.1 is realised provided the total number of such eigenvalues for A^+ and A^- is even. In view of Proposition 5.3, if the origin is stable fixed point for $\mu = 0$, we obtain the regular case. On the contrary, if the total number of eigenvalues greater than unity for A^+ and A^- is odd, the system undergoes one-side bifurcation, remainding the fold bifurcation in smooth systems. We refer to [4-6] for more details and restrict ourselves with the following example.

Example. Consider once more the system depicted in Fig. 11, but in more general case of connection between the two masses: it may include non-linear spring and dashpot. The equations of motion are

$$\ddot{x}_1 = -g - \ddot{h}(t) + G(x_1 - x_2, \dot{x}_1 - \dot{x}_2),$$
$$\ddot{x}_2 = -g - \ddot{h}(t) - G(x_1 - x_2, \dot{x}_1 - \dot{x}_2),$$
(5.37)

where x_1 and x_2 are the distances from the balls to the "pronges" and G is the interaction force, $G(0,0) = 0$. As previously, we consider symmetric motions $(x_1(t) = x_2(t) \equiv x^*(t))$ with one (multiple) collision per period. It is convenient to introduce new variables $z_1 = (x_1 + x_2)/2$ and $z_2 = (x_1 - x_2)/2$ to simplify equations (5.37): new variables describe motion of the centre of gravity and relative motion. We use formulas of Sect. 5.2 to obtain finally the following result:

$$\mathbf{J_2 J_1} = \begin{pmatrix} -e & 0 & 0 & 0 \\ -\Phi & -e & -\dfrac{(1+e)G_+}{V} & 0 \\ 0 & 0 & -e & 0 \\ 0 & 0 & -\Phi + \dfrac{(e-1)G_+}{V} & -e \end{pmatrix},$$

$$\mathbf{J_1' J_2'} = \begin{pmatrix} -e & 0 & 0 & 0 \\ -\Phi & -e & -\dfrac{(1+e)G_-}{V} & 0 \\ 0 & 0 & -e & 0 \\ 0 & 0 & -\Phi - \dfrac{(e-1)G_-}{V} & -e \end{pmatrix},$$
(5.38)

where $\Phi = (1+e)(g + \ddot{h}(t'))/V$, $G_\pm = G(0, \pm(1+e)V$. In case of symmetric connection $(G_+ = -G_-)$ the periodic solution may be stable or not; anyway, it is *structurally stable*, i.e. small parameter changes do not destroy it. On the contrary, in case where $G_+ \neq -G_-$) a possibility of structural instability exists in accordance with Proposition 4.5.

Acknowledgements

The work is supported in part by INTAS 96-2138 and by the Russian Foundation for Basic Researches 99-01-00281.

References

1. Guckenheimer J., Holmes P.J. (1997) Nonlinear Oscillations, Dynamical Systems, and Bifurcations of Vector Fields, corrected 5-th edn. Springer, New York
2. Kobrinsky A.A., Kobrinsky A.E. (1981) Two-dimensional Vibroimpact Systems. Nauka, Moscow (in Russian)
3. Babitsky V.I. (1998) Theory of Vibro-Impact Systems and Applications. Springer, Berlin (Rev. transl. from Russian, Nauka, Moscow, 1978)
4. Feigin M.I. (1970) Period Doubling in C-Bifurcations in Piecewise-Conti nuous Systems. J.Appl.Maths Mechs 34(5): 861–869
5. Feigin M.I. (1974) Appearance of Subharmonic Families in a Piecewise-Continuous System. J.Appl.Maths Mechs 38(5): 810–818
6. Feigin M.I. (1977) Behavior of Dynamical Dystems Close to the Boundary of Existece Region for Periodic Motions. J.Appl.Maths Mechs 41(4): 628–636
7. Whiston G.S. (1987) Global Dynamics of Vibro-Impacting Linear Oscilla tor. J.Sound Vibr 118: 395–429
8. Whiston G.S. (1992) Singularities in Vibro-Impact Dynamics. J. Sound Vibr 152(3): 427–460
9. Nordmark A.B. (1991) Non-Periodic Motion Caused by Grazing Incidence in an Impact Oscillator. J.Sound Vibr 145(2): 279–297
10. Nordmark A.B. (1992) Effects due to Low Velocity Impact in Mechanical Oscillators. Int. J. Bifurc Chaos 2: 597–605
11. Nordmark A.B. (1997) Universal Limit Mapping in Grazing Bifurcation. Phys.Review E, 55(1): 266–270
12. Foale S., Bishop S.R. (1992) Dynamical Complexities of Forced Impacting Systems. Proc. Roy. Soc. Lond. A, 338: 547–556
13. Foale S. (1994) Analytical Determination of Bifurcations in an Impact Oscillator. Proc. Roy. Soc. Lond. A, 347: 353–364
14. Foale S., Bishop S.R. (1994) Bifurcations in Impact Oscillator. Nonlin Dynamics 6: 285
15. Ivanov A.P. (1995) Bifurcations Associated with Grazing Collisions. In: Awrejcewicz J. (Ed.) Nonlinear Dynamics: New Theoretical and Applied Results. Akademie Verlag, Berlin, 67–91
16. Fredriksson M.H., Nordmark A.B. (1997) Bifurcations Caused by Grazing Incidence in a Many Degrees of Freedom Impact Oscillator. Proc. R. Soc. Lond. A, 453: 1261–1276
17. Fredriksson M.H. (1997) Grazing Bifurcation in Multibody Systems. Nonlin Analysis, 30(7): 4475-4483
18. Babitsky V.I., Veprik A.M., Krupenin V.L. (1988) Vibroimpact Effects in the Chain of Elastically Linked Point Masses. DAN, 300(3)
19. Astashev V.K., Krupenin V.L., Tresvjatsky A.N. (1996) On the Experimental Study of the Impact Synchronism in Distributed Systems with Paral- lel Impact Pairs. DAN, 351(1)

20. Ivanov A.P (1987) Impacts in Systems with Several Unilateral Constraints J.Appl.Maths Mechs 51(4): 559–566
21. Goldsmith W. (1960) Impact: The Theory and Physical Behaviour of Colliding Solids. Edward Arnolds, London
22. Zukas J. A. (1982) Impact Dynamics. John Wiley, New York
23. Ivanov A.P. (1995) On Multiple Impact. J.Appl.Maths Mechs 59(6): 930–946
24. Ivanov A.P. (1997) The Problem of Constrained Impact. J.Appl.Maths Mechs 61(3): 341–353
25. Ivanov A.P. (1994) Dynamic of Systems in a Vicinity of Grazing Impact. J.Appl.Maths Mechs 58(3): 63–70
26. Pfeiffer F., Glocker Ch. (1996) Multibody Dynamics with Unilateral Contacts. Wiley, New York
27. Rouche N., Habets P., Laloy M. (1977) Stability Theory by Liapunov's Direct Method. Springer, New-York
28. Holmes P.J. (1982) The Dynamics of Repeated Impacts with a Sinusoidally Vibrating Table. J. Sound Vibr 84(2) 173–189
29. Gajewsky K., Radziszevsky B. (1987) On the Stability of Impact Systems Bull. Polish Acad Sci, Tech Sci 35(3-4) 183–188
30. Ivanov A.P. (1998) The Stability of Periodic Solutions of Discontinuous Systems that Intersect Several Surfaces of Discontinuity. J.Appl.Maths Mechs 62(5): 677–685
31. Farkas M. (1994) Periodic Motions. Springer-Verlag, New York
32. Appell P. (1953) Traité de Méchanique Rationelle. T.2. Gauthier Villars, Paris
33. Akhmetov M.U., Perestyuk N.A. (1992) Differential Properties of Solutions and Integral Surfaces of Nonlinear Impulsive Systems. Diff Eqs, 28(4): 555-566
34. Ivanov A.P. (1997) Dynamics of Systems with Mechanical Collisions. Int. Programm of Education, Moscow
35. Peterka F. (1992) Dynamics of Impact Motions of a Beam Between Stops and of a Spherical Pendulum in a Channel with a Flow of Liquid. Advanc in Mech. 15(3-4): 93–121
36. Ivanov A.P. (1993) Stabilization of an Impact Oscillator Near Grazing Incidence Owing to Resonance. J. Sound Vibr, 162(3): 562–565
37. Ivanov A.P. (1994) Impact Oscillations: Linear Theory of Stability and Bifurcations. J.Sound Vibr 178(3): 361-378
38. Nusse H.E., Ott E., Yorke J.A. (1994) Border-Collision Bifurcations: A Possible Explanation for Observed Bifurcation Phenomena. Phys. Review E, 49(2): 1073–1076
39. Budd C., Dux F. (1995) The Effect of Frequency and Clearance Variations on Single-Degree-of-Freedom Impact Oscillator. J.Sound Vibr 184(3): 475–502
40. Chin W., Ott E., Nusse H.E., Grebogi C. (1994) Grazing Bifurcation in Impact Oscillators. Phys. Review E, 50(6): 4427–4444

41. Chin W., Ott E., Nusse H.E., Grebogi C. (1995) Universal Behaviour of Impact Oscillators near Grazing Incidence. Phys. Letters A, 201: 197–204

42. Ivanov A.P. Bifurcations in Impact Systems. (1996) Chaos, Solitons Fractals, 7(10): 1615–1634

43. Bishop S.R., Thompson M.G., Foale S. (1996) Prediction of Period-1 Impacts in a Driven Beam. Proc. R. Soc. London A, 452: 2579-2592

44. Fredriksson M.H., Borglund D., Nordmark A.B. (1999) Experiments on the Onset of Impacting Motion Using a Pipe Conveying Fluid. To be issued

45. Ivanov A.P. Stability of Periodic Motions in Systems with Unilateral Constraints. In: Babitsky V.I. (Ed.) Dynamics of Vibro-Impact Sys- tems. Springer: Berlin, 1999. 119-126.

Contact Problems for Elasto-Plastic Impact in Multi-Body Systems

W.J. Stronge

Department of Engineering
University of Cambridge
Cambridge CB2 1PZ, UK

Abstract. Low to moderate speed collision between elastic-plastic bodies results in imperceptible permanent indentation of the contact surfaces if the bodies are hard. Nevertheless in these collisions the contact force–indentation relation is irreversible since internal energy gained from work done by the contact force during the compression is partially trapped in elastic waves, work done in plastic deformation and work done to overcome friction during sliding. These sources of kinetic energy loss in collisions depend on relative velocity between the bodies at the contact point, material properties, inertia properties related to the impact configuration and geometric constraints on the deformation field.

Colliding bodies that can be represented as elastic-plastic or visco-plastic solids have been analyzed using specific models of contact compliance to calculate the energy transformed into irrecoverable forms. These calculations show how elastic waves, plastic deformation and friction affect the energetic coefficient of restitution — a coefficient that is a measure of impact energy loss from internal sources. The calculations indicate that there is considerable difference in the sources of energy loss for 2D and 3D deformations. Also, that friction little affects the plastic energy losses unless the impact speed is large enough to cause uncontained plastic deformation beneath the contact area. Consequently, for low-speed impact $\rho v_3^2(0)/Y < 10^{-3}$, the energetic coefficient of friction e_* is insensitive to friction.

In analysing impact of multi-body systems (e.g. mechanisms, kinematic chains or agglomerates of granules), it is crucial to employ these ideas by specifically modelling the contact compliance. During impact on a mechanism, the velocity changes in one area of contact induce small displacements that develop at other compliant contacts where the impacted body is supported or connected to other elements; the force that develops at these secondary contacts (as a consequence of relative displacement at each individual contact) is the means of transmitting the impact process through system. Although the *global compliance* of elements of the system may be small enough so that vibration energy is negligible, the dynamics of multi-body collisions requires consideration of *local compliance* in each contact regions.

Previously impact on a system of interconnected bodies has been analyzed as either a sequence of separate collisions or a set of simultaneous collisions. In general neither of these assumptions gives an accurate representation of the dynamic behaviour of multi-body systems. Rather, it is necessary to model the compliance of each contact and consider the contact forces which develop since it is these forces which prevent interpenetration or overlap. When applied to impact on multi-body systems such as mechanisms, the impulse-momentum methods used in rigid body dynamics give but one limit of the range of response — a range that depends on the distribution of local compliance at each contact between bodies in the system. An accurate analysis of multi-body system response to impact generally requires consideration of the time-dependent contact forces in a wave of reaction that propagates away from the initial site of impact.

1. Introduction

Impact occurs when two bodies come together with a normal component of relative velocity at an initial point of contact. Thereafter there would be interpenetration of the bodies were it not for local deformations in a small region around the point of initial contact. For solids composed of materials with large elastic moduli, an integral of these deformations (strains) over the small depth of the deforming region gives a small normal deflection of the contact surface — during the contact period this deflection is required for compatibility of displacements in the common area of contact which develops around the point of initial contact. The contact surface deflections are consistent with a very large interface force acting in the contact area. This force opposes interpenetration and tends to drive the bodies apart. Since the contact force is very large it causes large accelerations which rapidly change the velocity of the bodies outside the contact area; the contact force and these accelerations terminate at separation. The changes in velocity during the very brief period of contact result in some residual or rebound velocity at separation.

For problems where the main interest is the dynamics of impact (i.e. changes in velocity occurring during impact), the effort and computation time required to obtain predictions of the terminal state are substantially reduced by adopting some version of rigid-body impact theory. This theory assumes that the contact duration is infinitesimal so that changes in velocity occur instantaneously. The major analytical advantage of this assumption is that the displacements of the bodies during impact are negligible. With negligible displacements, the only active forces arc impulsive reactions at points of displacement constraint. For planar or nonfrictional problems this assumption gives equations of motion that are trivially integrable so that changes in velocity can be expressed as algebraic functions of the normal impulse p at the contact point.

On closer inspection however, it becomes clear that the factors which are neglected by rigid-body impact theory in order to obtain an analytically simple result can be important in some problems. Here my aim is to clearly identify the conditions where rigid-body theory is applicable by analysing some generic problems and comparing results with those from more detailed and/or complete analyses. The conclusions are that for impact between a pair of rough bodies, rigid-body theory can give accurate results if there is continuous unidirectional slip at the contact point; for multi-body impact however, the range of system parameters where rigid-body impact represents the system response is restricted to the limiting cases of very large or very small gradient of contact compliance through the set of contacts. Thus *local compliance* is significant wherever there are interaction forces that are coupled and the forces are functions of relative displacement at points of contact. Stereo-mechanical rigid body impact theory requires in addition that the bodies be so inflexible that structural vibration energy can be neglected; i.e. that the colliding elements have negligible *global compliance*.

2. Impact Process

Impact initiates when two colliding bodies B and B' first come into contact at C, an initial point of contact. Each body has a point of contact, C or C', and at incidence these points have velocities $\mathbf{V_C}(0)$ and $\mathbf{V_{C'}}(0)$, respectively. Between the contact points there is a relative velocity $\mathbf{v}(t)$ defined as $\mathbf{v}(t) \equiv \mathbf{V_C}(t) - \mathbf{V_{C'}}(t)$. If at least one of the bodies is smooth in a neighbourhood of C, there is a common tangent plane (c.t.p.) and perpendicular to this plane there is a normal direction $\mathbf{n_3}$. At incidence the bodies come together with a negative relative velocity at C; i.e. $\mathbf{n_3} \cdot \mathbf{v}(0) < 0$.

Following incidence, in a region around C the surfaces are indented by equal but opposite normal reaction forces F_3 — these forces produce the local deflections required for compatibility of displacements on the common contact surface that develops between the bodies. During an initial period of *compression*, elements just outside the contact area continue to approach each other so that the contact area spreads, the indentation increases and the normal contact force increases as shown in Fig. 1c. The change in relative velocity at C caused by the normal reaction force brings the normal component of relative velocity of particles just outside the contact regions to rest at a time of maximum compression t_c ; at this instant when $\mathbf{n_3} \cdot \mathbf{v}(t_c) = 0$ there is a transition from the period of compression to a subsequent period of *restitution* wherein the bodies are separating. During restitution the normal contact force drives the bodies apart as the normal relative velocity continues to increase. Finally at time t_f the normal reaction force drops to zero as the bodies separate (see Fig 1d).

Solid bodies that are composed of stiff (i.e. high elastic modulus) materials give a large reaction force at a very small normal deflection δ (see Fig. 1b). The large force rapidly accelerates the colliding bodies and changes the direction of the normal component of relative velocity at C from one of approach to one of separation. For hard solid bodies, the contact period ranges from $t_f \approx 0.2$ ms for steel spheres to $t_f \approx 1.5$ ms for golf balls [1]. This brief contact period is a consequence of the small normal contact compliance (or high stiffness); likewise, if the contacting surfaces are non-conforming, the small normal compliance results in a contact area that remains small in comparison with the size of the colliding bodies.

During contact the reaction force causes stresses and strains in the colliding bodies. The 3D elastic stress field arising from a normal force on the surface of a smooth solid body gives a stress field that decreases as r^{-2} with increasing radial distance r from the point of application. This very rapid decrease in stress (and strain) results in a deforming region in each of the colliding bodies which is not much larger than the size of the contact area — the remainder of these bodies essentially are rigid or undeforming.

3. 'Rigid' Body Impact Theory for Smooth Hard Bodies

Analyses of 'rigid' body impact can follow the process of velocity change at the contact point by introducing an infinitesimal deformable particle between the bodies at C — this particle simulates an infinitesimal deforming region. The deforming region is assumed to have negligible mass because it is small in comparison with the size of either body; consequently the contact forces on either side of the particle will be equal but opposite. Noting that the normal contact force F_3 is always compressive we recognize that the normal component of impulse $p_3 \equiv \int_0^t F_3(t)\,dt$ is a monotonously increasing function of time; thus the normal impulse $p \equiv p_3$ can replace time as an independent variable.

Let colliding bodies B and B' have masses M and M' respectively. Each body has a center of mass with a velocity $\hat{V}_B$ or $\hat{V}_{B'}$ respectively while each body has an angular velocity ω_B or $\omega_{B'}$. From each center of mass to the contact point C there is a position vector r or r'. In a collision where there is negligible displacement during contact, the only active force is the reaction at the contact point. This reaction gives equal but opposite differentials of impulse $d\,P_B$ or $d\,P_{B'}$ acting on each respective body,

$$d\,P_B = F_B\,dt, \quad d\,P_{B'} = F_{B'}\,dt. \tag{1}$$

Newton's second law of motion states that

$$d\hat{V}_B = M^{-1}d\,P_B, \quad d\hat{V}_{B'} = M'^{-1}d\,P_{B'}. \tag{2}$$

For planar motion of rigid bodies with radii of gyration $\hat{\rho}$ and $\hat{\rho}'$ about the out-of-plane axis through the respective centers of mass, the laws of rotational motion are written as

$$d\,\omega_B = (M\hat{\rho}^2)^{-1}r\times d\,P_B, \quad d\,\omega_{B'} = (M'\hat{\rho}'^2)^{-1}r'\times d\,P_{B'}. \tag{3}$$

A transformation to equations of motion for the respective contact points C or C' is made by noting that

$$V_C = \hat{V}_B + \omega_B \times r, \quad V_{C'} = \hat{V}_{B'} + \omega_{B'} \times r'. \tag{4}$$

Across the deformable particle at C there is a relative velocity given by $v \equiv V_C - V_{C'}$.

Assembling equations (2)-(4) and letting $d\,p \equiv d\,P_B = -d\,P_{B'}$ results in equations of planar relative motion for the pair of bodies colliding at C as a function of the contact impulse $p \equiv (p_1, p_3)$.

193

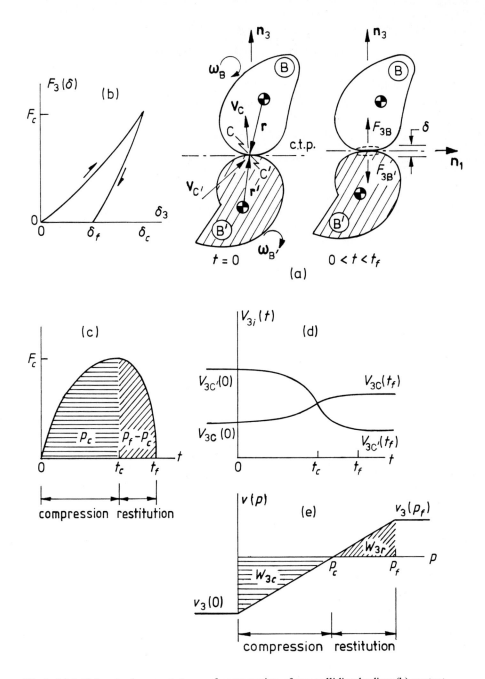

Fig 1 (a) Initial and subsequent stages of compression of two colliding bodies; (b) contact force as function of indentation; (c) contact force as function of time; (d) velocity changes as function time; (e) relative velocity changes vs normal impulse.

$$\left\{ \begin{matrix} d\,v_1 \\ d\,v_3 \end{matrix} \right\} = m^{-1} \left[\begin{matrix} \beta_1 & -\beta_2 \\ -\beta_2 & \beta_3 \end{matrix} \right] \left\{ \begin{matrix} d\,p_1 \\ d\,p_3 \end{matrix} \right\} \tag{5}$$

where an effective mass $m = MM'/(M + M')$. In this expression the inertia coefficients β_i depend on the system configuration,

$$\begin{aligned}
\beta_1 &= 1 + mr_3^2 / M\hat{\rho}^2 + mr_3'^2 / M'\hat{\rho}'^2, \\
\beta_2 &= mr_1 r_3 / M\hat{\rho}^2 + mr_1' r_3' / M'\hat{\rho}'^2, \\
\beta_3 &= 1 + mr_1^2 / M\hat{\rho}^2 + mr_1'^2 / M'\hat{\rho}'^2.
\end{aligned}$$

This form of equation of planar motion was presented by Wang & Mason [2] and Stronge [3]. It is useful to recognize that $\beta_1 > 0$, $\beta_3 > 0$ and $\beta_1\beta_3 > \beta_2^2$.

For smooth bodies the tangential impulse vanishes $dp_1 = 0$; in this case the equations (5) can be integrated to obtain,

$$\begin{aligned}
v_1(p) &= v_1(0) - m^{-1}\beta_2 p \\
v_3(p) &= v_3(0) + m^{-1}\beta_3 p
\end{aligned} \tag{6}$$

where the normal impulse $p \equiv p_3$.

A characteristic impulse p_c for the period of compression is defined as the normal impulse that brings the normal component of relative velocity at C to rest; i.e.

$$p_c = -mv_3(0)/\beta_3. \tag{7}$$

Hence the components of relative velocity in (6) can be expressed as

$$\frac{v_1(p)}{v_3(0)} = \frac{v_1(0)}{v_3(0)} - \frac{\beta_2}{\beta_3}\frac{p}{p_c},$$

$$\frac{v_3(p)}{v_3(0)} = 1 - \frac{p}{p_c}.$$

This expression for the normal component of relative velocity is shown in Fig. 1e.

The partial work done by the normal component of impulse $W_3(p)$ during contact is calculated from

$$W_3(p) = \int_0^p v_3(p)\,dp = v_3(0)p + \frac{\beta_3}{m}\frac{p^2}{2}. \tag{8a}$$

where the integrand was obtained from (6b). The partial work for the period of compression $W_{3c} \equiv W_3(p_c)$ is cross-hatched in Fig. 1e; this work done to bring the initial normal relative velocity at the contact point to rest, is given by substituting (7) into (8a),

$$W_{3c} = 0.5\beta_3^{-1} m v_3^2(0). \tag{8b}$$

For a terminal impulse p_f the terminal relative velocity at separation $\mathbf{v}(p_f) = (v_1(p_f), v_3(p_f))$ is given by (6). If tangential impulse is negligible, the terminal relative velocity has components

$$v_1(p_f) = v_1(0) - \beta_2\beta_3^{-1}v_3(0)\big(p_f / p_c\big), \tag{9a}$$

$$v_3(p_f) = v_3(0)\big(1 - p_f / p_c\big). \tag{9b}$$

The question that remains — how to obtain the ratio of normal impulse at termination to that at the end of compression p_f / p_c — has been answered in several different ways. Both the kinematic coefficient of restitution $e \equiv -v_3(p_f)/v_3(0)$ and the kinetic coefficient of restitution $e_0 \equiv (p_f - p_c)/p_c$ are empirical coefficients that can be used to estimate this ratio; they are based on measurements from experiments that must be physically similar to the impact conditions where they are applied. On the other hand, the energetic coefficient e_* is based on work done during contact by the normal component of contact force [4]; the energetic coefficient of restitution provides an opportunity to calculate the impulse ratio p_f / p_c based on material properties, the impact configuration and the relative incident velocity at the impact point. This calculation is presented in Sect. 4.

Assumptions of 'Rigid' Body Theory

i. Small contact area (deforming region small in comparison with all dimensions of bodies)
ii. Compliance of contact region small so that contact period is infinitesimal (negligible changes in configuration during impact).
iii. Compliance of contact region large in comparison with compliance of all other sections (i.e. all sections of colliding bodies are rigid in comparison with contact region).

4. Extended Hertz Theory for Elastic-Plastic Impact

Elasticity provides a continuum field theory for the stress and displacement distributions near the contact area of two bodies which are pressed together. This theory will be representative of the local stresses, strains and displacements if the stresses are within the yield limit. The elastic analysis can be used to evaluate the relative displacement or indentation which equates with the elastic limit state.

4.1 Elastic Indentation from Normal Contact Force

Hertz [5] proposed a theory of contact which neglects any local effect of inertia; i.e. this theory is quasi-static and neglects effects of stress waves. Let non-conforming bodies B and B′ come into contact at an initial contact point C, as shown in Fig. 1a. Suppose that in a neighbourhood of C the surfaces of the bodies have radii of curvature R_B and $R_{B'}$ respectively; i.e. both bodies are smooth in the contact region. At the initial contact point C the surfaces have a common tangent plane and normal to this plane is the common normal direction n_3.

Let bodies with initial surface curvatures R_B^{-1} and $R_{B'}^{-1}$ be compressed by a normal force F_3 as shown in Fig. 2; as the compressing force increases the common area of contact spreads from C to include a small region around C. The perimeter of this contact area has radius a. Hertz showed that for compatible normal displacements within the contact area, there is an elliptic distribution of contact pressure $p(r)$ between the bodies,

$$\frac{p}{p_0} = \left(1 - \frac{r^2}{a^2}\right)^{1/2} , \quad r \le a . \tag{10}$$

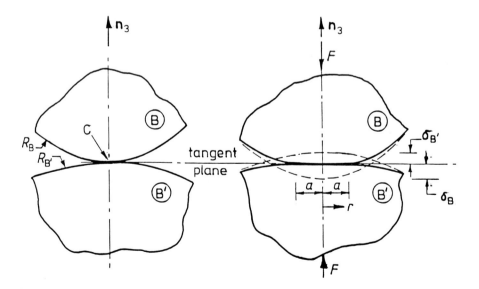

Fig. 2 Normal indentation $\delta = \delta_B + \delta_{B'}$ of spherical contact surfaces.

where r is a radial coordinate originating at C and $p_0 = p(0)$ is the maximum pressure which occurs at the center of the contact area. This contact pressure results in a normal contact force F_3 on each body,

$$F_3 = \int_0^a p(r) 2\pi r \, dr = \left(\frac{2\pi}{3}\right) p_0 a^2.$$

(11)

In the contact area the mean pressure $\bar{p}$ is 2/3 of the maximum pressure, $\bar{p} = 2p_0/3$.

The pressure distribution on the surface causes a normal component of displacement at the surface $w_i(r)$, $i = $ B, B'; an expression for this surface displacement can be obtained from the Boussinesq solution by integration of the strains arising from a distribution of normal force on the surface of an elastic half-space [6],

$$w_i(r) = \frac{\pi}{4} \frac{1-v_i^2}{E_i} \left(2 - \frac{r^2}{a^2}\right) a p_0, \quad r \le a,$$

(12)

where body i has elastic moduli given as Poisson's ratio v_i and Young's modulus E_i. Each body has an indentation δ_i defined as the maximum normal displacement (relative to the center of mass of the body); i.e. $\delta_i \equiv w_i(0)$. Defining the total indentation δ as the sum of the indentations of each body in the colliding pair,

$$\delta = \delta_B + \delta_{B'} = \frac{3\pi}{4} \frac{a\bar{p}}{E_*},$$

(13)

where an effective elastic modulus E_* and an effective initial contact surface radius of curvature R_* are defined as,

$$E_*^{-1} \equiv \left[\frac{1-v_B^2}{E_B} + \frac{1-v_{B'}^2}{E_{B'}}\right]^{-1}, \quad R_*^{-1} = \left(R_B^{-1} + R_{B'}^{-1}\right)^{-1}.$$

By rearranging expressions (11)–(13), the elastic indentation, initial contact radius and normal contact force are obtained as,

$$\frac{\delta}{R_*} = \frac{a^2}{R_*^2},$$

(14a)

$$\frac{a}{R_*} = \left(\frac{3F_3}{4E_*R_*^2}\right)^{1/3}, \tag{14b}$$

$$\frac{F_3}{E_*R_*^2} = \frac{4}{3}\left(\frac{\delta}{R_*}\right)^{3/2}. \tag{14c}$$

For an initial period of increasing magnitude of force (compression), the work W_3 done by the normal component of contact force F_3 in indenting the colliding bodies can be calculated;

$$\frac{W_3}{E_*R_*^3} = \int_0^{\delta/R_*} \frac{F_3(\delta/R_*)}{E_*R_*^2} \, d\left(\frac{\delta}{R_*}\right) = \frac{8}{15}\left(\frac{\delta}{R_*}\right)^{5/2}. \tag{14d}$$

This work that increases the strain energy in the deforming region simultaneously reduces the kinetic energy of relative motion at C, the kinetic energy is transformed into internal energy of deformation.

4.2 Indentation at Yield of Elastic-Plastic Bodies

With increasing indentation, elastic behaviour continues until yield occurs in one or the other of the bodies that are pressed together at C. Suppose that plasticity (i.e. irreversible rate-independent deformation) initiates at a uniaxial yield stress Y. For an axisymmetric stress field beneath the elliptic interface pressure distribution, either the Tresca or von Mises yield criterion gives a mean pressure to initiate yield $\bar{p}_Y$ as $\bar{p}_Y = 1.1Y$, [7]. Plastic yielding initiates substantially below the contact surface at a depth $z \approx 0.45a_Y$. In the present formulation the ratio of mean pressure for initiation of yield to uniaxial yield stress is a parameter ϑ_Y that can be specified to obtain continuity of mean pressure at the transition between elastic and fully plastic indentation. Thus

$$\frac{\bar{p}_Y}{Y} \equiv \vartheta_Y = \frac{4}{3\pi}\frac{E_*}{Y}\left(\frac{\delta_Y}{R_*}\right)^{1/2}. \tag{15}$$

Nondimensional indentation, normal force and indentation work at yield are material properties;

$$\frac{\delta_Y}{R_*} = \left(\frac{3\pi}{4}\right)^2\left(\frac{\vartheta_Y Y}{E_*}\right)^2, \tag{16a}$$

$$\frac{F_{3Y}}{YR_*^2} = \pi\vartheta_Y \left(\frac{3\pi}{4}\right)^2 \left(\frac{\vartheta_Y Y}{E_*}\right)^2, \tag{16b}$$

$$\frac{W_{3Y}}{YR_*^3} = \frac{2\pi\vartheta_Y}{5} \left(\frac{3\pi}{4}\right)^4 \left(\frac{\vartheta_Y Y}{E_*}\right)^4. \tag{16c}$$

Elastic indentation has a nondimensional contact area, normal force and indentation work that can be described as

$$\frac{a}{a_Y} = \left(\frac{\delta}{\delta_Y}\right)^{1/2}, \quad \frac{F_3}{F_{3Y}} = \left(\frac{\delta}{\delta_Y}\right)^{3/2}, \quad \frac{W_3}{W_{3Y}} = \left(\frac{\delta}{\delta_Y}\right)^{5/2}.$$

Contours of maximum shear stress in an elastic solid compressed by a spherical indentor are shown in Fig. 3.

Partial work done by the normal component of contact force in collinear impact is used to define a characteristic normal component of relative velocity v_Y that is sufficient to initiate yield. Letting $W_{3Y} \equiv 0.5mv_Y^2$ we obtain

$$v_Y^2 = \frac{4\pi\vartheta_Y}{5} \left(\frac{3\pi}{4}\right)^4 \left(\frac{\vartheta_Y Y}{E_*}\right)^4 \frac{YR_*^3}{m}. \tag{17}$$

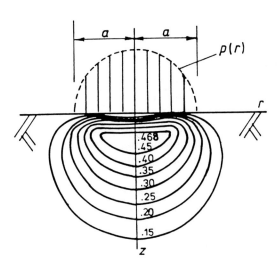

Fig. 3 Contours of equal shear stress in elastic half-space indented by sphere.

Table 1 Material properties from indentation and impact tests

Material	Density ρ (gcm^{-3})	Young's Modulus E_i (GPa)	Static Yield Y_s (MPa)	Dynamic Yield Y_d (MPa)	Impact Speed to Initiate Yield v_Y (ms^{-1})	Source
Mild Steel (as received)	7.8	210	600	583–780	.049–.101	[8]
Mild Steel (work hard.)	7.8	210	650	1160	.055	[9]
Brass (drawn)	8.5	100	200	250	.007	[10]
Alum. 1180-H14	2.7	69	110	130	.004	
Alum. 2014-T6	2.8	69	410	410	.076	

The normal incident speed for yield v_Y is a function of material properties, the curvature of the contact region and the effective mass of the colliding bodies; it is however, independent of the impact configuration and friction [8]. With yield at $\vartheta_Y = 1.1$, characteristic normal speeds for yield v_Y and the elastic-plastic transition v_p are approximately equal to

$$v_Y^2 \cong 125 \left(\frac{Y}{E_*} \right)^4 \frac{YR_*^3}{m}, \quad v_p^2 = v_Y^2 \left(\frac{2.8}{\vartheta_Y} \right)^5 \cong 13,362 \left(\frac{Y}{E_*} \right)^4 \frac{YR_*^3}{m}.$$

4.3 Fully Plastic Indentation

As indentation increases beyond that required to initiate yield there is a range of indentation where the region of plastic deformation expands but does not break through to the contact interface. A stress and deformation analysis of this intermediate elastic-plastic range of indentation was presented by Stronge [11]. Elastic-plastic indentation results in a smooth transition between the elastic and a fully plastic range of behaviour.

After the plastically deforming region intersects the contact surface, further indentation causes uncontained fully plastic flow. In the fully plastic range the interface pressure is constant throughout the contact area. For spherical contact surfaces and negligible strain-hardening the fully plastic indentation pressure equals $\bar{p} = 2.8Y$; this pressure is the same as that measured in a Brinell hardness test [10]. Fully plastic indentation initiates at an indentation δ_p, contact radius a_p, contact force F_p and indentation work W_p that are somewhat larger than the values required to initiate yield,

$$\frac{\delta_p}{\delta_Y} = \left(\frac{2.8}{1.1}\right)^2 = 6.48, \quad \frac{a_p}{a_Y} = 2.54, \quad \frac{F_{3p}}{F_{3Y}} = 16.5, \quad \frac{W_{3p}}{W_{3Y}} = 106.9. \quad (18)$$

Numerical simulations shown in Fig. 4 by Hardy, Baronet and Tordion [12] indicate that for spherical contact surfaces, fully plastic indentation begins at $F_3 / F_{3p} \cong 20$.

Fully plastic indentation is related to the contact radius by a condition that at the edge of the contact area the surface neither sinks in nor piles up; this condition ensures continuity of mass,

$$\frac{\delta}{R_*} = \frac{1}{2}\left(\frac{a^2}{R_*^2} + \frac{a_p^2}{R_*^2}\right) \quad \text{or} \quad \frac{\delta}{\delta_p} = \frac{1}{2}\left(\frac{a^2}{a_p^2} + 1\right), \quad \delta \geq \delta_p. \quad (19)$$

In this range of behaviour the mean contact pressure $\bar{p}$ is independent of indentation so that the contact force increases with the contact area,

$$\frac{F_3}{F_{3p}} = \frac{2\delta}{\delta_p} - 1, \quad \delta \geq \delta_p. \quad (20)$$

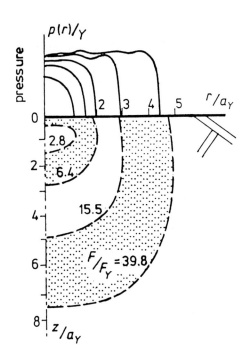

Fig. 4 Region of plastic deformation expands with increasing normal force.

With this force-indentation relation, the total work done on both bodies during compression can be expressed as

$$\frac{W_3}{W_{3p}} = 1 + \frac{5}{2} \int_1^{\delta/\delta_p} \left(\frac{2\delta}{\delta_p} - 1 \right) d\left(\frac{\delta}{\delta_p} \right) = 1 + \frac{5}{2} \left(\frac{\delta^2}{\delta_p^2} - \frac{\delta}{\delta_p} \right), \quad \delta \geq \delta_p. \quad (21)$$

4.4 Elastic Unloading from Fully Plastic Indentation

For rate-independent materials, maximum force occurs simultaneously with maximum indentation at the end of the compression period. As the bodies start to separate or unload, the normal contact force decreases. For plastically deformed bodies, the unloading path for separation is distinct from the force-deflection relation during compression.

Denote the maximum indentation by δ_c and the maximum force by $F_{3c} \equiv F_3(\delta_c)$. Unloading from maximum indentation occurs elastically. During unloading there is a change in indentation δ_r, so that when the compressing force vanishes there is a residual final deflection $\delta_f = \delta_c - \delta_r$. In calculating the compliance relation for unloading it is important to recognize that as a result of plastic deformation during compression, the initial contact curvature R_*^{-1} has been changed to a new unloaded curvature $\bar{R}_*^{-1}$. We assume this transition occurs at maximum indentation. The curvature of the deformed surface depends on whether both arc convex or whether one has become concave,

$$\bar{R}_* = \begin{cases} \bar{R}_B \bar{R}_{B'} / (\bar{R}_B + \bar{R}_{B'}) & \bar{R}_B \geq 0, \ \bar{R}_{B'} \geq 0, \\ \bar{R}_B \bar{R}_{B'} / (\bar{R}_B - \bar{R}_{B'}) & -\bar{R}_B > \bar{R}_{B'} \geq 0. \end{cases}$$

During elastic unloading the deformations in the contact region are geometrically similar to the changes that occur during elastic compression; i.e.

$$\frac{\delta_p}{R_*} = \frac{\delta_r}{\bar{R}_*}. \quad (22)$$

The relationships between indentation and contact radius at maximum elastic-plastic compression and during elastic separation are as follows:

$$\frac{\delta_c}{R_*} = \frac{1}{2}\left(\frac{a_c^2}{R_*^2} + \frac{\delta_p}{R_*} \right), \quad \frac{\delta_r}{\bar{R}_*} = \frac{a_r^2}{\bar{R}_*^2}.$$

Letting the change in contact radius during unloading a_r equal the change in radius during compression a_c gives $\delta_r / \delta_c = 2R_*\bar{R}_*(R_*^2 + \bar{R}_*^2)^{-1}$. Together with

Eq. (22) this gives a relation for indentation recovered during separation as a function of the maximum indentation,

$$\frac{\bar{R}_*}{R_*} = \frac{\delta_r}{\delta_p} = \left(\frac{2\delta_c}{\delta_p} - 1\right)^{1/2}. \tag{23}$$

During the period of restitution the normal contact force $F_3(p)$ is unloading elastically until it vanishes at a terminal indentation δ_f. For unloading the normal contact force can be expressed as

$$\frac{F_3}{E_*\bar{R}_*^2} = \frac{4}{3}\left(\frac{\delta - \delta_f}{\bar{R}_*}\right)^{3/2}.$$

This extended Hertz theory gives a force-deflection relation for indentation of elastic-perfectly plastic spherical surfaces which is illustrated in Fig. 5.

During compression the normal contact force does work on the bodies that transforms kinetic energy into internal energy of deformation — both elastic strain energy and plastic work. The elastic strain energy is the source of the normal (unloading) force that drives the bodies apart during restitution. During unloading the normal force is in the opposite direction from the change in the normal component of relative displacement so the partial work done on the bodies W_{3r} is negative (the deforming region loses strain energy and the bodies regain part of the initial kinetic energy of relative motion);

$$\frac{W_{3r}}{YR_*^3} = -\int_{\delta_f/\bar{R}_*}^{\delta_f/\bar{R}_* + \delta_r/\bar{R}_*} \left(\frac{4E_*}{3Y}\right)\left(\frac{\bar{R}_*}{R_*}\right)^3 \left(\frac{\delta - \delta_f}{\bar{R}_*}\right)^{3/2} \mathrm{d}\left(\frac{\delta}{\bar{R}_*}\right)$$

$$= -\frac{8}{15}\frac{E_*}{Y}\left(\frac{\delta_r}{\delta_p}\right)^3 \left(\frac{\delta_p}{R_*}\right)^{5/2}. \tag{24}$$

For elastic-fully plastic behaviour, the indentation that initiates uncontained plastic deformation is $\delta_p/R_* = (3\pi/4)^2 (2.8Y/E_*)^2$ so that (23) and (24) result in an expression for ratio of partial work recovered during restitution to partial work for fully plastic indentation

$$\frac{W_{3r}}{W_{3p}} = -\left(\frac{\delta_r}{\delta_p}\right)^3 = -\left(\frac{2\delta_c}{\delta_p} - 1\right)^{3/2}, \qquad \delta_c > \delta_p.$$

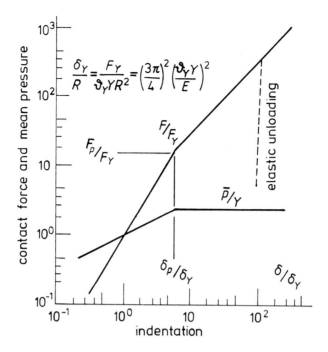

Fig. 5 Nondimensional mean pressure $\bar{p}/Y$ and normal force F_3/F_Y for indentation of elastic-perfectly plastic body.

4.5 Energetic Coefficient of Restitution

The partial work done by the normal component of contact force during compression W_{3c} and that recovered during restitution W_{3r} are given as

$$\frac{W_{3c}}{W_{3p}} = 1 + \frac{5}{2}\left(\frac{\delta^2}{\delta_p^2} - \frac{\delta}{\delta_p}\right), \qquad \frac{W_{3r}}{W_{3p}} = -\left(\frac{2\delta_c}{\delta_p} - 1\right)^{3/2}, \qquad \delta_c > \delta_p \qquad (25)$$

For energy losses in collision that are related to irreversible internal deformations in the contact region, the *energetic coefficient of restitution* e_* relates the partial work recovered during restitution W_{3r} to the partial work done during compression W_{3c} [4].

The square of the coefficient of restitution e_*^2 is negative of the ratio of the elastic strain energy released during restitution to the internal energy of deformation absorbed during compression,

$$e_*^2 \doteq -\frac{W_3(p_f) - W_3(p_c)}{W_3(p_c)} = -\frac{W_{3r}}{W_{3c}} \qquad (26)$$

Equation (26) assumes that the internal energy absorbed during compression is solely due to the normal component of contact force; this neglects any interaction between the stresses generated by normal and tangential components of contact force. Strictly, the assumption is valid if the tangential compliance is negligible. Ordinarily, for a half-space, the normal compliance is somewhat larger than the tangential compliance but the success of this assumption is more due to the stress field around the contact area and the fact that the plastically deforming region is contained within a subsurface inclusion located beneath the contact area. So long as the contact pressure does not increase to the extent required to produce uncontained plastic deformation, the tangential force-deflection relation remains elastic although the normal force-deflection relation is elastic-plastic. Consequently, at separation the partial work done by tangential forces vanishes, see Lim & Stronge [13].

From (8b) and (25a) the maximum normal indentation δ_c can be related to the normal component of incident relative velocity $v_3(0)$,

$$\frac{\delta_c}{\delta_p} = \frac{1}{2}\left[1 + \sqrt{\frac{8}{5\beta_3}\left(\frac{\vartheta_Y}{2.8}\right)^5 \frac{v_3^2(0)}{v_Y^2} - \frac{3}{5}}\right]. \qquad (27)$$

This maximum indentation can be substituted into relations (25) and (26) to obtain a expression for the energetic coefficient of restitution,

$$e_*^2 = \frac{0.4\left(2\delta_c/\delta_p - 1\right)^{3/2}}{\left(\delta_c/\delta_p\right)^2 - \delta_c/\delta_p + 0.4}. \qquad (28)$$

The ratio of terminal impulse to normal impulse for compression p_f/p_c, is obtained from (8a) together with (26);

$$\frac{W_{3r}}{W_{3c}} = -\left(\frac{p_f}{p_c} - 1\right)^2 = -e_*^2. \qquad (29)$$

Hence if the impact configuration is collinear, the terminal normal impulse p_f and the normal impulse for compression p_c are related by the energetic coefficient of restitution.

$$\frac{p_f}{p_c} = 1 + e_* .$$

This expression is the same as the definition of the kinetic coefficient of restitution e_0. In general the kinematic, kinetic and energetic coefficients of restitution are identical unless there is friction, the configuration is eccentric and slip changes direction before separation — a change in the direction of slip can occur only for low initial speeds of sliding.

The energetic coefficient of restitution shown in Fig. 6 was calculated from (27) and (28) for inertia properties representative of an end of a uniform rigid rod striking an elastic-plastic half-space. This coefficient has some dependence on orientation of the rod since this affects the maximum force, and hence maximum indentation [14]. For normal incident velocities that are large in comparison with the velocity at yield v_Y the coefficient of restitution $e_* \to \left(v_3(0)/v_Y \right)^{-1/4}$; this asymptotic limit is consistent with measurements on many hard metals reported by Goldsmith [15]. Generally larger normal incident speeds result in larger indentation and more energy loss due to plastic deformation; thus higher impact speeds give a smaller energetic coefficient of restitution.

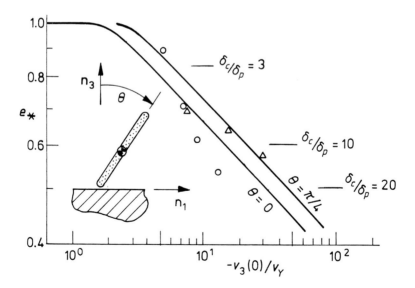

Fig. 6 Theoretical coefficient of restitution for elastic-perfectly plastic slender rod at two angles of inclination at incidence, $\theta = 0, \pi/4$.

While the previous analysis of contact deformations employs elastic and elastic-plastic field theory, this level of detail is frequently not necessary. The following analyses use discrete or lumped parameter models that represent the contact compliance in the contact region. For any particular application, the contact stiffness must be evaluated either experimentally or from a continuum analysis. The introduction of interface elements which give contact force as a specific function of displacement has a cost — the analysis must employ time t rather than the normal component of impulse p as an independent variable.

5. Effect of Tangential Compliance Between Colliding Bodies

A "superball" is a highly elastic (low hysteresis) sphere made of a polymer that gives a large coefficient of friction. A rotating "superball" bouncing back and forth on a hard level surface suffers a reversal in direction of rotation at each bounce. The spherical ball gives an approximately collinear impact configuration; thus the reversal in direction is not possible if tangential compliance is negligible. The motion of a "superball" is an example of behaviour that results directly from tangential compliance [16].

Effects of tangential compliance for collinear impact of rough (frictional) elastic continua were investigated by Maw, Barber and Fawcett [17]. To extend that investigation to considerations of inelastic collisions, a method based on discrete modelling of contact compliance has been developed by Stronge [18]. Figure 7 illustrates the modelling of both normal and tangential compliance at the contact point C. The infinitesimal particle C is subject to both normal and tangential contact forces which then are transmitted through the springs to the respective rigid bodies. Across each spring the displacement δ of the particle from the initial equilibrium position is denoted by $\delta = (\delta_1, \delta_3)$. Compressive displacement is defined as positive.

The normal stiffness k for the period of compression is obtained from material properties $k = (4\pi/5)\vartheta_Y Y R_*;^1$ hysteresis causes the stiffness for the period of restitution to increase to k/e_*^2 where the coefficient of restitution $e_* \leq 1$. This hysteresis loop gives an energy loss in a cycle of loading and unloading that is consistent with the energetic coefficient of restitution e_*. The hysteresis relation leaves a residual indentation $\delta_{3r} = (1 - e_*^2)\delta_3(t_c)$ after complete unloading as shown in Fig. 8. Ordinarily the tangential stiffness k/η^2 is somewhat smaller than the normal stiffness, $k/\eta^2 = 2k(1 - v^2)/(2 - v)(1 + v)$ (see Johnson [7], p221). Hence a ratio of stiffness gives $\eta^2 = (2 - v)/2(1 - v)$ or $\eta^2 = 1.25$ for Poisson's ratio $v = 0.33$.

---

[1] The linear spring constant k is an approximation for the nonlinear stiffness k_s from the relation for a spherical contact region $F_3 = k_s \delta^{3/2}$. Equivalence was obtained by equating the partial work of the normal force F_3 during indentation up to the yield limit δ_Y.

208

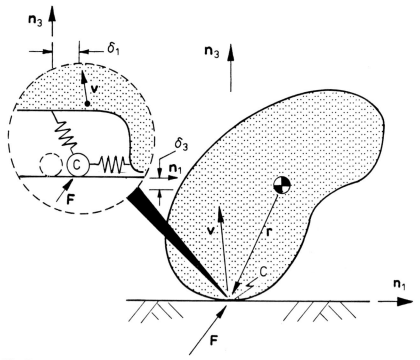

Fig. 7 Discrete element model with both normal and tangential compliance at the contact point C.

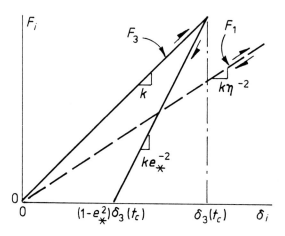

Fig. 8 Force-deflection relations for normal and tangential stiffness of smooth solids.

209

With these bilinear stiffness relations, the equations of planar motion can be written in terms of the relative displacement (δ_1, δ_3),

$$\begin{Bmatrix} \ddot{\delta}_1 \\ \ddot{\delta}_3 \end{Bmatrix} = -\omega_0^2 \begin{bmatrix} \beta_1 \eta^{-2} & -\beta_2 \\ -\beta_2 \eta^{-2} & \beta_3 \end{bmatrix} \begin{Bmatrix} \delta_1 \\ \delta_2 \end{Bmatrix}, \qquad \omega_0^2 = \frac{k}{m}. \tag{30}$$

These equations represent a system with two natural frequencies ω and Ω which are eigenvalues of the frequency equation,

$$\begin{Bmatrix} \omega^2 / \omega_0^2 \\ \Omega^2 / \omega_0^2 \end{Bmatrix} = \frac{(\beta_1 \eta^{-2} - \beta_3)}{2} \left\{ 1 \pm \sqrt{1 - \frac{4(\beta_1 \beta_3 - \beta_2^2)}{(\beta_1 \eta^{-2} + \beta_3)^2}} \right\}. \tag{31}$$

If the collision configuration is collinear ($\beta_2 = 0$) the tangential and normal equations of motion are decoupled. In this case the contact point undergoes simple harmonic motion (s.h.m.) in the normal direction. When the contact point is sticking rather than sliding, there is s.h.m. in the tangential direction also. The circular frequencies of normal relative motion Ω and tangential motion ω are defined as

$$\Omega \equiv \sqrt{\frac{\beta_3 k}{m}} = \frac{\pi}{2 t_c}, \qquad \omega \equiv \sqrt{\frac{\beta_1 k}{\eta^2 m}} = \frac{\pi}{2\eta t_c} \sqrt{\frac{\beta_1}{\beta_3}}. \tag{32}$$

where the time of maximum compression t_c occurs when the normal relative velocity $\dot{\delta}_3(t_c) = 0$.

The collision terminates and separation occurs at a final time t_f. For collinear collisions the time of separation and the time of maximum compression are related, $t_f = (1 + e_*) t_c$. Between these times during the period of restitution $t_c \le t \le t_f$ the normal stiffness k is replaced by k / e_*^2 in the expressions above; i.e. if there are energy losses in collision $e_* < 1$, the frequency of normal relative motion is larger during restitution than that during compression. This approximation assumes that separation occurs at a final normal indentation $\delta_3(t_f) = (1 - e_*^2)\delta_3(t_c)$.

5.1 Normal Relative Velocity for Collinear Impact

Where the impact configuration is collinear, $\beta_2 = 0$ the equations for normal and tangential relative motion are decoupled; i.e. the normal component of motion is unaffected by slip and friction at the contact point. With piecewise linear normal compliance, the normal relative velocity undergoes simple harmonic motion (s.h.m.) as a function of time t with a frequency Ω for the period of compression and a frequency $e_*^{-1}\Omega$ for the period of restitution. Thus the normal component of relative velocity is expressed as,

$$v_3(t) = v_3(0)\cos\Omega t, \qquad\qquad 0 \le t \le t_c, \qquad (33a)$$

$$v_3(t) = e_* v_3(0)\cos e_*^{-1}\Big(\Omega t - (1 - e_*)\pi/2\Big), \quad t_c < t \le t_f. \quad (33b)$$

At maximum compression there is a change in contact stiffness; after maximum compression the frequency of normal relative motion is changed from Ω to $e_*^{-1}\Omega$. Corresponding to this variation in relative velocity, one can calculate the variation in normal displacement and normal contact force during the contact period occurring between incidence and separation, see Table 2. Notice that because normal indentation is positive in compression, $v(t) = -\dot\delta_3(t)$.

Table 2 Normal relative displacement, relative velocity and contact force at C.

	compression, $0 \le t \le t_c$	restitution, $t_c \le t \le t_f$
disp.	$\delta_3(t) = -\Omega^{-1}v_3(0)\sin\Omega t$	$\delta_3(t) = -\dfrac{e_*^2 v_3(0)}{\Omega}\sin e_*^{-1}\left(\Omega t - \dfrac{\pi}{2}(1 - e_*)\right)$
vel.	$v_3(t) = v_3(0)\cos\Omega t$	$v_3(t) = e_* v_3(0)\cos e_*^{-1}\left(\Omega t - \dfrac{\pi}{2}(1 - e_*)\right)$
force	$F_3(t) = -\dfrac{m\Omega v_3(0)}{\beta_3}\sin\Omega t$	$F_3(t) = -\dfrac{m\Omega v_3(0)}{\beta_3}\sin e_*^{-1}\left(\Omega t - \dfrac{\pi}{2}(1 - e_*)\right)$

5.2 Tangential Relative Velocity for Collinear Impact and Dry Friction

Tangential contact forces can arise from dry friction or from interlocking of rugosities on the contact surface. Where the source of the force can be represented as dry friction, the tangential and normal contact forces are related through the *Amontons-Coulomb law of friction*. This law states that the contact point sticks $0 = v_1 + \dot\delta_1$ if the ratio of magnitude of tangential to normal force $|F_1|/F_3$ is less than a static *coefficient of friction*, μ_s. If the contact point slips (or slides) however, the ratio of magnitude of tangential to normal force equals the dynamic coefficient of friction, μ_d;

stick:	$|F_1|/F_3 < \mu_s,$	$v_1 + \dot\delta_1 = 0,$
slip:	$|F_1|/F_3 = \mu_d,$	$v_1 + \dot\delta_1 \ne 0.$

The friction force acts in a direction opposed to sliding.

Measurements typically show that $\mu_s > \mu_d$ so that with increasing tangential relative displacement, the tangential force increases until sliding initiates and then the force decreases somewhat. During sliding the tangential force is almost constant irrespective of the speed of sliding. The subsequent analysis presented here neglects this detail and assumes that $\mu_s = \mu_d \equiv \mu$.

Tangential Displacement and Force During Stick

Consider the case where initial slip at the contact point is brought to a halt at a time t_s where $t_s < t_f$. For the period of time $t > t_s$ suppose that the contact point sticks. Hence the contact relative displacement, relative velocity and contact force can be written as

$$\delta_1(t) = \delta_1(t_s)\cos\omega(t - t_s) - \omega^{-1}v_1(t_s)\sin\omega(t - t_s), \tag{34a}$$

$$v_1(t) = \omega\delta_1(t_s)\sin\omega(t - t_s) + v_1(t_s)\cos\omega(t - t_s), \tag{34b}$$

$$F_1(t) = \frac{m\omega^2\delta_1(t_s)}{\beta_1}\cos\omega(t - t_s) - \frac{m\omega v_1(t_s)}{\beta_1}\sin\omega(t - t_s), \quad t \geq t_s. \tag{34c}$$

The state of stick continues until the ratio of tangential to normal force has a magnitude that increases to a value equal to that of the coefficient of friction; i.e. $|F_1|/F_3 = \mu$. When the components of force first satisfy this equality, the contact point begins to slip.

Tangential Velocity Changes During Slip

If the force ratio equals the coefficient of friction $|F_1|/F_3 = \mu$, the particle at the contact point slips at a speed $v_1 + \dot{\delta}_1$ in direction $s(t)$. During any period of slip, because friction is opposed to the direction of slip s, the differential of impulse can be expressed in terms of the normal component of impulse $dp \equiv dp_3$; i.e.

$$\begin{Bmatrix} dp_1 \\ dp_3 \end{Bmatrix} = \begin{Bmatrix} -\mu s \\ 1 \end{Bmatrix} dp.$$

This gives differential equations for any period of slip — differential equations for the velocity changes as a function of the normal component of impulse dp rather than time,

$$\begin{Bmatrix} dv_1/dp \\ dv_3/dp \end{Bmatrix} = m^{-1}\begin{bmatrix} \beta_1\eta^{-2} & 0 \\ 0 & \beta_3 \end{bmatrix}\begin{Bmatrix} -\mu s \\ 1 \end{Bmatrix}. \tag{35}$$

Angles of Incidence giving Initial Stick

During stick the tangential force F_3 and the normal force F_1 are given by Eq (34c) and an equation from Table 2. At the contact point there is stick if the ratio of these forces is inside the cone of friction. If stick initiates at incidence, $t = 0$, then the period of stick persists so long as the force ratio satisfies

$$\frac{|F_1(t)|}{F_3(t)} = \frac{1}{\mu\eta^2}\frac{v_1(0)}{v_3(0)}\frac{\Omega\sin\omega t}{\omega\sin\Omega t} < 1, \qquad 0 \le t < t_c. \qquad (36)$$

Hence, at the contact point the sliding process begins with initial stick if and only if the relative velocity at the contact point is inside a friction-compliance ratio cone

$$\left|\frac{v_1(0)}{v_3(0)}\right| < \mu\eta^2. \qquad (37)$$

At the point of contact the relative velocity ratio is related to the angle of the incident relative velocity ϕ (measured from normal), $\phi = \tan^{-1}[v_1(0)/v_3(0)]$. Hence initial stick at C occurs only if the angle of incidence ϕ is less than the product of the coefficient of friction and the compliance ratio η^2. For larger angles of incidence $\phi > \tan^{-1}(\mu\eta^2)$ the contact point slides at incidence.

Transition from Initial Slip to Intermediate Period of Stick

Sliding begins at incidence if the angle of incidence is not small $\phi > \tan^{-1}(\mu\eta^2)$. Subsequently, during the contact period, the speed of sliding is slowed by friction. If the speed of sliding vanishes during impact, the contact points then begin a period of stick. The time of transition from sliding to stick t_s is calculated from a condition of smoothness that results from the local tangential deformation $\delta_1(t)$ being a continuous function at the transition time t_s. Hence,

$$\lim_{\varepsilon \to 0}\left[\mu\frac{dF_3(t_s-\varepsilon)}{d\varepsilon} = \left|\frac{dF_1(t_s+\varepsilon)}{d\varepsilon}\right|\right]. \qquad (38)$$

For stick that begins during restitution $t_c \le t_s < t_f$, the derivatives of tangential force at instants of sliding before and stick after the transition time t_s are given by

$$\frac{dF_1}{dt} = -\frac{m\omega^2 v_1(t_s)}{\beta_1},$$

$$\mu\frac{dF_3(t_s)}{dt} = -\frac{\mu m\Omega^2 v_3(0)}{e_*\beta_3}\cos e_*^{-1}\left(\Omega t_s - \frac{\pi}{2}(1-e_*)\right),$$

where the kinetic equations during an initial period of sliding $t < t_s$ result at the time of transition t_s in small displacement and sliding at the contact point,

$$\delta_1 = \mu \frac{\beta_1 \Omega v_3(0)}{\beta_3 \omega^2} \sin e_*^{-1}\left(\Omega t_s - \frac{\pi}{2}(1-e_*)\right), \quad t_c \leq t_s < t_f,$$

$$v_1(t_s) = v_1(0) - \mu \frac{\beta_1}{\beta_3} v_3(0)\left[1 - e_* \cos e_*^{-1}\left(\Omega t_s - \frac{\pi}{2}(1-e_*)\right)\right], \quad t_c \leq t_s < t_f.$$

Substituting the rate-of-change of forces into (38) we obtain an equation for the time t_s when initial slip halts and stick begins,

$$\Omega t_s = \frac{\pi}{2}(1-e_*) + e_* \cos e_*\left(\frac{v_1(0)/\mu v_3(0) - \beta_1/\beta_3}{\eta^2 - e_*^2 \beta_1/\beta_3}\right), \quad t_c \leq t_s < t_f. \tag{39}$$

This gives an initial velocity ratio where initial slip halts at the end of the period of compression; i.e. $\Omega t_s = \Omega t_c = \pi/2$ implies

$$\left|\frac{v_1(0)}{v_3(0)}\right| = \mu \frac{\beta_1}{\beta_3}. \tag{40}$$

Initial sliding does not halt during the contact period if the initial velocity ratio is sufficiently large. The relative velocity ratio (angle of incidence) that results in initial slip halting at the time of separation t_f is obtained by letting $\Omega t_s = \Omega t_f = (\pi/2)(1+e_*)$; hence,

$$\left|\frac{v_1(0)}{v_3(0)}\right| = \mu \frac{\beta_1}{\beta_3}(1+e_*), \qquad s_0 < 0. \tag{41}$$

5.3 Change of Relative Velocity for Different Slip Processes

Small Angle of Incidence, $|v_1(0)/v_3(0)| < \mu \eta^2$

If the angle of incidence is small, at the contact point there is initial stick and this persists until late in the restitution period when there is a transition to a final stage of sliding. This final stage of sliding is always present; it occurs because the magnitude of the normal compressive force decreases during the final phase of the contact period. If the initial direction of sliding $s(0) \equiv s_0$ is in the negative tangential direction $s_0 < 0$, at the time of transition from stick to sliding t_s the direction of sliding has reversed and the tangential relative velocity of the bodies can be calculated as

$$v_1(t_s) = v_1(0)\cos \omega t_s.$$

The terminal tangential relative velocity depends on the impulse during the final period of sliding

$$v_1(t_f) = v_1(0)\cos \omega t_s - \frac{\mu s \beta_1}{m}\left[p(t_f) - p(t_s)\right], \qquad (42a)$$

where s is the current direction of sliding and the sliding impulse is expressed as

$$p(t_f) - p(t_s) = -\frac{m v_3(0)}{\beta_3} e_* \left\{1 + \cos e_*^{-1}\left(\Omega t_s + \frac{\pi}{2}(1 - e_*)\right)\right\}. \qquad (42b)$$

The variation with time of normal and tangential forces for an impact initiating with a small tangential relative velocity is shown in Fig. 9a while the variation of the tangential component of relative velocity is shown in Fig. 9b. Notice that colliding bodies composed of isotropic elastic solids have a compliance ratio in the range $1 < \eta^2 < 1.5$ so that the frequency of tangential motion is larger than the frequency of normal motion $\omega > \Omega$.

Intermediate Angle of Incidence, $\mu \eta^2 \leq |v_1(0)/v_3(0)| < \mu(1 + e_*)\beta_1/\beta_3$

Intermediate angles of incidence result in initial slip that halts at a time t_s which is before separation; i.e. $t_s < t_f$. After slip halts the contact sticks and elastic energy stored during sliding retards the relative tangential motion; subsequently the tangential elastic element draws the bodies together until the tangential relative displacement vanishes and the direction of sliding reverses. Finally as the normal contact force decreases during restitution, the tangential force undergoes a second transition and enters a final stage of sliding.

The contact point has an intermediate angle of incidence ϕ if initial slip halts during the contact period, $\tan^{-1}(\mu \eta^2) < \phi \leq \tan^{-1}[\mu(1 + e_*)\beta_1/\beta_3]$. For an intermediate angle of incidence, the tangential velocity during the initial period of slip is obtained as

$$v_1(t) = v_1(0) - \frac{\mu s \beta_1}{m} p(t).$$

This initial slip terminates and stick begins at a time t_s when subsequent slip or stick give the same rate of change for the tangential force; this transition from slip to stick occurs when

$$\lim_{\varepsilon \to 0}\left[\left|\frac{dF_1(t_s + \varepsilon)}{d\varepsilon}\right| = \mu \frac{dF_3(t_s - \varepsilon)}{d\varepsilon}\right]. \qquad (43)$$

For the period of stick, $t > t_s$, Eq. (34c) gives

$$\frac{dF_1(t)}{dt} = -\frac{m\omega^3 \delta_1(t_s)}{\beta_1} \sin \omega(t - t_s) - \frac{m\omega^2 v_1(t_s)}{\beta_1} \cos \omega(t - t_s).$$

If $v_1(0) < 0$ the dynamics of sliding during $t \le t_s$ give transition values

$$\delta_1(t_s) = \frac{\eta^2 F_1(t_s)}{\kappa} = \mu \frac{\beta_1 \Omega v_3(0)}{\beta_2 \omega^2} \sin \Omega t_s, \qquad t_s \le t_c,$$

$$v_1(t_s) = v_1(0) - \mu \frac{\beta_1 v_3(0)}{\beta_2} \left[1 - \cos \Omega t_s \right], \qquad t_s \le t_c,$$

$$\delta_1(t_s) = \mu \frac{\beta_1 \Omega v_3(0)}{\beta_2 \omega^2} \sin e_*^{-1} \left[\Omega t_s - \frac{\pi}{2}(1 - e_*) \right], \qquad t_c < t_s \le t_f,$$

$$v_1(t_s) = v_1(0) - \mu \frac{\beta_1 v_3(0)}{\beta_2} \left\{ 1 - e_* \cos e_*^{-1} \left[\Omega t_s - \frac{\pi}{2}(1 - e_*) \right] \right\}, \quad t_c < t_s \le t_f.$$

The normal force is obtained from Table 2; hence after differentiation one obtains

$$\frac{dF_3(t)}{dt} = -\frac{m\Omega^2 v_3(0)}{\beta_3} \cos \Omega t, \qquad t_s \le t \le t_c$$

$$\frac{dF_3(t)}{dt} = -\frac{m\Omega^2 v_3(0)}{e_* \beta_3} \cos e_*^{-1} \left[\Omega t - \frac{\pi}{2}(1 - e_*) \right], \qquad t_c \le t_s \le t$$

After equating rates of change for components of force in the limit as $t \to t_s$ and noting that the time when slip halts equals the time of compression $t_s = t_c$ if $v_1(0)/v_3(0) = \mu\beta_1/\beta_3$, we obtain an expression to be solved for the non-dimensional transition time Ωt_s,

$$\frac{v_1(0)}{\mu v_3(0)} = \frac{\beta_1}{\beta_3} \{ 1 - \cos \Omega t_s \} + \eta^2 \cos \Omega t_s, \qquad \frac{v_1(0)}{v_3(0)} \le \frac{\beta_1}{\mu \beta_3}, \quad (44a)$$

$$\frac{v_1(0)}{\mu v_3(0)} = \frac{\beta_1}{\beta_3} \left\{ 1 - e_* \cos e_*^{-1} \left[\Omega t_s - \frac{\pi}{2}(1 - e_*) \right] \right\}$$

$$+ \frac{\eta^2}{e_*} \cos e_*^{-1} \left[\Omega t_s - \frac{\pi}{2}(1 - e_*) \right], \qquad \frac{v_1(0)}{v_3(0)} > \frac{\beta_1}{\mu \beta_3}. \quad (44b)$$

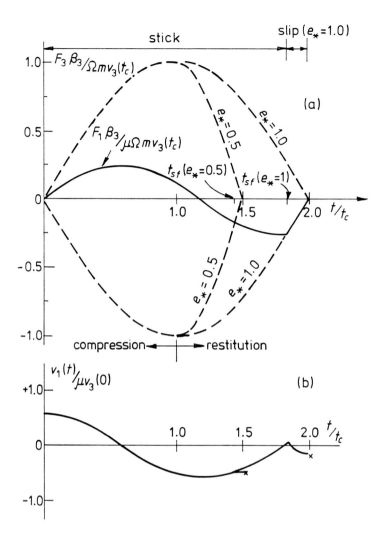

Fig. 9 For small angle of incidence (a) variation of normal and tangential force during stick-slip motion, and (b) slip velocity during collision.

In the equations above, the form of expression depends on whether slip is brought to a halt during compression or restitution.

After slip halts at time t_s the contact point sticks and the tangential compliant element (linear) begins a period of simple harmonic motion (s.h.m.). The tangential components of velocity and force for this period of stick are given by Eq. (34). The intermediate period of stick continues until time t_{sf} when a terminal period of slip begins. Terminal slip begins because the ratio of tangential to normal

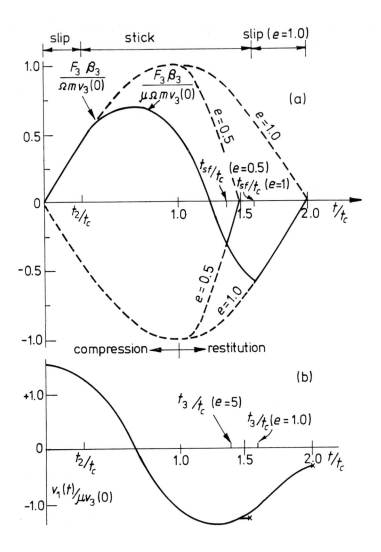

Fig. 10 For intermediate angle of incidence (a) variation of normal and tangential force during slip-stick-slip motion, and (b) slip velocity during collision.

force has increased to the friction limit, $|F_1|/F_3 = \mu$. Hence, the terminal period of slip begins at a time t_{sf} that satisfies

$$\left| \frac{\Omega \delta_1(t_s)}{\mu v_3(0)} \cos \omega(t_{sf} - t_s) - \frac{\Omega v_1(t_s)}{\mu \omega v_3(0)} \sin \omega(t_{sf} - t_s) \right|$$

$$= \eta^2 \sin e_*^{-1} \left[\Omega t_s - \frac{\pi}{2}(1 - e_*) \right]. \qquad (45)$$

Changes in velocity and force during the terminal period of slip are given by expressions similar to (34),

$$v_1(t) = \omega \delta_1(t_{sf}) \sin \omega(t - t_{sf}) + v_1(t_{sf}) \cos \omega(t - t_{sf}),$$ (46a)

$$F_1(t) = \frac{m\omega^2 \delta_1(t_{sf})}{\beta_1} \cos \omega(t - t_{sf}) - \frac{m\omega v_1(t_{sf})}{\beta_1} \sin \omega(t - t_{sf}).$$ (46b)

During the second phase of slip, as the force reduces the tangential compliant element relaxes   The second phase of slip terminates at separation.

The second phase of slip terminates at separation. For a collinear collision, the time of separation t_f is related to the period of compression t_c by the coefficient of restitution e_*; i.e. $t_f = (1 + e_*)t_c$. At separation the tangential component of relative velocity at the contact point is

$$v_1(t_f) = v_1(t_{sf}) - \frac{\mu\beta_1}{m}\{p(t_f) - p(t_{sf})\}, \quad s_0 < 0.$$ (47)

Figure 10a shows the variation of normal and tangential force as a function of time for impact at an intermediate angle of obliquity. The corresponding variation in tangential velocity is illustrated in Fig. 10b.

Large Angle of Incidence, $\mu(1 + e_*)\beta_1 / \beta_3 \le |v_1(0)/v_3(0)|$

There is continuous or gross slip at the contact point if the angle of incidence is large enough so that slip does not cease before separation; i.e. friction merely slows the speed of sliding but is not large enough to halt sliding during the contact period. Initial conditions that result in continuous sliding can be identified from the condition that slip is brought to a halt only at a time larger than the time of separation, $t_{sf} \ge (1 + e_*)t_c$. If slip is continuous, the terminal (separation) impulse $p(t_f)$ is related to the impulse for compression by $p(t_f) = (1 + e_*)p(t_c)$ and the tangential component of relative velocity at the contact point is obtained as

$$v_1(t_f) = v_1(0) + \frac{\mu\beta_1}{m}(1 + e_*)p(t_c), \quad s_0 < 0.$$ (48)

5.4 Example: Oblique Impact Of Sphere

The effects of friction and tangential compliance on dynamics of a sphere hitting a nonlinear elastic half-space at an angle of incidence have been evaluated and compared with an analytical elasticity solution by Maw, Barber & Fawcett [17]. At impact the sphere has tangential and normal components of velocity $v_1(0)$ and $v_2(0)$ respectively and it has no initial rotation. The elasticity solution is based on quasi-static behaviour of an elastic solid with Coulomb friction at the interface; this continuum formulation is used to develop the local stress and deformation fields throughout the contact region.

An alternative analysis based on the present discrete element modelling uses a Poisson's ratio $v = 0.3$ and for an initially spherical contact surface, a ratio of normal to tangential stiffness $\eta^2 = 1.21$ (see Johnson [7]). For a spherical solid striking a half-space we obtain a ratio of inertia coefficients $\beta_1 / \beta_3 = 3.5$ which gives the ratio of frequencies, $\omega / \Omega = 1.7$.

Figure 11 is a plot of the terminal tangential velocity resulting from oblique impact of an elastic sphere striking at an angle of incidence ϕ where $\phi = \tan^{-1}(v_1(0)/v_3(0))$; the terminal tangential velocity has been plotted as a function of the initial ratio of tangential to normal velocity $v_1(0)/v_3(0)$ (directly related to the initial angle of incidence). These results were calculated with both the elastic continuum theory and discrete element modelling; these different analytical methods gave almost identical results. For a small angle of incidence, there is stick throughout most of the contact period and only a late stage of terminal slip as the normal contact force becomes small. If the coefficient of restitution is large, small angles of incidence give a final tangential velocity in the same direction as that at incidence; for a somewhat smaller coefficient of restitution $e_* < 0.7$ the direction of terminal slip reverses from that at incidence. For an intermediate value of the angle of incidence, the direction of terminal slip reverses from that at incidence for all values of the coefficient of restitution; slip reversal occurs for angles of incidence $v_1(0)/v_3(0) < \mu(1+e_*)\beta_1 / \beta_3$. For a large angle of incidence

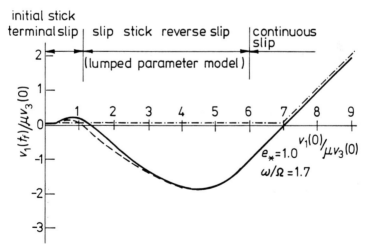

Fig. 11 Terminal sliding speed as function of incident sliding velocity for edge impact of disk on half-space: discrete element model ———— ; Maw, Barber & Fawcett elastic solution — — — — ; rigid body with negligible tangential compliance — · — · — .

$v_1(0)/v_3(0) \ge \mu(1+e_*)\beta_1/\beta_3$ there is gross slip throughout the contact period and sliding that continues in the same direction throughout the contact period.

In Fig. 11 the result of assuming that tangential compliance is negligible is illustrated by the chain-dot line. With this assumption, the principles of mechanics result in either terminal stick at zero tangential velocity or else gross slip. Throughout most of the range of gross slip the assumption of negligible tangential compliance gives velocity changes that are identical with those of the elastic continuum and discrete element analytical models.

Similar results for various coefficients of restitution have been obtained by Stronge [18]. These results, shown in Fig. 12, indicate that increasing losses of energy at impact have the effect of compressing the small and intermediate range of angles of incidence to a smaller range of values. Thus for perfectly plastic impact ($e_* = 0$) the small and intermediate ranges are roughly half the size of those for a coefficient of restitution $e_* = 1$.

5.5 Maximum Force From Oblique Impact Of Sphere

The effects of friction and tangential compliance on the maximum normal and tangential contact force were investigated in experiments by Lewis and Rodgers [19]. They mounted a 25.4 mm diameter steel sphere at the end of a 1.8 m long pendulum which controlled the direction of oblique impact of the sphere against a heavy steel plate. This apparatus allowed the angle of incidence ϕ to vary between $0°$ and $85°$ from normal. In the experiments of Lewis and Rodgers the normal impact speed was small, $0.01\text{-}0.05 \text{ ms}^{-1}$ so that the impact was effectively elastic. Piezoelectric force transducers were used too make separate measurements of tangential and normal components of contact force during impact. The coefficient of friction measured in experiments giving gross sliding was $\mu = 0.179$.

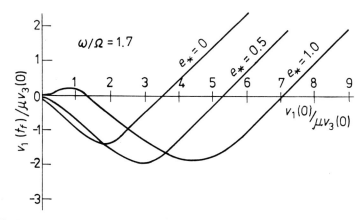

Fig. 12 Terminal sliding speed as function of incident sliding velocity for edge impact of disk on half-space with coefficient of restitution $e_* = 0.0, 0.5, 1.0$.

Experimental data from impact of the pendulum at an incident speed $V_0 = 0.048$ ms^{-1} is compared with calculations obtained from the present discrete element modelling in Fig. 13. The calculations depend on an estimate of the mass of the pendulum arm that supports the sphere. The agreement shown in Fig. 13 was achieved by increasing the mass of the sphere by 50% to account for the inertia of the pendulum arm. Also, these calculations assume that the compliance ratio $\eta^2 = 1.21$ and the inertia ratio $\beta_1 / \beta_3 = 3.5$ — values that are representative of a solid sphere. These values gave quantitative agreement between the experiments and the calculated values throughout the full range of angles of incidence. The largest friction force occurred at an angle of incidence of about $40°$ whereas with $\mu = 0.18$ and $e_* = 1$, the calculations gave a value of maximum force at $35°$ irrespective of impact speed. Notice that the largest ratio of tangential to normal force equals the coefficient of friction μ only if the contact is sliding at the transition from compression to restitution when the normal force is a maximum. This occurs if the angle of incidence is intermediate of large; i.e. $\phi \geq 32°$ for a rough solid sphere with a coefficient of friction $\mu = 0.18$.

5.6 Discrete Modelling of Tangential Compliance

For elastic bodies with a large coefficient of friction, impact experiments show that at small angles of obliquity there can be reversal in the direction of slip during collision; it is not possible however, to obtain this reversal from a theory that neglects tangential compliance. The discrete element model presented here is the simplest formulation where an analysis of a collinear collision can give a solution which shows a reversal in direction of tangential relative velocity during a collision. Again the penalty of incorporating the detail of local compliance at the contact

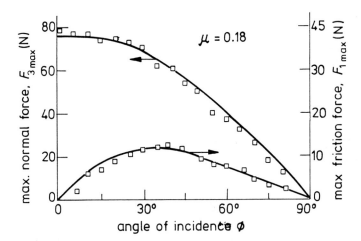

Fig. 13 Comparisons of calculation of maximum normal and tangential components of force during impact of rough sphere on half-space with measurements by Lewis & Rodgers [19].

point is that the analysis becomes time dependent. In a problem involving Coulomb friction and slip-stick relative motion however, this level of detail is necessary in order to represent the relationship between components of contact force.

6. Chain Reactions From Impact In Multi-Body System

To analyze impact response of a system of 'rigid' bodies using impulse-momentum methods, some authors employ an assumption regarding the timing of reactions that occur at contact points where relative displacements of the bodies are related by constraints. Consistent with a strict interpretation of rigid body theory, Wittenburg [20], Pfeiffer & Glockner [21] and Pereira & Nikravesh [22] assume that impulses are applied simultaneously at the points of constraint; simultaneity implies that the period of compression is identical irrespective of local properties around each point of contact (or constraint). On the other hand, Johnson [23], Hurmuzlu & Marghitu [24] and Ivanov [25] assume that reaction impulses at the contact points occur sequentially, usually in a sequence akin to Moreau's sweeping process [26]. This assumption is motivated by experience of simple experiments using a tightly-packed chain of identical spheres — a 'Newton cradle' apparatus.

A different method of approach is to incorporate the additional detail of a force-displacement analysis in place of impulse-momentum; consideration of displacements requires integration of time-dependent equations of motion rather than the algebraic analysis appropriate for planar problems that involve two-body collisions. For impact in multi-body systems, Stronge [27] has argued that very small relative displacements at the contact points must be considered because these are the source of the reaction forces that prevent interpenetration at points of constraint. Previous applications of this methodology include Cundall & Strack [28], Lankarani & Nikravesh [29] and Wang, Kumar & Abel [30].

With consideration of very small deformations that remain local to the contact region, the analyst can explicitly represent the local compliance by means of discrete elements; e.g. springs and dashpots. The coefficients for these elements can be selected in accord with the extended Hertz theory presented in Sect. 4; i.e. these coefficients are representative of the local compliance and they depend on the geometry of the contact region, the material properties and the contact pressure. Incorporating these explicit force-displacement relations in a rigid-body impact analysis implicitly involves two different displacement scales. Very small relative displacements must be considered in order to obtain the time-dependent interaction forces at points of constraint — the displacements, however, must be so small that they have negligible effect on the inertia properties of the system.

Here the effect of impact in a multi-body system is analyzed using discrete elements at the contacts between individual 'rigid' bodies. For linear local compliance the result is independent of the elastic stiffness of the contacts; i.e. this stiffness affects only the speed of propagation through the system. Two different examples are used to elucidate the nature of propagation of a pulse of reaction traveling away from the impact point. To focus on the qualitative features of propagation, each of these examples considers direct axial impact in a collinear system where friction is irrelevant.

6.1 Example: Collinear Impact In Row Of Identical Spheres

We consider an axially aligned row of many identical spheres which are initially in diametrical contact but not compressed. The jth element or sphere in the row has mass M, diameter D and an axial displacement u_j. The local compressive stiffness of each contact region is denoted by k, while in extension the stiffness vanishes. Between the contacts each body is assumed to be rigid. This one-dimensional system is illustrated in Fig. 14.

Wave Propagation In Linear Coaxial Periodic System

As long as the contact forces remain compressive, a typical element of this periodic system has an equation of motion

$$\ddot{u}_j + 2\omega_0^2 u_j = \omega_0^2(u_{j-1} + u_{j+1}), \qquad \omega_0^2 = k/M. \tag{49}$$

For a wave of slowly varying amplitude the displacement u_j varies according to

$$u_j = Ue^{i(\kappa x - \omega t)} = Ue^{i(\kappa jD - \omega t)},$$

where the wave-number κ is related to the wave-length λ by $\kappa = 2\pi/\lambda$ and $i = \sqrt{-1}$. By substituting this solution into (49) a dispersion relation is obtained,

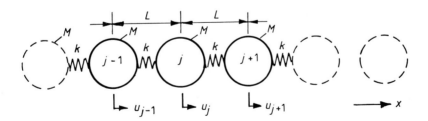

Fig. 14 Direct impact on collinear system with interface springs that represent the local elastic stiffness at each point of contact.

$$\frac{\omega}{\omega_0} = \pm 2\sin\left(\frac{\kappa D}{2}\right), \quad \frac{\omega}{\omega_0} < 2.$$

In this system a particular wave-length component of a compressive pulse propagates with the phase velocity c_p where

$$c_p = \frac{\omega}{\kappa} = \pm\omega_0 \frac{\sin(\kappa D/2)}{\kappa D/2} = \pm\frac{\omega_0\lambda}{\pi}\sin\left(\frac{\pi D}{\lambda}\right). \tag{50}$$

For a compressive pulse containing a range of wave-lengths however, the speed of energy propagation is given by the group velocity,

$$c_g = \frac{d\omega}{d\kappa} = \omega_0 D\cos\left(\frac{\kappa D}{2}\right). \tag{51}$$

Figure 15 shows the phase and group velocities for this periodic system. The speed of energy propagation decreases with increasing wave number (decreasing wavelength) until cut-off at $\kappa = \pi/D$. This cut-off wave number gives a lower bound for the wavelength of a propagating disturbance; i.e. propagation occurs only if $\lambda > 2D$. A slightly longer minimum wavelength has been measured in photoelastic experiments in a periodic series of collinear disks driven axially by a small explosive charge [31].

In general, the primary reaction force generated by impact can be resolved into components with various frequencies. High frequency components $\omega > 2\omega_0$ result in a non-propagating disturbance which corresponds to evanescence; i.e. a

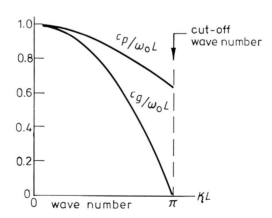

Fig. 15 Phase velocity c_p and group velocity c_g of periodic system as functions of wave number κ.

standing wave with vibration amplitude that decreases exponentially as distance from the impact site increases [27].

Impact Response Of Linear Coaxial Periodic System

Numerical simulations have been used to investigate the response in an aligned series of five initially touching, identical elastic spheres when one end is subject to coaxial impact by an identical sphere — the simplest case that can be tested using Newton's cradle. Figure 16 shows the velocity of individual spheres as a function of time for (a) bilinear stiffness $F_j = k\delta_j$ and (b) Hertz stiffness $F_j = k_s\delta_j^{3/2}$.

The single mass colliding against one end of a chain of identical spheres results in a pulse of compression that propagates through the system in a direction away from the impact point. Near the impact point the length of the pulse is between $2D$ and $3D$ but as the pulse propagates, dispersion broadens this pattern. Because momentum is conserved, the pulse amplitude decreases as the breadth increases. The increase in breadth of the pulse with increasing distance is noticeably more for the bilinear contact compliance rather than the nonlinear (Hertzian) compliance. The front of this compression pulse travels at a speed roughly equal to $\approx \omega_0 D$; for the linear stiffness relation, this equals the upper limit for group velocity which corresponds to an indefinitely long wavelength, $\lambda \to \infty$. Using the stiffness relation from Sect. 5 for k, this speed of propagation is $\approx \sqrt{1.32 Y / \rho}$ in contrast to the one-dimensional wave speed for uniaxial stress, $c_0 = \sqrt{E / \rho}$; for typical metals this difference in speeds is a factor of ten.

Figure 16 also shows that near the impact end of the chain, the spheres have a terminal velocity that is slightly negative. The magnitude of this residual velocity appears to decrease exponentially with increasing distance from the impact point, prompting the association with evanescent waves. In the case of a system with a Hertzian contact relation, the maximum amplitude of this residual velocity is roughly half that of the bilinear system.

6.2 Example: Impact Response Of Multi-body System With Graduated Properties

Here, we consider a row of 3 spheres B_1, B_2, B_3 as shown in Fig. 17; in this collinear system the bodies have distinct masses, M_1, M_2, M_3 respectively. Also, at contact points C_1 and C_2 the local contact stiffness are distinct. Initially the spheres are in contact but not compressed.

The displacement of the ith sphere from its initial position $u_i(t)$ is a function of time t. Each sphere has a velocity in the axial direction $\dot{u}_i(t)$. During the contact period $0 < t \le t_f$ these velocities change as a consequence of contact forces — forces that arise from small local indentation of the bodies δ_i $i = 1, 2$ in the regions immediately surrounding C_1 and C_2. Other than the small contact regions, the bodies are assumed to be rigid.

At any time t this system has a kinetic energy T where

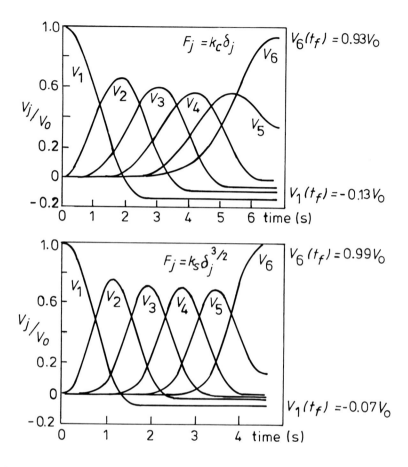

Fig. 16 Variation in velocity as function of time for elements in collinear system of 5 identical spheres struck axially at one end by single sphere, B_1: (a) bilinear interaction stiffness (light lines) and (b) Hertz nonlinear interaction stiffness (heavy lines).

$$2T = \sum_{i=1}^{3} M_i \dot{u}_i^2. \tag{52}$$

If there are no external active forces acting on the system during the contact period, momentum is conserved so that the velocity of the center of mass $\hat{V}$ is obtained from,

$$\hat{V} = M^{-1} \sum_{i=1}^{3} M_i \dot{u}_i, \quad M \equiv \sum_{i=1}^{3} M_i. \tag{53}$$

Relative to the steadily moving center of mass the translational momentum vanishes so that,

$$0 = \sum_{i=1}^{3} M_i \left(\dot{u}_i - \hat{V} \right). \tag{54}$$

Since momentum is conserved, the kinetic energy can be separated into part that is constant during the contact period and a partial kinetic energy of relative motion T_{rel}; i.e. $T = T_{rel} + 0.5 M \hat{V}^2$. The contact forces do work during contact so that some of T_{rel} is transformed into strain energy of internal deformation; in an elastic collision, this transformed kinetic energy is recovered. The kinetic energy of relative motion T_{rel} is defined as

$$T_{rel} = \sum_{i=1}^{3} M_i \left(\dot{u}_i^2 - \hat{V}^2 \right). \tag{55}$$

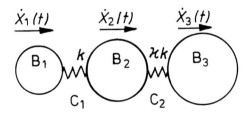

Fig. 17 Collinear system of 3 spheres with smoothly varying properties for mass and local contact stiffness.

Recognizing that the indentation $\delta_i \equiv u_{i+1} - u_i$ and using (52), (54) and (55) we obtain an expression for the kinetic energy of relative displacement in terms of indentation,

$$2T_{rel} = M^{-1}\left[M_1(M_2 + M_3)\dot{\delta}_1^2 + 2M_1M_2\dot{\delta}_1\dot{\delta}_2 + M_3(M_1 + M_2)\dot{\delta}_2^2\right]$$
$$= \dot{z}_1^2 + 2\dot{z}_1\dot{z}_2\cos\alpha + \dot{z}_2^2. \tag{56}$$

where nondimensional indentations that symmetrize the equations were suggested by Ivanov and Larina [32].

$$z_1 = \delta_1\left[M^{-1}M_1(M_2 + M_3)\right]^{1/2}, \quad z_2 = \delta_2\left[M^{-1}M_3(M_1 + M_2)\right]^{1/2}.$$

Normal contact forces $F_j, j = 1, 2$ arise between the bodies as a result of the relative displacements δ_i. These forces are solely compressive; they have a relationship to contact indentation which depends on material properties and the geometry of the contact region. If the bodies are elastic and the contacting surfaces are cylindrical with parallel axes, the stress field is 2-D and $F_j = k_j\delta_j H(\delta_j)$ where $H(\delta_j) = 0$ if $\delta_j < 0$, and $H(\delta_j) = 1$ if $\delta_j > 0$. On the other hand, if the contacting surfaces are spherical, the stress field is 3-D and Hertz theory gives a normal contact force $F_j = k_{sj}\delta_j^{3/2}H(\delta_j)$. The stiffness coefficients k_j or k_{sj} can be obtained from expressions in Sect. 4. The interaction forces are used to develop potential functions U_c or U_s for cylindrical or spherical contacts respectively,

$$2U_c = k_1\delta_1^2 H(\delta_1) + k_2\delta_2^2 H(\delta_2),$$

$$(5/2)U_s = k_{s1}\delta_1^{5/2}H(\delta_1) + k_{s2}\delta_2^{5/2}H(\delta_2). \tag{57}$$

For spherical contacts, the equations of relative motion are obtained from (56) and (57) by means of Lagrange's equations,

$$\begin{Bmatrix}0\\0\end{Bmatrix} = \begin{bmatrix}1 & \cos\alpha\\\cos\alpha & 1\end{bmatrix}\begin{Bmatrix}d^2z_1/d\tau^2\\d^2z_2/d\tau^2\end{Bmatrix} + \sin^2\alpha\begin{Bmatrix}z_1^{3/2}H(z_1)\\\gamma^2z_2^{3/2}H(z_2)\end{Bmatrix}. \tag{58}$$

where a non-dimensional time τ and ratio of stiffness to mass gradients γ^2 are defined as

$$\tau \equiv t\sqrt{\frac{k_{s1}(M_1 + M_2)}{M_1M_2}}, \quad \gamma^2 \equiv \frac{k_{s2}}{k_{s1}}\frac{M_1(M_2 + M_3)}{M_3(M_1 + M_2)},$$

$$\cos\alpha \equiv \sqrt{\frac{M_1 M_2}{(M_1 + M_2)(M_2 + M_3)}}.$$

The present example is concerned with effects of a smooth gradation of properties; e.g. let η be a gradient of mass and ς be a gradient of stiffness. Hence, the individual masses are $M_1 = \eta^{-1}M_2$, M_2, $M_3 = \eta M_2$ where the central mass $M_2 = \eta M/(1+\eta+\eta^2)$. The stiffness of the compliant elements are in the ratio $k_{s2}/k_{s1} = \varsigma$. These inertia and stiffness gradients give

$$\cos\alpha = \frac{1}{1+\eta}, \qquad \gamma^2 = \frac{\varsigma}{\eta}.$$

Now γ is recognizable as the gradient of natural frequency (or more precisely, a gradient of speed of propagation through this system of 'rigid' masses linked by discrete compliant elements). The characteristic frequencies for this system can be obtained from the eigenvalues of (57) with $\omega_0^2 = (k_1 + k_2)/2M_2$.

To retrieve the velocities in an inertial frame from the non-dimensional relative velocities, we use the transformation,

$$\begin{Bmatrix} \dot{u}_1 \\ \dot{u}_2 \\ \dot{u}_3 \end{Bmatrix} = \begin{Bmatrix} \hat{v} \\ \hat{v} \\ \hat{v} \end{Bmatrix} + \frac{1}{\sqrt{(1+\eta)(1+\eta+\eta^2)}} \begin{bmatrix} \eta(1+\eta) & \eta^{3/2} \\ -1 & \eta^{3/2} \\ -1 & -\eta^{1/2}(1+\eta^2) \end{bmatrix} \begin{Bmatrix} dz_1/d\tau \\ dz_2/d\tau \end{Bmatrix}. \tag{59}$$

Discussion of Solution

The nonlinear differential equations have been solved numerically for the terminal relative velocities, then these solution were transformed to obtain the final velocities in an inertial reference frame. The equations of motion were solved beginning from the following initial condition,

I.C. $\qquad \dot{u}_1 = V_0, \ \dot{u}_2 = \dot{u}_3 = 0 \quad$ or $\quad \dot{\delta}_1/V_0 = 1, \ \dot{\delta}_2/V_0 = 0.$

Figure 18 shows the distribution of final velocity in a 3 sphere chain subject to unilateral constraints at contact points. The results are plotted as a function of the gradient of natural frequency γ at two different values of the mass gradient η; either uniform mass density $\eta = 1$ or an increasing mass density $\eta = 4$. Principally, if $\gamma < 1$ the result asymptotically approaches that for sequential collisions whereas if $\gamma \gg 1$ the result asymptotically approaches that for simultaneous collisions. The distribution of final velocities depends almost solely on the gradient of natural frequency γ while the mass gradient affects only the absolute value of the final velocities (not the relative velocities). The results for

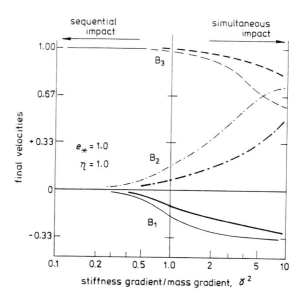

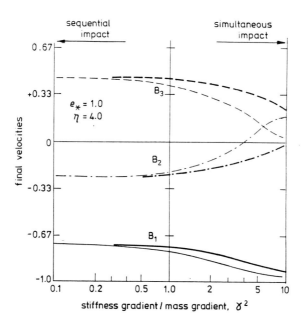

Fig. 18 Variation of terminal velocities in 3 sphere chain as functions of the gradient of natural frequency γ: bilinear compliance (light lines) and Hertz compliance (heavy lines).

231

cylindrical contacts converge to asymptotic limits slightly faster than the results for spherical contacts — qualitatively however, the behaviour is the same.

The timing of reactions at different contact points is a function of the gradient of natural frequency (or gradient of wave propagation speed). If speed of propagation through the first contacting pair is larger than that of the second pair, the first pair will complete a cycle of compression and restitution before the cycle at the second contact point has proceeded very far; i.e. the compression cycles occur sequentially. On the other hand if the speed of propagation increases with distance, the compression cycles at the 1st and 2nd contacts, C_1 and C_2, can occur simultaneously. Figure 19 shows the variation of the ratio of times of maximum compression t_c at contacts C_1 and C_2 as functions of the gradient of natural frequency. If $\gamma < 1$ maximum compression at C_2 occurs substantially later than that at C_1. Alternatively if $\gamma \gg 1$ the times of maximum compression at C_1 and C_2 are almost simultaneous. With linear compliance, maximum compression occurs simultaneously at $\gamma \cong 6.7$ and larger values of γ result in double hits between spheres B_2 and B_3.

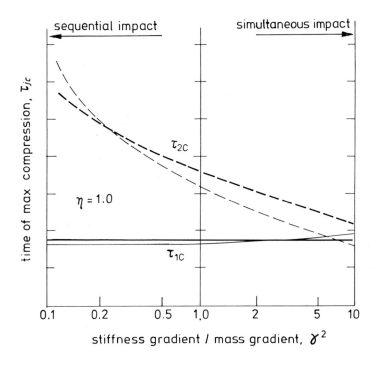

Fig. 19 Variation of ratio of time of maximum compression at contacts C_1 and C_2 as function of the gradient of natural frequency γ: bilinear compliance (light lines) and Hertz compliance (heavy lines).

In a collinear system of compact hard bodies with non-conforming contact regions, impact results in a wave of compression that travels away from the point of impact. This wave travels at a speed that depends on the compliance of local deformation in each contact region and the inertia of the bodies adjacent to each contact. For a chain of elastic spherical bodies, this speed is of the order of $\sqrt{\vartheta_Y Y / \rho}$; for metals, this speed is roughly 0.1 of the elastic wave speed through the material $\sqrt{E/\rho}$. This difference in speeds occurs because of the variation in stiffness in the direction of propagation along the diametrical line between contact points. For hard bodies the compliance of contact regions is small, nevertheless, between these regions the compliance of the bodies is very much smaller.

The principal factor affecting simultaneity of compression cycles in these collisions is the gradient of natural frequency γ. If $\gamma \gg 1$ the contact reactions are almost simultaneous while if $\gamma < 1$ the contact reactions occur sequentially in order of increasing distance from the impact point. The contact reactions result from very small relative displacements that prevent overlap or interference during the contact period. This was foreseen by Brogliato [33] who stated that "neglecting the bodies displacements during the collision process may yield erroneous conclusions."

7. CONCLUSION

Feynman characterized the Newtonian program by the phrase, "Pay attention to the forces." I repeat that clarion call. In most cases solutions of impact problems which are based on impulse-momentum relations turn out to be a limiting case; i.e. they accurately represent the system dynamics only if initial slip is large, the gradient of natural frequency in the system is either very large or small, etc. In general, contact problems require consideration of forces and the force-displacement relations in order to accurately represent the contact dynamics for a broad range of initial conditions and system parameters.

The penalty associated with incorporating in the analysis the detail of force-deflection relations is that the calculation becomes time-dependent. This can be computationally expensive but it seems inevitable in order to represent phenomena which depend directly on the magnitude of relative displacements at the contacts between hard bodies.

The analyses presented here are appropriate for collisions of hard bodies (or multi-body systems) where small areas of contact develop during collision. These conditions insure that the total contact period is small so that effects can be neglected that are associated with changes in inertia and/or the size of the contact area.

The results presented here imply that for dynamics of collision, the analyst must explicitly model the force-deflection relation for each localized region or cross-section where the compliance is not much smaller that at the point of external impact.

Acknowledgement

This research was partly supported by INTAS grant 96-2138, Mechanical Impact in Multi-body Systems.

References

[1] H. Deresiewicz, "A note on Hertz's theory of impact", *ACTA Mechanica* **6**, 110.

[2] Y. Wang & M.T. Mason, Two-dimensional rigid-body collisions with friction," *ASME J. Appl. Mech.* **59**, 635-642.

[3] W.J. Stronge, "Two-dimensional rigid-body collisions with friction – Discussion," *ASME J. Appl. Mech.* **60**, 564–566 (1993).

[4] W.J. Stronge, "Rigid body collisions with friction," *Proc. R. Soc. Lond.* A**431**, 169–181 (1990).

[5] H. Hertz, "Uber die beruhhrung fester elastischer korper", *J. Reine und Angewundte Mathematik* **92**, 156-171.

[6] S. Timoshenko & N. Goodier, *Theory of Elasticity,* McGraw-Hill, New York (1970).

[7] K.L. Johnson, *Contact Mechanics,* Cambridge University Press, 1985.

[8] W.J. Stronge, "Theoretical coefficient of friction for planar impact of rough clastoplastic bodies," *Impact, Waves and Fracture,* ASME AMD-205 (ed. R.C. Batra, A.K. Mal and G.P. MacSithigh) 351-362 (1995).

[9] D. Tabor, *The Hardness of Metals,* Oxford University Press, UK, pp. 128–138 (1951).

[10] D. Tabor, "A simple theory of static and dynamic hardness," *Proc. Royal Soc. Lond.* A**192**, 247-274 (1948).

[11] W.J. Stronge, *Impact Mechanics*, Cambridge University Press, 2000.

[12] C. Hardy, C.N. Baronet &G.V. Tordion , "Elastoplastic indentation of a half-space by a rigid sphere", *J. Num. Methods in Engineering* **3**, 451 (1971).

[13] C.T. Lim & W.J . Stronge, "Oblique elastic-plastic impact between rough cylinders in plane strain," *Int. J. Engr. Sci* **37**, 97-122 (1998).

[14] G.G. Adams, "The coefficient of restitution for a planar two-body eccentric impact," *J. Appl. Mech.* **60**, 1058-1060 (1993).

[15] W. Goldsmith, *Impact: the Theory and Physical Behaviour of Colliding Solids*, Edward Arnold Pub. London (1960).

[16] K.L. Johnson, "The bounce of 'superball'," *Int. J. Mechanical Engng Education* 111, 57-63 (1983).

[17] N. Maw, J.R. Barber & J.N. Fawcett, "The role of elastic tangential compliance in oblique impact", *ASME J. Lub. Technology* **103**(74), 74-80 (1981).

[18] W.J. Stronge, "Planar impact of rough compliant bodies," *Int. J. Impact Engng.* **15**, 435-450.

[19] A.D. Lewis & R.J. Rodgers, "Experimental and numerical study f forces during oblique impact", *J. Sound & Vibration* **125**(3), 403-412 (1988).

[20] J. Wittenburg *Dynamics of Systems of Rigid Bodies*, B.G. Teubner, Stuttgart (1977).

[21] Glockner, Ch. and Pfeiffer, F. (1995) "Multiple impacts with friction in rigid multibody systems", *Advances in Nonlinear Dynamics* (ed. A.K. Bajaj & S.W. Shaw), Kluwer Academic Pub.

[22] M.S. Pereira and P. Nikravesh, "Impact dynamics of nultibody systems with frictional contact using joint coordinates and canonical equations of motion", *Nonlinear Dynamics* **9**, 53-71 (1996).

[23] W. Johnson, "'Simple' linear impact", *Int J. Mech. Engng. Ed.* **4**, 167-181 (1976).

[24] Y. Hurmuzlu and D.B. Marghitu, "Rigid body collisions of planar kinematic chains with multiple contact points", *Int. J. Robotics Research* **13,** 82-92 (1994).

[25] A.P. Ivanov, "On multiple impact", *Prikl. Mat. Mekh.* **59**(6), 887-902 (1995).

[26] J.J. Moreau, "Standard inelastic shocks and the dynamics of unilateral constraints'" CISM courses and lectures, no 288, Springer-Verlag, 173-221.

[27] W.J. Stronge, "Mechanics of impact for compliant multi-body systems", *IUTAM Symp. on Unilateral Multibody Dynamics* (ed. Glockner, Ch. and Pfeiffer, F.), Munich (1998).

[28] P.A. Cundall and O.D.L. Strack, "A discrete numerical model for granular assemblies", *Geotechnique* **29**, 47-65.

[29] H.M. Lankarani & P.E. Nikravesh, "Contact force model with hysteresis damping for impact analysis of multibody systems", *ASME J. Mech Design* **112,** 360-376 (1990).

[30] Y.T. Wang, V. Kumar & J. Abel, "Dynamics of rigid bodies with multiple frictional contacts, Proceedings IEEE Int'l Conf. on Robotics and Automation II, 442-448 (1992).

[31] R. Singh, A. Shukla &H. Zervas, "Explosively generated pulse propagation through particles containing natural cracks", *Mechanics of Materials* **23**, 255-270 (1996).

[32] A.P. Ivanov and T.V. Lavina, *"On the problem of collinear triple collision"*, Proceedings of EUROMECH 397, Grenoble (1999).

[33] B. Brogliato, *Nonsmooth Impact Mechanics: Models, Dynamics and Control*, Springer-Verlag LNCIS 220 (1996).

Impulse Correlation Ratio in Solving Multiple Impact Problems

Yildirim Hurmuzlu and Viorel Ceanga

Southern Methodist University, Dallas, TX 75275

Abstract. In this chapter we consider first the most basic multi-impact system, the so called "Newton's Cradle". The task of developing an analytical method to predict the post impact velocities of the balls in the cradle has baffled investigators in the field of impact research for many years. The impulse based rigid body as well as the alternative time based approaches have failed to produce valid solutions to this problem.

A new method that produces energetically consistent solutions to the problem is presented. The method is based on the traditional impulse-momentum based rigid body approach. We do, however, resolve the non-uniqueness difficulty in the rigid body approach by introducing a new constant called the Impulse Correlation Ratio. Finally, we verify our method by conducting a set of experiments and comparing the theoretical predictions with the experimental outcomes.

The second part of the chapter deals with a second class of multi-impact problems. Such problems arise when a rigid body strikes an external surface at a point when it is resting on the surface at another point. The impulse correlation ratio is also applied to solve this problem. The method produces physically valid and energetically consistent solutions to the problem.

1 Introduction

Multi-impact problems pose many difficulties and unanswered questions (Marghitu and Hurmuzlu, 1995). The simplest of these problems, the linear chain or the "Newton's Cradle" represents the most basic problem of this type. This classical problem involves a collision problem where one ball strikes one end of a linear chain of stationary balls in contact with each other. The system represents the simplest of the multi-body impact problems that one may consider. Yet, it encapsulates the difficulties that are present in more complex systems. Many investigators attempted to develop analytical solutions that produce post impact velocities in this class of problems. The fact remains, however, that the methods that have been proposed possess deficiencies and inconsistencies that are yet to be resolved.

One approach to the solution of the problem has been the consideration of sequential impacts and use of impulse-momentum rules and coefficients of restitution. Johnson (1976), introduced the notion of sequential impacts. The method was based on a succession of simple impacts that occur one at a time. Han and Gilmore (1993), proposed a solution algorithm that accommodated multiple impacts, but their methods resulted in multiple sets of feasible post impact velocities that were valid for the same initial conditions. Brogliato (1996) also considered the three ball impact problem. The conclusion drawn by this author was that the rigid body rules did not possess sufficient information to yield a unique post impact solution. He proceeded by stating that compliance based methods can be useful in providing physical information that can be used in rigid body approach.

The second approach is the compliance based method that introduces linear springs between consecutive balls in the chain. The approach results in a set of second order, linear differential equations that can be cast as functions of the ratios of the spring constants. One of the earliest studies of this type was presented in Smith (1955). Walkiewicz and Newby (1972), considered the possible solutions of the three ball chain that simultaneously satisfied momentum and energy equations. They showed that there were infinitely many solutions that fit this description. Newby (1979), studied the three ball impacts by placing linear springs to model the contacts. He used the spring stiffness ratio as a parameter, and analyzed its effect on the velocity outcomes. He demonstrated that it was not always possible to determine the values of the necessary stiffness ratio that yielded a specific velocity outcome. The compliance based methods are significantly more difficult to apply that the impulse-momentum based ones. To the best of our knowledge, the investigators that applied these methods, never included damping in their analysis, and have always assumed perfectly elastic collisions. In addition, complexity of such methods makes it very difficult to experimentally estimate the model parameters, specifically for chains with large number of balls.

Cholet (1998), uses the methods of convex analysis that is inspired by an adaptation of Moreau's sweeping process (Moreau, 1994; Fremond, 1995) to analyze the multiple impacts in a three ball cradle. This work represents the most advanced up to date and produces unique and energetically consistent results. The only drawback of the approach is that the solution is formulated in terms of three parameters that do not have obvious physical meanings. For example, with this method, it becomes very difficult to identify the parameter values that lead to purely elastic impacts. In addition, the post impact velocities are nonlinear functions of the parameters. Thus, one may encounter difficulties in estimating the parameter values from experiments. Often, there is not a one to one correspondence between a particular post impact velocity set and parameter set. One may obtain the same outcome for different set of parameters. We believe that, a solution method that is based on physically meaningful parameters such as the coefficient of restitution will be more effective in dealing with the multiple impact problem at hand.

The first objective of this chapter is to develop an impulse-momentum based method to determine the post impact velocities of the general N-Ball chain. The method should produce unique and energetically consistent solutions. The predicted outcomes should be physically consistent and experimentally verifiable (experimental verification of the previous methods is almost nonexistent). For this purpose we present a new methodology that uses the energetic coefficient of restitution, Stronge(1990). We propose a new constant, that we call the "Impulse

Correlation Ratio". We test our method by conducting a set of experiments, and comparing the theoretical outcomes with the experimental ones.

As second class of multi-impact problems occur when a rigid body is impacted from a certain point while it has passive contacts with other external objects. Applications such as robotics or systems geared mechanisms, are confronted with such impacts that are transmitted to the system's constraints. Accurate modeling of such impacts requires solution of multiple impact problems in rigid body mechanics. The problem of distributing the impact force from the initial point of collision, where the system receives an impulse, to other passive contacts is not trivial. The conventional approaches to this type of problems may not produce unique solution, or may not produce a solution at all.

The second objective of the present chapter is to develop a method to determine the post-impact velocities of the block undergoing an impact with a massive external surface at one edge, while the other edge is stationary on the surface. The method should produce unique and energetically consistent solutions. The predicted outcomes should be physically consistent. For this purpose we present a new methodology that uses the energetic coefficient of restitution, Stronge (1990).

2 Multiple collisions in an N-Ball Chain

2.1 Three ball chain

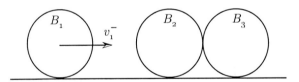

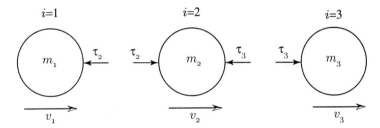

Fig. 1. Impulses

Consider the three balls that are depicted in Fig. (1). Ball B_1 strikes the other two balls (with initial velocities of v_2^- and v_3^-) that are in contact at time

$t = t^-$ with a velocity of v_1^- subject to $v_1^- \geq v_2^- \geq v_3^-$. The collision causes the two normal impulses τ_2 and τ_3 as shown in the figure. The problem at hand is to determine the post impact velocities v_1^+, v_2^+, and v_3^+. For this purpose one can write the conservation of linear momentum equations for the three balls, this yields:

$$m_1 \Delta v_1 = -\Delta \tau_2 \tag{1}$$

$$m_2 \Delta v_2 = \Delta \tau_2 - \Delta \tau_3 \tag{2}$$

$$m_3 \Delta v_3 = \Delta \tau_3 \tag{3}$$

where, Δv_i and $\Delta \tau_i$ are the changes in velocities and impulses as a result of the collision. Here, we have three equations in terms of the three post impact velocities and the two changes that occur in normal impulses. Additional assumptions are needed to obtain the two additional equations that are necessary to solve the problem. One can use coefficients of restitution between pairs of balls to resolve this problem. Since the balls can be treated as particles, the kinematic coefficient of restitution e_2^k between B_2 and B_3 can be used to obtain one additional velocity relationship:

$$v_1^+ - v_2^+ = -e_2^k v_1^- \tag{4}$$

where the superscript "+" denotes the quantity at the end of the collision. One encounters problems in applying the restitution law between B_2 and B_3 because two assumptions regarding their contact situation at the instant of collision yield different solutions. Accordingly, Han and Gilmore (1993), report the following solutions for three equal mass balls and $e = 1$, $v_1^- = 1$, $v_2^- = 0$, and $v_3^- = 0$:

$$v_1^+ = -\frac{1}{3} \quad v_2^+ = v_3^+ = \frac{2}{3} \tag{5}$$

$$v_1^+ = v_2^+ = 0 \quad v_3^+ = 1 \tag{6}$$

The former solution is obtained by assuming that B_2 and B_3 are in contact when B_1 strikes and they can be treated as a single mass ($v_2^+ = v_3^+$). The second solution, on the other hand, is obtained by assuming that B_2 and B_3 are not in contact when B_1 strikes, which leads to the impulse condition $\tau_3 = 0$. Having two equally possible solutions poses a serious difficulty in accepting this approach as a valid way of solving this problem.

2.2 Impulse Correlation Ratio

We now consider the compliant model that is presented in Fig. (2), which is the example considered in Brogliato (1996). For simplicity, we will choose $m_1 = m_2 = m_3 = v_1^- = 1$ and $v_2^- = v_3^- = 0$. When all the balls are in contact, their displacements can be obtained as follows:

$$q_1 = \frac{t}{3} - \frac{k(2\gamma - \gamma_1)\sin\left(\sqrt{k\gamma_1}t\right)}{(k\gamma_1)^{3/2}(\gamma_1 - \gamma_2)} + \frac{k(2\gamma - \gamma_2)\sin\left(\sqrt{k\gamma_2}t\right)}{(k\gamma_2)^{3/2}(\gamma_1 - \gamma_2)} \tag{7}$$

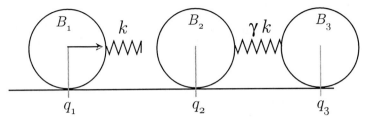

Fig. 2. The compliant, three ball chain.

$$q_2 = \frac{t}{3} + \frac{k(\gamma - \gamma_1)\sin\left(\sqrt{k\gamma_1}t\right)}{(k\gamma_1)^{3/2}(\gamma_1 - \gamma_2)} - \frac{k(\gamma - \gamma_2)\sin\left(\sqrt{k\gamma_2}t\right)}{(k\gamma_2)^{3/2}(\gamma_1 - \gamma_2)} \tag{8}$$

$$q_3 = \frac{t}{3} + \frac{k\gamma \sin\left(\sqrt{k\gamma_1}t\right)}{(k\gamma_1)^{3/2}(\gamma_1 - \gamma_2)} - \frac{k\gamma \sin\left(\sqrt{k\gamma_2}t\right)}{(k\gamma_2)^{3/2}(\gamma_1 - \gamma_2)} \tag{9}$$

where $\gamma_1 = 1 + \gamma + \sqrt{1 - \gamma + \gamma^2}$ and $\gamma_2 = 1 + \gamma - \sqrt{1 - \gamma + \gamma^2}$. The left and right impulses acting on B_2 can be written as follows:

$$\Delta\tau_2 = \int_0^t k(q_2 - q_1)dt$$
$$= \frac{(3\gamma - 2\gamma_1)\gamma_2^2 \sin^2\left(\frac{1}{2}\sqrt{k\gamma_1}t\right) - (3\gamma - 2\gamma_2)\gamma_1^2 \sin^2\left(\frac{1}{2}\sqrt{k\gamma_2}t\right)}{\frac{1}{2}\gamma_1^2\gamma_2^2(\gamma_1 - \gamma_2)} \tag{10}$$

$$\Delta\tau_3 = \int_0^t k\gamma(q_3 - q_2)dt = \frac{\left[1 - \cos\left(\sqrt{k\gamma_1}t\right)\right]\gamma\gamma_2 - \left[1 - \cos\left(\sqrt{k\gamma_2}t\right)\right]\gamma\gamma_1}{\gamma_1\gamma_2(\gamma_1 - \gamma_2)} \tag{11}$$

Now, we investigate the relationship between $\Delta\tau_2$ and $\Delta\tau_3$. First we form the following linear relationship between the impulses:

$$\delta = \alpha_2 \Delta\tau_2 + \Delta\tau_3 \tag{12}$$

Substituting Eqs. (10) and (11) into (12) and simplifying we obtain:

$$\delta = \frac{2\left[3\alpha_2\gamma + (\gamma - 2\alpha_2)\gamma_1\right]}{\gamma_1^2(\gamma_1 - \gamma_2)}\sin^2\left(\frac{1}{2}\sqrt{k\gamma_1}t\right)$$
$$- \frac{2\left[3\alpha_2\gamma + (\gamma - 2\alpha_2)\gamma_2\right]}{\gamma_2^2(\gamma_1 - \gamma_2)}\sin^2\left(\frac{1}{2}\sqrt{k\gamma_2}t\right) \tag{13}$$

Considering the first extreme case when the first spring is much stiffer than the second ($\gamma \ll 1$) in Eq. (13) yields:

$$\delta_1 \approx -\frac{(\alpha_2 + \gamma)\left[1 - \cos\left(\sqrt{\alpha kt}\right)\right]}{2\gamma} = 0 \quad \text{for } \alpha_2 = -\gamma \tag{14}$$

Thus, the impulses can be related as $\Delta\tau_3 = \gamma\Delta\tau_2$ when the first spring is much stiffer than the second. Next, we consider the other extreme, $\gamma \gg 1$ in Eq. (13)

yields:

$$\delta_2 \approx -\frac{2\gamma(1+2\alpha_2)\sin^2\left(\sqrt{3k/8t}\right)}{3\gamma} = 0 \quad \text{for } \alpha_2 = -1/2 \tag{15}$$

Once again we have a proportional relationship in the form of $\Delta\tau_3 = 1/2\Delta\tau_2$.

Based on the two extreme trends that we have shown, we form the following hypothesis that we assume is valid for triplets of balls:

> Consider a linear sequence of three balls: B_{i-1}, B_i, and B_{i+1}, where B_{i-1} impacts B_i while it is in contact with B_{i+1}. Then, the impulses that develop subsequent to the impact between the pair B_{i-1}-B_i and the pair B_i-B_{i+1} are proportional, and related through a constant $\alpha_i \geq 0$. This constant, we term as the Impulse Correlation Ratio, depends on the specific sequence, mass and material properties of the three adjacent balls.

We can apply this hypothesis to the following two possible cases that may arise during the multi-body collisions of a linear chain:

1. *Forward Impact:* This situation arises when B_i establishes contact and initiates a collisions with B_{i+1} during the impact of B_{i-1} and B_i. Then, for the forward impact we have:

$$\Delta\tau_{i+1} = \alpha_i \Delta\tau_i \tag{16}$$

2. *Backward Impact:* This situation arises when B_i establishes contact and initiates a collision with B_{i-1} during the impact of B_i and B_{i+1}. Thus, for the backward impact we have:

$$\Delta\tau_i = \alpha_i \Delta\tau_{i+1} \tag{17}$$

The two definitions allow us to establish causality among the impulses according to the impact direction. In addition, the definitions should be understood in the context of the impact direction, as in this chapter we assume that the impact propagates in the increasing direction of ball indices. Adaptation of these definitions to cases with reverse impact direction is a matter of reversing the indexing.

In the next subsection, we use the momentum based approach and the Impulse Correlation Ratio, to formulate an energetically consistent solution to obtain the post impact velocities.

2.3 The solution method

Solving Eqs. (1), (2), (3), and (16) for the velocity changes Δv_1, Δv_2, in terms of $\Delta\tau_2$ and Δv_2, Δv_3 in terms of $\Delta\tau_3$ yields:

$$\Delta v_1 = -\frac{1}{m_1}\Delta\tau_2 \tag{18}$$

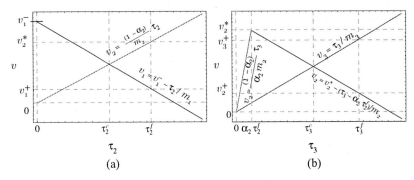

Fig. 3. Velocity-Impulse graphs

$$\Delta v_2 = \frac{1-\alpha_2}{m_2}\Delta\tau_2 = \frac{1-\alpha_2}{\alpha_2 m_2}\Delta\tau_3 \tag{19}$$

$$\Delta v_3 = \frac{1}{m_3}\Delta\tau_3 \tag{20}$$

Figure (3.a) and (3.b) depict the velocity-impulse diagrams for the collisions between the balls B_1, B_2 and B_2, B_3 respectively. In this study we use the energetic coefficient of restitution (Stronge, 1990), to determine the terminal impulse for each collision. Accordingly, for the first diagram we have (see Figure (3.a)):

$$v_1 = v_1^- - \frac{1}{m_1}\tau_2 \tag{21}$$

$$v_2 = v_2^- + \frac{1-\alpha_2}{m_2}\tau_2 \tag{22}$$

Now, setting $v_2 = v_1$ and solving for the maximum compression impulse τ_2^c, results in the following expression:

$$\tau_2^c = \frac{m_1 m_2(v_1^- - v_2^-)}{(1-\alpha_2)m_1 + m_2} \tag{23}$$

Next, we compute the work done during the compression and restitution phases and use the definition of the energetic coefficient of restitution to get the following equation:

$$e_2^2 \int_0^{\tau_2^c}(v_1 - v_2)d\tau_2 + \int_{\tau_2^c}^{\tau_2^f}(v_1 - v_2)d\tau_2 = 0 \tag{24}$$

where e_2 is the coefficient of restitution between B_1 and B_2. Solving Eq. (24) for τ_2^f yields:

$$\tau_2^f = \frac{(1+e_2)m_1 m_2(v_1^- - v_2^-)}{(1-\alpha_2)m_1 + m_2} \tag{25}$$

Moving on to the second diagram (see Figure (3.b)) where we have:

$$v_2 = \begin{cases} v_2^- + \frac{1-\alpha_2}{\alpha_2 m_2}\tau_3 & \text{if } 0 \leq \tau_3 \leq \alpha_2 \tau_2^f \\ v_2^* - \frac{\tau_3 - \alpha_2 \tau_2^f}{m_2} & \text{if } \tau_3 \geq \alpha_2 \tau_2^f \end{cases} \tag{26}$$

where,

$$v_2^* = v_2^- + \frac{(1-\alpha_2)(1+e_2)m_1(v_1^- - v_2^-)}{(1-\alpha_2)m_1 + m_2} \tag{27}$$

Note that, when $\tau_3 \geq \alpha_2 \tau_2^f$ the impact between B_1 and B_2 is over. Thus, only the right impulse, τ_3 acts on the ball B_2 $(-m_2(v_2 - v_2^*) = \tau_3 - \alpha_2 \tau_2^f)$. This leads to the second part of the expression given for v_2 in Eq. (26). Finally the expression for v_3 can be written as follows:

$$v_3 = v_3^- + \frac{1}{m_3}\tau_3 \tag{28}$$

Once again, we compute the maximum compression impulse between B_2 and B_3 by setting $v_3 = v_2$, which yields:

$$\tau_3^c = \frac{\left\{(v_1^- - v_2^-)(1+e_2)m_1 + (v_2^- - v_3^-)[m_1(1-\alpha_2) + m_2]\right\} m_2 m_3}{[(1-\alpha_2)m_1 + m_2](m_2 + m_3)} \tag{29}$$

Now, set up the energy equation as:

$$e_3^2 \int_0^{\tau_3^c} (v_2 - v_3)d\tau_3 + \int_{\tau_3^c}^{\tau_3^f} (v_2 - v_3)d\tau_3 = 0 \tag{30}$$

where e_3 is the coefficient of restitution between B_2 and B_3. Solving Eq. (30) for τ_3^f yields:

$$\tau_3^f = \left\{(v_1^- - v_2^-)(1+e_2)m_1 + (v_2^- - v_3^-)[m_1(1-\alpha_2) + m_2]\right\} m_2 m_3$$

$$\frac{1 + e_3\sqrt{1 - \alpha_2(\frac{m_2}{m_3} + 1)\left[1 + \frac{(v_2^- - v_3^-)}{(v_1^- - v_2^-)}\frac{(1-\alpha_2)m_1 + m_2}{(1+e_2)m_1}\right]^{-2}}}{[(1-\alpha_2)m_1 + m_2](m_2 + m_3)} \tag{31}$$

The post impact velocities can now be computed as follows:

$$v_1^+ = v_1^- - \frac{1}{m_1}\tau_2^f \tag{32}$$

$$v_2^+ = v_2^* - \frac{\tau_3^f - \alpha_2 \tau_2^f}{m_2} \tag{33}$$

$$v_3^+ = v_3^- + \frac{1}{m_3}\tau_3^f \tag{34}$$

We note that the positiveness of the inside of the radical in Eq. (31) can be used to establish an upper bound on the correlation ratio α_2. If we consider

$\alpha_2 \le 1$, and impose the two underlying conditions $v_1^- - v_2^- \ge 0$ and $v_2^- - v_3^- \ge 0$ we can write the upper limit of the correlation ratio as follows:

$$0 \le \alpha_2 \le \frac{m_3}{m_2 + m_3} \tag{35}$$

Figure (4) depicts the three post impact velocities for $m_1 = m_2 = m_3 = e_2 = e_3 = v_1^- = 1$ as α_2 is varied in its limiting interval $0 \le \alpha_2 \le 0.5$. We note that the two outcomes that are produced by the solution method of Han and Gilmore (1993), corresponds to the two limiting values of the Impulse Correlation Ratio. Our method exposes a spectrum of solutions that bridge the gap between the two limiting outcomes. By specifying the value of the Impulse Correlation Ratio, one can obtain a unique solution for the problem. The question of the validity of the ratio as a material constant remains to be answered. In the latter part of this section we will present the results of an experimental study that addresses this issue.

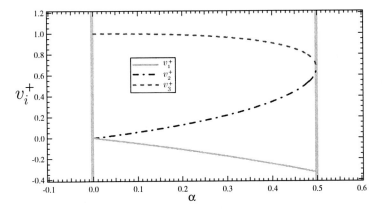

Fig. 4. Effect of α on the post impact velocities

2.4 Multiple impacts

To explain the multiple impacts that may arise during the present problem, we consider a specific example that leads to the velocity impulse diagrams that are depicted in Fig. (5). The example corresponds to a three ball case with initial velocities of $v_1^- = 1$ m/s, $v_2^- = 0.8$ m/s, and $v_3^- = 0$ m/s. At the onset of the collision, we have the ball B_2 having simultaneous impacts with B_1 and B_3. The first B_1-B_2 collision takes place during the impulse intervals $0 \le \tau_2 \le \tau_{2,1}^f$ ($0 \le \tau_3 \le \tau_{3,1}^* = \alpha_2 \tau_{2,1}^f$). Meanwhile, the first B_2-B_3 collision takes place during the impulse intervals $0 \le \tau_2 \le \tau_{2,1}^*$ ($0 \le \tau_3 \le \tau_{3,1}^f$). When the B_1-B_2 impact ends, the velocity of B_2 continues to decrease as result of the continuing collision between B_2 and B_3 during the interval $\tau_{3,1}^* \leq \tau_3 \leq \tau_{3,2}^*$. The decrease in v_2 during this interval can be observed as the sudden change in the velocity on the diagram that corresponds to τ_2. This is true because during this interval B_1 and B_2 do not interact, and thus the impulse τ_2 between these balls remains constant. Subsequently, the slowdown in the velocity v_2 initiates a backward impact between B_1 and B_2 when $v_2 = v_1 = v_{1,1}^+$ at $\tau_3 = \tau_{3,2}^*$. During the next interval $\tau_{2,1}^f \leq \tau_2 \leq \tau_{2,1}^*$ ($\tau_{3,2}^* \leq \tau_3 \leq \tau_{3,1}^f$) we have simultaneous collisions between B_1-B_2 and B_2-B_3. Next, the first B_2-B_3 collision ends at $\tau_2 = \tau_{2,1}^*$ ($\tau_3 = \tau_{3,1}^f$). But, when this collision ends the velocity of B_2 continues to increase as a result of the collisions between B_1 and B_2. Once again, the increase in this velocity can be observed as the vertical jump at $\tau_3 = \tau_{3,1}^f$ on the τ_3 diagram. Subsequently, the increase in v_2 leads to a forward collision between B_2 and B_3 at $v_2 = v_3 = v_{3,1}^+$ and $\tau_2 = \tau_{2,2}^*$. Then, we have simultaneous B_1-B_2, B_2-B_3 impacts during the impulse interval $\tau_{2,2}^* \leq \tau_2 \leq \tau_{2,2}^f$ ($\tau_{3,1}^f \leq \tau_3 \leq \tau_{3,3}^*$). Finally, the B_2-B_3 collision continues during the interval $\tau_{3,3}^* \leq \tau_3 \leq \tau_{3,2}^f$, which is also reflected as the sudden jump in v_2 on the τ_2 diagram. The overall process ends at $\tau_2 = \tau_{2,2}^f$ ($\tau_3 = \tau_{3,2}^f$) since $v_{2,4}^+ < v_{1,2}^+$.

Now, using the Impulse Correlation Ratio according to the rules that were enumerated in section 2.2 along with Eqs. (1), (2), and (3) we may express the velocities as follows:

$$v_1 = \begin{cases} v_1^- - \tau_2/m_1 & \text{if } 0 \leq \tau_2 < \tau_{2,2}^f \\ v_{1,2}^+ & \text{if } \tau_2 \geq \tau_{2,2}^f \end{cases} \tag{36}$$

$$v_2 = \begin{cases} v_2^- + (1 - \alpha_2)\frac{\tau_2}{m_2} & \text{if } 0 \leq \tau_2 < \tau_{2,1}^f \\ v_{1,1}^+ + (1 - \frac{1}{\alpha_2})\frac{\tau_2 - \tau_{2,1}^f}{m_2} & \text{if } \tau_{2,1}^f \leq \tau_2 \leq \tau_{2,1}^* \\ v_{2,2}^+ + \frac{\tau_2 - \tau_{2,1}^*}{m_2} & \text{if } \tau_{2,1}^* \leq \tau_2 \leq \tau_{2,2}^* \\ v_{3,1}^+ + \frac{\tau_2 - \tau_{2,2}^*}{m_2} & \text{if } \tau_{2,2}^* \leq \tau_2 < \tau_{2,2}^f \\ v_{2,4}^+ & \text{if } \tau_2 \geq \tau_{2,2}^f \end{cases} \tag{37}$$

$$v_2 = \begin{cases} v_2^- + (\frac{1}{\alpha_2} - 1)\frac{\tau_3}{m_2} & \text{if } 0 \leq \tau_3 \leq \tau_{3,1}^* \\ v_{2,1}^+ - \frac{\tau_3 - \tau_{3,1}^*}{m_2} & \text{if } \tau_{3,1}^* \leq \tau_3 \leq \tau_{3,2}^* \\ v_{1,1}^+ + (\alpha_2 - 1)\frac{\tau_3 - \tau_{3,2}^*}{m_2} & \text{if } \tau_{3,2}^* \leq \tau_3 < \tau_{3,1}^f \\ v_{3,1}^+ + (\frac{1}{\alpha_2} - 1)\frac{\tau_3 - \tau_{3,1}^f}{m_2} & \text{if } \tau_{3,1}^f \leq \tau_3 < \tau_{3,3}^* \\ v_{2,3}^+ - \frac{\tau_3 - \tau_{3,3}^*}{m_2} & \text{if } \tau_{3,3}^* \leq \tau_3 < \tau_{3,2}^f \\ v_{2,4}^+ & \text{if } \tau_3 \geq \tau_{3,2}^f \end{cases} \tag{38}$$

$$v_3 = \begin{cases} v_3^- + \tau_3/m_3 & \text{if } 0 \leq \tau_3 < \tau_{3,2}^f \\ v_{3,2}^+ & \text{if } \tau_3 \geq \tau_{3,2}^f \end{cases} \tag{39}$$

The specific computation of final impulses is carried out using the energetic definition of the coefficient of restitution. The respective final impulses and velocities can be found by sequentially solving the following equations for $\tau_{2,1}^f$, $\tau_{3,1}^f$,

$\tau_{2,2}^f$, $\tau_{3,2}^f$:

$$e_2^2 \int_0^{\tau_{2,1}^c} (v_1 - v_2)d\tau_2 + \int_{\tau_{2,1}^c}^{\tau_{2,1}^f} (v_1 - v_2)d\tau_2 = 0 \tag{40}$$

$$e_3^2 \int_0^{\tau_{3,1}^c} (v_2 - v_3)d\tau_3 + \int_{\tau_{3,1}^c}^{\tau_{3,1}^f} (v_2 - v_3)d\tau_3 = 0 \tag{41}$$

$$e_2^2 \int_{\tau_{2,1}^f}^{\tau_{2,2}^c} (v_1 - v_2)d\tau_2 + \int_{\tau_{2,2}^c}^{\tau_{2,2}^f} (v_1 - v_2)d\tau_2 = 0 \tag{42}$$

$$e_3^2 \int_{\tau_{3,1}^f}^{\tau_{3,2}^c} (v_2 - v_3)d\tau_3 + \int_{\tau_{3,2}^c}^{\tau_{3,2}^f} (v_2 - v_3)d\tau_3 = 0 \tag{43}$$

where $\tau_{2,1}^c$, $\tau_{2,2}^c$, $\tau_{3,1}^c$, and $\tau_{3,2}^c$ are the maximum compression impulses as shown in Fig. (5). It should be obvious from this example that one may have to go through a complex set of computations in order to solve even this simple example. For this purpose, we have written a computer program using the software package Mathematica that automatically performs the computations required to compute the final impulses, the maximum compression impulses, etc.

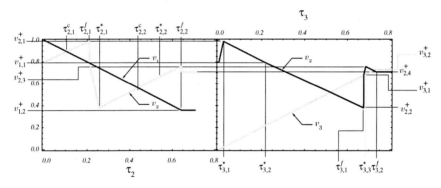

Fig. 5. Multiple impacts on the velocity-impulse diagram

2.5 Post impact bouncing patterns of a three ball cradle

Having developed our solution method, now we study the effect of various parameters on the possible bouncing patterns that can be exhibited by the three ball case. Here, we let $v_2^- = v_3^- = 0$, and used $\alpha_2 = 0.15$, $e_2 = e_3 = 0.5$ in order to obtain the regions in Fig. (6). Yet, we should point out that the regions would be qualitatively preserved if one chooses to use different values of impulse correlation ratio and coefficients of restitution.

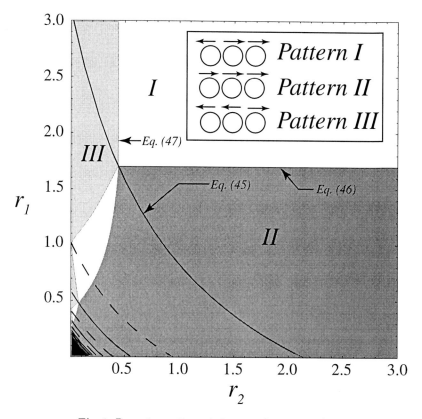

Fig. 6. Bouncing pattern regions on the $r_1 - r_2$ plane

Now, we divide the post impact patterns according to occurrence of multiple collisions. Multiple collisions exist if we have at least one back impact between B_2 and B_1. This occurs if $v_1^+ = v_2$ during the B_2-B_3 impact. The impulse condition that would lead to this situation can be obtained by setting $v_1^+ = v_2$ and solving for τ_3, which yields:

$$\tau_{3,1}^* = \frac{m_2(\alpha_2 m_1 + e_2(m_1 + m_2))v_1^-}{(1 - \alpha_2)m_1 + m_2} \qquad (44)$$

Multiple impacts take place only when $\alpha_2 \tau_{2,1}^f \leq \tau_{3,1}^* \leq \tau_{3,1}^f$, which leads to the following condition:

$$r_1 \leq \frac{(1 + e_2)(1 + e_3\sqrt{1 - \alpha_2(1 + r_2)})}{e_2(1 + r_2)} - (1 + \frac{\alpha_2}{e_2}) \qquad (45)$$

where $r_1 = \frac{m_2}{m_1}$ and $r_2 = \frac{m_2}{m_3}$. Thus, we may divide the $r_1 - r_2$ plane into two regions that are below and above the upper-most curve shown in Fig. (6). The region above this curve is the distinct collision area, while in the lower region we have at least one back impact. We may continue in the same fashion to obtain the boundary for the region where we have at least one forward impact between B_2 and B_3 as a result of the back impact between B_2 and B_1. We obtain a condition for the forward impact in a similar manner to the one we obtained Eq. (45). This results in the dashed curve that is situated below the topmost curve. Then, we can continue in the same manner to obtain further alternating boundaries that lead to more pairs of back and forward impacts.

We may also partition the $r_1 - r_2$ plane into three regions that correspond to post-impact velocity directions. Each of these regions corresponds to a unique post-impact bouncing pattern (see Fig. (6)). The expression for the line that separates regions I and II in the distinct-collision region can be obtained by setting v_1^+ in Eq. (32) equal to zero and solving for r_1 as:

$$r_1 = \frac{1 - \alpha_2}{e_2} \tag{46}$$

The line that separates regions I and III in the distinct-collision region can be obtained by setting v_2^+ in Eq. (33) equal to zero and solving for r_2 as:

$$r_2 = e_3 \sqrt{1 - \alpha_2 + \left(\frac{1}{2}\alpha_2 e_3\right)^2} - \frac{1}{2}\alpha_2 e_3^2 \tag{47}$$

We may obtain similar partitioning in the multiple collisions zones by setting respective velocity pairs equal to one another and obtaining conditions.

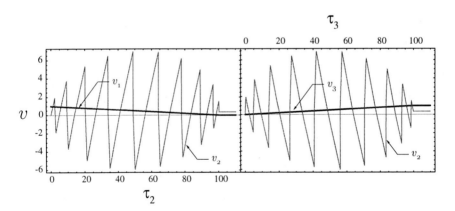

Fig. 7. Velocity-impulse diagram of a small center ball example

Figure (6) depicts all possible solutions and demonstrates the consistency and uniqueness of the solutions that are obtained by the method proposed in this chapter. One particularly interesting region of the parameter plane is the lower left corner where we have a dense set of multiple collisions. This region corresponds to configurations where the middle ball (B_2) is significantly lighter than the other two balls. Figure (7) depicts the impulse velocity diagrams of a three ball cradle with $v_1^- = 1$ m/s, $e_2 = e_3 = 1$, $\alpha_2 = 0.15$, and $r_1 = r_2 = 0.01$. The post impact velocities computed for this case are $v_1^+ = -0.003$ m/s, $v_2^+ = 0.348$ m/s, and $v_3^+ = 0.999$ m/s. In this example, we have a very small central ball and two very large peripheral balls. If we have neglected the central ball then the resulting post impact velocities of the large balls would have been 0 and 1 m/s respectively. The presence of the central ball does not effect this outcome, yet it introduces several micro-collisions that transmit the momentum from the incident large ball to the other. As the central ball becomes smaller we may expect more multiple collisions to arise, which is why we see a dense region of separator curves in the lower left of Fig. (6).

2.6 Generalization of the three ball approach to N-balls

The system under consideration consists of N balls of given masses m_i. The last $N-1$ balls are arranged in a chain such that all consecutive pairs of balls are in contact. One ball strikes one end of the chain with a non-zero pre-impact velocity of v_1^-, while the other balls are in contact with($v_i^- \geq v_{i+1}^-$ for $i = 2, N$). As in the previous section, our objective is to determine the post impact velocities v_i^+ of the balls.

We start by writing the conservation of momentum equations for any consecutive triplets of balls, this yields the following N equations:

$$m_i \Delta v_i = \Delta \tau_i - \Delta \tau_{i+1} \quad \text{for } i = 1, N \tag{48}$$

with

$$\Delta \tau_1 = \Delta \tau_{N+1} = 0 \tag{49}$$

where, Δv_i and $\Delta \tau_i$ are the changes in velocities and impulses as a result of the collision. The Impulse Correlation Ratios for each ball B_i can be used to obtain additional $N-2$ equations in the following form:

$$\Delta \tau_i = \begin{cases} \alpha_{i-1} \Delta \tau_{i-1} \text{ for some } i \in \mathcal{I}(q) \subseteq \{1, \dots, m\} \\ \alpha_i \Delta \tau_{i+1} \quad \text{for } i \notin \mathcal{I}(q) \end{cases} \tag{50}$$

Solving for Δv_i yields solutions of the form:

$$\Delta v_1 = -\frac{1}{m_1} \Delta \tau_2 \tag{51}$$

$$\Delta v_i = \begin{cases} (1-\alpha_i)/m_i \Delta \tau_i = (1-\alpha_i)/(\alpha_i m_i) \Delta \tau_{i+1} \text{ for } i \in \mathcal{I}(q) \subseteq \{1, \dots, m\} \\ (\alpha_i - 1)/(\alpha_i m_i) \Delta \tau_i = (\alpha_i - 1)/m_i \Delta \tau_{i+1} \text{ for } i \notin \mathcal{I}(q) \end{cases} \tag{52}$$

$$\Delta v_N = \frac{1}{m_N} \Delta \tau_N \tag{53}$$

The next step in solving the impact problem is to proceed as we have done for the three ball case. That is, the velocities should be tracked on the impulse diagrams. The points where the balls lose and re-establish contact should be calculated and the related impulse expressions should be adjusted properly. As we have mentioned in the three ball case, this is a very complex procedure and we have developed a Mathematica package that will automatically set up the calculations for N number of balls.

Finally, a symbolic analysis of the velocity outcomes similar to the three ball case yields the following bounds on the Impulse Correlation Ratios for the N ball case:

$$0 \leq \alpha_i \leq \frac{m_{i+1}}{m_{i+1} + m_i(1 - \alpha_{i+1})} \quad \text{for } i = 2, N-1 \quad \text{with } \alpha_N = 0 \qquad (54)$$

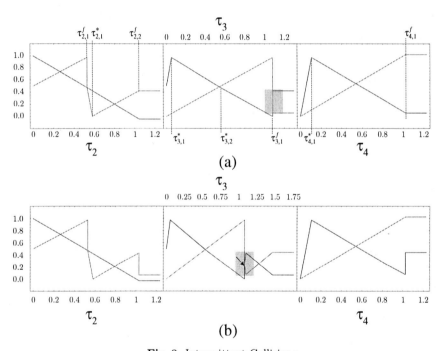

Fig. 8. Intermittent Collisions

2.7 Intermittent collisions

There is a special situation that may arise during collisions for chains with $N \geq 4$. Figure (8) depicts a 4 ball collision with $m_1 = m_2 = m_3 = m_4 = e_2 = e_3 = e_4 = 1$, $\alpha_2 = \alpha_3 = 0.1$, and initial velocities of $v_1 = 1, v_2 = 0.5, v_3 = v_4 = 0$

m/s. At the onset of the impact we have simultaneous collisions at all three contact points. The first B_1-B_2 collision takes place during the impulse intervals $0 \leq \tau_2 \leq \tau_{2,1}^f$ ($0 \leq \tau_3 \leq \tau_{3,1}^*$). Meanwhile, the first B_2-B_3 and B_3-B_4 collisions occur during $0 \leq \tau_3 \leq \tau_{3,1}^f$ ($0 \leq \tau_4 \leq \tau_{4,1}^*$) and $0 \leq \tau_4 \leq \tau_{4,1}^f$ respectively. Yet, the continuing B_2-B_3 impact leads to a second B_1-B_2 collision that causes the slope change at $\tau_{3,2}^*$ on the τ_3 diagram. The special case arises when the B_2-B_3 impact ends at $\tau_{3,1}^f$. At this point, we encounter an unusual case on the τ_3 diagram (see Fig. (8.a)). We observe an upward vertical jump in v_2 due to the continuing second B_1-B_2 collision and downward jump in v_3 caused by the ongoing B_3-B_4 impact. The problem that is encountered here is to determine the point where we have the onset of the second B_2-B_3 impact. If we continue the B_1-B_2 and B_3-B_4 collisions by ignoring the possible contact between B_2-B_3 we obtain the shaded region that is shown in Fig. (8.a)). The onset of the second B_2-B_3 collision will be somewhere in the shaded region where the two velocities overlap. When this situation arises, we assume that the two balls meet in the midpoint of the overlap region. This assumption leads to the final diagrams that are depicted in Fig.(8.b)).

2.8 Special Examples

In this section we will consider several interesting examples that will be very difficult to solve without the solution method presented in this chapter. Most of these cases can be very easily observed by using a simple Newton's cradle that can be purchased as toy from commercial vendors

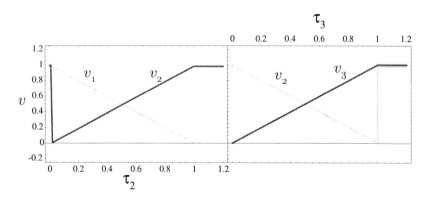

Fig. 9. Three ball separation

I. Three ball separation: The first examples concerns a chain with three, equal mass balls such that $m_1 = m_2 = m_3 = 1$, $e_2 = e_3 = 1$, and $\alpha_2 = 0.01$ (these are typical restitution coefficients and Impulse Correlation Ratios for steel balls). The initial conditions are chosen such that $v_1^- = v_2^- = 1$ and $v_3^- = -1$ m/s. This means that the first two balls strike the last ball with an equal velocity. Figure (9) depicts the impulse diagrams. One may easily observe the separation of $B_1 - B_2$ and attachment of $B_2 - B_3$ during the collision. At the onset, there is a triple impact between all balls. First, $B_2 - B_3$ collision ends with the last ball having a velocity of 1 m/s and the middle ball having a zero velocity. Thus, the linear momentum of the middle ball is transferred to the last one. Subsequently, during the remainder of the impact period, the linear momentum of the first ball is transferred to the middle one. The resulting final velocities that are computed as: $v_1^+ = 0$ and $v_2^+ = v_3^+ = 1$ m/s. Effectively, the separation occurs through a forward and then backward impacts of the middle ball.

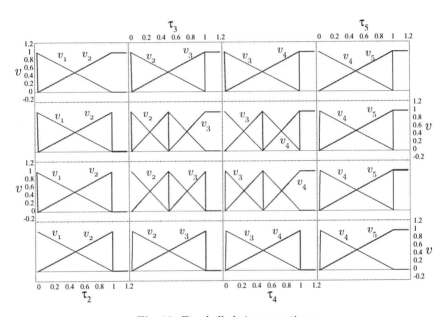

Fig. 10. Five ball chain separations

II. Five ball separations: The second example is similar to the first one we considered, but it involves five equal mass balls with coefficients of restitution and Impulse Correlation ratios identical to the previous case. Figure (10) depicts impulse diagrams for four cases. In the first case the first four balls of the chain strike the last ball with a common velocity of 1 m/s. As we can see from the figure, we have one collision for each ball pair. The order of linear momentum transfer is from B_4 to B_5, B_3 to B_4, B_2 to B_3, and B_1 to B_2. This results in the the following post impact velocities: $v_1^+ = 0$ and $v_2^+ = v_3^+ = v_4^+ = v_5^+ = 1$ m/s. The second case corresponds to the release of the first three balls with a common pre-impact velocity of 1 m/s while the last to balls being at rest. In this case, the inner balls have two impacts while the outer balls have one impact with one another. The order of linear momentum transfer for this case is not a straightforward as the previous one. One can, however can easily track the momentum transfers by studying Fig. (10), which yield in the following post-impact velocities: $v_1^+ = v_2^+ = 0$ and $v_3^+ = v_4^+ = v_5^+ = 1$ m/s. The third case corresponds to the first two balls striking the resting last three balls with a pre-impact velocity of 1 m/s. In this case one obtains $v_1^+ = v_2^+ = v_3^+ = 0$ and $v_4^+ = v_5^+ = 1$ m/s. In final case the first ball has a pre-impact velocity of 1 m/s, while the last four are at rest. One can observe that the linear momentum transfer is from B_1 to B_2, B_2 to B_3, B_3 to B_4, and B_4 to B_5. Then, the resulting post impact velocities are: $v_1^+ = v_2^+ = v_3^+ = v_4^+ = 0$ and $v_5^+ = 1$ m/s.

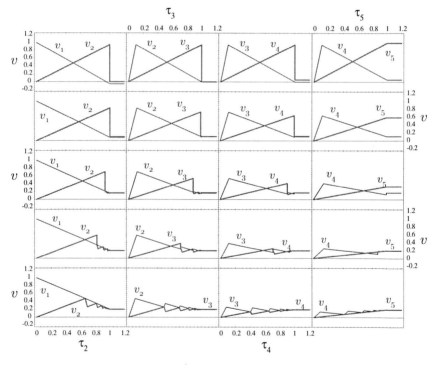

Fig. 11. Inelastic collisions of a Five ball chain

III. Inelastic collisions: The third example is conducted with five unit mass balls such that $e_2 = e_3 = e_4 = e_5 = 0.25n$ $(n = 0, \ldots, 4)$ and $\alpha_2 = \alpha_3 = \alpha_4 = 0.1$. For all cases the initial velocities are chose as $v_1^- = 1$ m/s and $v_2^- = v_3^- = v_4^- = v_5^- = 0$. Figure (11) depicts the impulse diagrams for

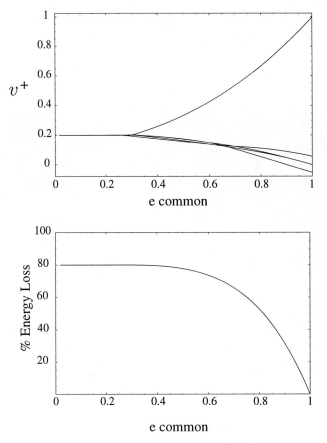

Fig. 12. Five ball chain. restitution effect

decreasing values of the uniform coefficient of restitution. Two characteristics of the post-impact results can be noted. First, as the collisions become more inelastic, the post impact velocities of the balls approach the same value. For a coefficient of restitution of 0.25 the post-impact velocities of the balls are given as $v_1^+ = v_2^+ = v_3^+ = v_4^+ = v_5^+ = 0.2$ m/s. In addition, as the coefficient of restitution is decreased, the energy loss approaches a constant level of 80%. These characteristics of Newton's cradle can be very easily observed in the real system. After a certain period of time, the balls start swing together. In addition, the impacts never deplete the energy of the system, since the motion always end as a collision free swinging of the balls. Figure (12) depicts the variation of the post impact velocities and the percent energy loss as a function of the uniform restitution coefficient. The second characteristic of the inelastic impacts is the emergence of a succession of micro-collisions between pairs of balls. (see Fig (11)). This was interesting to observe that the impulse diagrams can capture a vibrational event and exhibit the velocity oscillations during the impact process.

2.9 Experiments

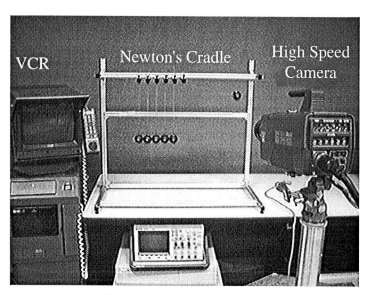

Fig. 13. The experimental set-up

The objective of the present study was to analyze multiple impacts in a multi-body system for the non-frictional case. For this purpose the classical collision experiment known as Newton's Cradle was set up (see Fig. (13)). Various chains with 3 to 6 balls of different masses and materials were arranged by suspending each ball from a frame using two threads. The strings are attached to the frame through sliding guides to ensure properly aligned balls before impact. Proper alignment of the mass centers ensured the elimination of the rotation of the balls and tangential forces at the points of impact. A chain of central impacts was generated by releasing the first ball in the chain from a pre-determined elevation.

The experimental data was captured by using a high speed video system capable of 1000 frames per second. Retro-reflective markers were used to mark the mass center of each ball. The relative positions of the balls before and after the impact were determined with respect to a fixed marker, used as reference. The acquired video images were transferred to a personal computer, where a specialized program was used to digitize the markers positions. The digitized positions were used to compute the dropping height of the first ball and the maximum post impact heights attained by each ball. Finally, the pre-impact velocity of the first ball and post impact velocities of all balls were computed from the calculated heights.

The experiments were conducted using three types of balls, designated as A, B, and C. The masses of the A, B, and C balls were 45, 53, and 53 grams respectively. An initial set of experiments with pairs of balls were conducted to determine the coefficients of restitution that are presented in Table. 1.

Ball Type	A	B	C
A	0.87	0.53	0.31
B	0.53	0.36	0.28
C	0.31	0.28	0.27

Table 1. Coefficients of restitution.

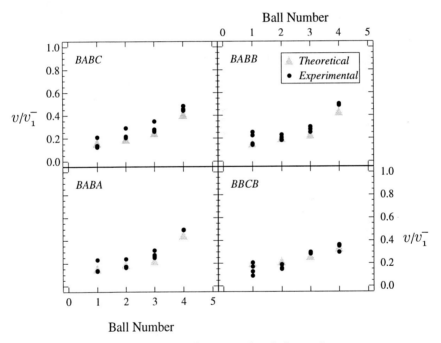

Fig. 14. Velocity predictions in four ball experiments

2.10   Experimental verification of the Impulse Correlation Ratio

The hypothesis of the Impulse Correlation Ratio is based on the fact that it is a constant which depends on the material and geometric properties of triplet of balls in contact. To check this hypothesis, we first determined the Impulse Correlation Ratios for all possible combinations of the three balls arranged in triplets. For this purpose, twenty seven experiments were conducted. Each experiment involved a chain of three balls. The analytical algorithm and the coefficients of restitution in Table 1 were used to compute the impulse correlation ratios that produced the best fit to the experimentally acquired post impact velocities. The resulting impulse correlation ratios are listed in Table 2.

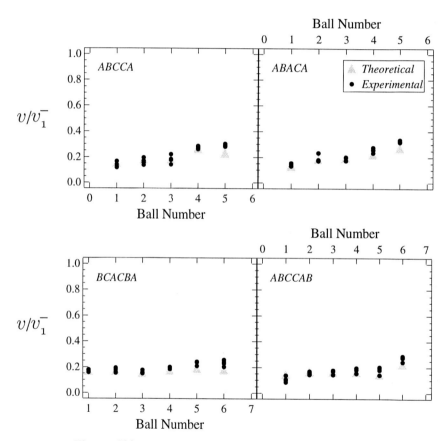

Fig. 15. Velocity predictions in five and six ball experiments.

Sequence	ICR	Sequence	ICR	Sequence	ICR
AAA	0.167	AAB	0.232	AAC	0.218
BAA	0.374	BAB	0.296	BAC	0.088
CAA	0.495	CAB	0.499	CAC	0.400
ABA	0.127	ABB	0.104	ABC	0.111
BBA	0.340	BBB	0.310	BBC	0.216
CBA	0.435	CBB	0.455	CBC	0.342
ACA	0.262	ACB	0.268	ACC	0.216
BCA	0.389	BCB	0.408	BCC	0.319
CCA	0.462	CCB	0.414	CCC	0.338

Table 2. Experimental Impulse Correlation Ratios.

Now, to test the hypothesis, we conducted eight sets of experiments: four with four ball, two with five ball, and two with six ball sequences. Each set of experiments included four individual experiments with four incident velocities of the first ball. The drop height of the first ball was adjusted such that we would have approximately 0.5, 1.0, 1.5, and 2.0 m/s pre impact velocities v_1^-. Next, we used the proposed analytical procedure and the experimental values listed in Tables 1 and 2 to compute the post impact velocities.

Fig. 16. Set-up of the small center ball experiment.

Figures (14) and (15)depict the experimental results. The figures depict the results of eight sets of experiments conducted with a specific sequences of balls as shown on individual graphs. The results clearly demonstrate that the theoretical outcomes are in agreement with the experimental results.

2.11 Experimental observation of the small center ball behavior

A final experiment was set up to observe a behavior that was considered in section 2.5. This was a three ball case with a relatively small central ball. The analytical solution resulted in velocity outcomes that are depicted in Fig. (7). The remarkable aspect of this case is the high number of impacts that were seen among the three balls. The present experiment was conducted to physically verify this observation. For this purpose, a three ball chain, with metallic balls was formed (see Fig. (16)). Then, a special electrical circuit was built in order mark periods of contact among the three balls. Figure (17) depicts the voltages that are measured for the collision of two balls (excluding the small center ball) and three balls. The voltage profiles clearly exhibit the multiple impacts that take place between the small center ball and the two larger outer balls.

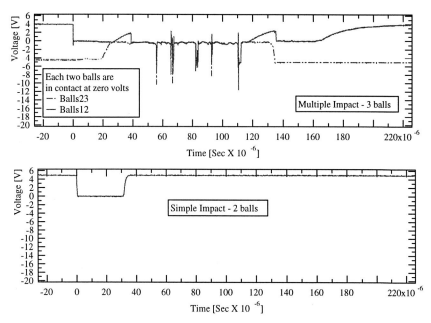

Fig. 17. Contact voltages of the small center ball experiment.

3 Multiple collisions of a rocking block

3.1 Problem Description

We consider the collision of a rigid block with a massive external surface that is schematically depicted in Fig. (18). The collision takes place at the right edge O_2 (see Fig. (18.a)) while the block is in contact with the the surface at the left edge O_1. Immediately before impact, we assume that the block is undergoing a non-centroidal rotation about O_1 with an angular velocity of ω^- (the superscripts − and + denote pre- and post-impact quantities respectively). Our goal is to compute the post-impact linear and angular velocities of the block in terms of the pre-impact angular velocity. To simplify the problem, we assume that the

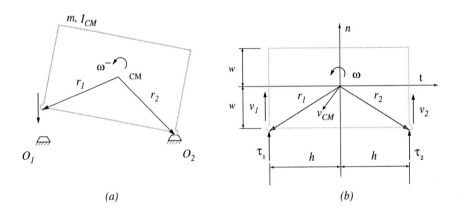

(a)   (b)

Fig. 18. Rigid block striking the external plane at O_1 (a) and Free body diagram of the block during the impact (b)

block and the external plane are smooth at the contact points such that we have frictionless impact. Accordingly, the tangential components of the impulses remain equal to zero throughout the collision process.

As shown in Fig. (18.b), the origin of the reference coordinate system is attached to center of mass of the block. The counter-clockwise direction of the angular velocity is considered to be positive. We assume that each the two edges where the impact occurs (at O_1 and O_2), are symmetrically displaced from the Center of Mass (CM) at a distance of h. At $t = t^-$ the edge O_1 collides with the surface with an initial velocity $v_1^- < 0$, while at edge O_2, $v_2^- = 0$. Using the laws of conservation of the linear and angular impulse and momentum yields:

$$mv_{CM} = \tau_1 + \tau_2 \tag{55}$$

$$I_{CM}\omega = -(\tau_1 - \tau_2)h \qquad (56)$$

where m and I_{CM} are the mass and centroidal moment of inertia of the block respectively and the subscript n denotes the normal direction. One can write the kinematic relations between the linear velocities at the two contact points and the velocity at the center of mass as follows:

$$\boldsymbol{v_1} = \boldsymbol{v}_{CM} + \boldsymbol{\omega} \times \boldsymbol{r_1} \qquad (57)$$

$$\boldsymbol{v_2} = \boldsymbol{v}_{CM} + \boldsymbol{\omega} \times \boldsymbol{r_2} \qquad (58)$$

The restitution law for the normal impact at O_1 can be written in the following general form:

$$R(\boldsymbol{v}_1^+, \boldsymbol{v}_1^-, e) = 0 \qquad (59)$$

where R is a certain restitution rule with a coefficient of e. A similar equation cannot be formed for the non-colliding edge O_2 because the pre-impact normal velocity there is equal to zero. The system formed by eight equations, Eq. (55) to Eq. (59) includes nine unknowns (i.e. the two impulses (τ_1 and τ_2), the post impact angular velocity (ω), and the three post impact velocity vectors: $\boldsymbol{v}_{CM}$, $\boldsymbol{v}_1$, and $\boldsymbol{v}_2$). Therefore, this multiple impact problem with two contact points cannot be solved using the classical approach.

3.2   Impulse Correlation Ratio

As in the case of three ball chain (section 2.2), now seek derive a relationship between the two impulses at the two contact points. The objective is to find out whether the impulses generated by the elastic forces at the two edges are related. For simplicity, a block of unit mass and dimensions are picked and it has been assumed that the edge O_1 strikes the external plane with an initial velocity $v_1^- = -1$ (see Fig. (19)).

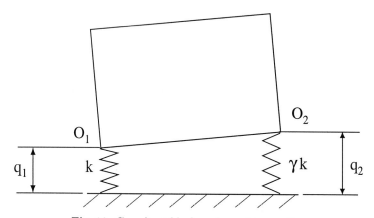

Fig. 19. Compliant block under rocking motion

The displacements at the two ends of the block can be computed as follows:

$$q_1 = \frac{1}{48\sqrt{k}(\gamma_1 - \gamma_2)}\left[\sin\left(\frac{1}{2}\sqrt{k\gamma_1}t\right)\sqrt{\gamma_1}(46 - 50\gamma + 5\gamma_1 - 5\gamma_2)\right.$$
$$\left. + \sin\left(\frac{1}{2}\sqrt{k\gamma_2}t\right)\sqrt{\gamma_2}(-46 + 50\gamma + 5\gamma_1 - 5\gamma_2)\right] \tag{60}$$

$$q_2 = \frac{1}{48\sqrt{k}(\gamma_1 - \gamma_2)}\left[\sin\left(\frac{1}{2}\sqrt{k\gamma_1}t\right)\sqrt{\gamma_1}(-10 - 10\gamma + \gamma_1 + \gamma_2)\right.$$
$$\left. - \sin\left(\frac{1}{2}\sqrt{k\gamma_2}t\right)\sqrt{\gamma_2}(10 + 10\gamma + \gamma_1 - \gamma_2)\right] \tag{61}$$

where $\gamma_1 = 5 + 5\gamma + \sqrt{25 - 46\gamma + 25\gamma^2}$ and $\gamma_2 = -5 - 5\gamma + \sqrt{25 - 46\gamma + 25\gamma^2}$. The impulses of the elastic forces at the two contact points can now be written as:

$$\Delta\tau_1 = \int_0^t kq_1 dt$$
$$= \frac{5}{12} + \frac{\cos\left(\frac{1}{2}\sqrt{k\gamma_1}t\right)(46 - 50\gamma - 5\gamma_1 + 5\gamma_2)}{24(\gamma_1 - \gamma_2)}$$
$$+ \frac{\cos\left(\frac{1}{2}\sqrt{k\gamma_2}t\right)(-46 + 50\gamma - 5\gamma_1 + 5\gamma_2)}{24(\gamma_1 - \gamma_2)} \tag{62}$$

$$\Delta\tau_2 = \int_0^t k\gamma q_2 dt$$
$$= \frac{1}{12} + \frac{\cos\left(\frac{1}{2}\sqrt{k\gamma_1}t\right)(10 + 10\gamma - \gamma_1 + \gamma_2)}{24(\gamma_1 - \gamma_2)}$$
$$- \frac{\cos\left(\frac{1}{2}\sqrt{k\gamma_2}t\right)(10 + 10\gamma + \gamma_1 - \gamma_2)}{24(\gamma_1 - \gamma_2)} \tag{63}$$

Next, in order to examine whether a relationship between the two normal impulses $\Delta\tau_1$ and $\Delta\tau_2$ exists, we propose the following linear expression:

$$\delta = \alpha\Delta\tau_1 + \Delta\tau_2 \tag{64}$$

where α is a proportionality constant. Substituting Eqs. (62) and (63) into (64) and simplifying yields:

$$\delta = -\frac{1}{24(\gamma_1 - \gamma_2)}\left\{\sin^2\left(\frac{1}{4}\sqrt{k\gamma_1}t\right)[10 - 46\alpha + (10\gamma - \gamma_1 + \gamma_2)(1 + 5\alpha)]\right.$$
$$\left. + \sin^2\left(\frac{1}{4}\sqrt{k\gamma_2}t\right)[-10 + 46\alpha + (-10\gamma - \gamma_1 + \gamma_2)(1 + 5\alpha)]\right\} \tag{65}$$

When the spring at O_1 is much stiffer than the one at O_2, i.e. for $\gamma \ll 1$ in Eq. (65) becomes:

$$\delta_1 \approx -3\left[5\gamma + \alpha(-32 + 25\gamma)\right]\sin^2\left(\sqrt{\frac{5k}{8}}t\right) = 0 \quad \text{for } \alpha = -\frac{5\gamma}{-32 + 25\gamma} \quad (66)$$

Thus, for this case Eq. (64) can be written as $\Delta\tau_2 = \dfrac{5\gamma}{-32 + 25\gamma}\Delta\tau_1$. Here, $\alpha \to 0$ for small values of the stiffness ratio γ.

For the second limiting case when the first spring at O_2 is much stiffer than the one at O_1 ($\gamma \gg 1$) Eq. (65) becomes:

$$\delta_2 \approx 40(1 + 5\alpha)\gamma\sin^2\left(\sqrt{\frac{3k}{5}}t\right) = 0 \quad \text{for } \alpha = -\frac{1}{5} \quad (67)$$

resulting in a second proportional relationship between the two impulses of the form $\Delta\tau_2 = 1/5\Delta\tau_1$.

Based on the two relations between the impulses resulted from the two extreme values for , we hypothesize that the impulse initiated at the colliding edge triggers an impulse at non-colliding edge and they are related by a proportional relation of the form:

$$\Delta\tau_c = \alpha\Delta\tau_i \quad \text{with} \quad \alpha > 0 \quad (68)$$

where we denoted with $\Delta\tau_i$ and $\Delta\tau_c$ are the impulse at the impacting and non-impacting edges respectively.

3.3 Velocity impulse relationships

The impulse correlation ratio that was formulated in section 3.2 is now used to cast the velocities in terms of the collision impulse. There are four possible contact situations for the block during collision:

1. *Impact at O_1 contact at O_2*:
   Using Eqs. (55), (56), (57), (58), and the impulse correlation relationship (Eq. (68), which becomes $\Delta\tau_2 = \alpha\Delta\tau_1$) one obtains the following expressions for the velocities:

$$\Delta v_{CM}^n = \frac{1 + \alpha}{m}\Delta\tau_1 \quad (69)$$

$$\Delta v_{CM}^t = 0 \quad (70)$$

$$\Delta v_1^n = \frac{[1 - \alpha + \beta(1 + \alpha)]\,\Delta\tau_1}{\beta m} \quad (71)$$

$$\Delta v_1^t = -\frac{w(1 - \alpha)\Delta\tau_1}{\beta h m} \quad (72)$$

$$\Delta v_2^n = \frac{[-1 + \alpha + \beta(1 + \alpha)]\,\Delta\tau_1}{\beta m} \quad (73)$$

$$\Delta v_2^t = -\frac{w(1-\alpha)\Delta\tau_1}{\beta hm} \tag{74}$$

$$\Delta w = -\frac{(1-\alpha)\Delta\tau_1}{\beta hm} \tag{75}$$

where,

$$\beta = \frac{I_{CM}}{mh^2} \tag{76}$$

2. *Impact at O_1 no contact at O_2*: Using Eqs. (55), (56), (57), (58), and no impulse condition at O_2 ($\Delta\tau_2 = 0$) one obtains the following expressions for the velocities:

$$\Delta v_{CM}^n = \frac{1}{m}\Delta\tau_1 \tag{77}$$

$$\Delta v_{CM}^t = 0 \tag{78}$$

$$\Delta v_1^n = \frac{(1+\beta)\Delta\tau_1}{\beta m} \tag{79}$$

$$\Delta v_1^t = -\frac{w\Delta\tau_1}{\beta hm} \tag{80}$$

$$\Delta v_2^n = \frac{(-1+\beta)\Delta\tau_1}{\beta m} \tag{81}$$

$$\Delta v_2^t = -\frac{w\Delta\tau_1}{\beta hm} \tag{82}$$

$$\Delta w = -\frac{\Delta\tau_1}{\beta hm} \tag{83}$$

3. *Contact at O_1 impact at O_2*: Using Eqs. (55), (56), (57), (58), and the impulse correlation relationship (Eq. (68), which becomes $\Delta\tau_1 = \alpha\Delta\tau_2$) one obtains the following expressions for the velocities:

$$\Delta v_{CM}^n = \frac{1+\alpha}{m}\Delta\tau_2 \tag{84}$$

$$\Delta v_{CM}^t = 0 \tag{85}$$

$$\Delta v_1^n = \frac{[-1+\alpha+\beta(1+\alpha)]\Delta\tau_2}{\beta m} \tag{86}$$

$$\Delta v_1^t = \frac{w(1-\alpha)\Delta\tau_2}{\beta hm} \tag{87}$$

$$\Delta v_2^n = \frac{[1-\alpha+\beta(1+\alpha)]\Delta\tau_2}{\beta m} \tag{88}$$

$$\Delta v_2^t = \frac{w(1-\alpha)\Delta\tau_2}{\beta hm} \tag{89}$$

$$\Delta w = \frac{(1-\alpha)\Delta\tau_2}{\beta hm} \tag{90}$$

4. *No contact at O_1 impact at O_2:*
   Using Eqs. (55), (56), (57), (58), and no impulse condition at O_2 ($\Delta\tau_1 = 0$) one obtains the following expressions for the velocities:

$$\Delta v^n_{CM} = \frac{1}{m}\Delta\tau_2 \tag{91}$$

$$\Delta v^t_{CM} = 0 \tag{92}$$

$$\Delta v^n_1 = \frac{(-1+\beta)\Delta\tau_2}{\beta m} \tag{93}$$

$$\Delta v^t_1 = \frac{w\Delta\tau_2}{\beta h m} \tag{94}$$

$$\Delta v^n_2 = \frac{(1+\beta)\Delta\tau_2}{\beta m} \tag{95}$$

$$\Delta v^t_2 = \frac{w\Delta\tau_2}{\beta h m} \tag{96}$$

$$\Delta w = \frac{\Delta\tau_2}{\beta h m} \tag{97}$$

3.4 The single impact case

Consider the case where the block strikes the external surface at O_1 while resting at O_2 ($v^{n-}_1 \neq 0$ and $v^{n-}_1 = 0$, or more specifically $w^- = w_0 \neq 0$). There are two possible bouncing patterns that result from the collision at O_1, they can be enumerated as follows:

i. The non-impacting end detaches at the onset of the collision:

The non-impacting end bounces at the onset of the collision if the impulse at O_2 (τ_2) causes a positive change in the normal velocity at this point (v^n_2). This condition can be obtained by considering the slope of τ_2 in Eq. (81), which yields:

$$\beta > 1 \tag{98}$$

Thus, one can obtain the post collision velocities by solving Eqs. (77) through (83. Figure (20) depicts the impulse diagram at O_1. Now, the maximum compression impulse can be found by setting $\Delta v^n_1 = 0$ in Eq. (79) and solving for τ_1 (note that $\Delta\tau_1 = \tau_1 - 0$) as:

$$\tau^c_1 = \frac{2hm\beta w_0}{1+\beta} \tag{99}$$

The impulse at the end of the collision can be found by using the energetic definition of the coefficient of restitution (here one may also use kinematic or kinetic definitions since the case frictionless) can be obtained as follows:

$$\tau^f_1 = \frac{2(1+e_1)hm\beta w_0}{1+\beta} \tag{100}$$

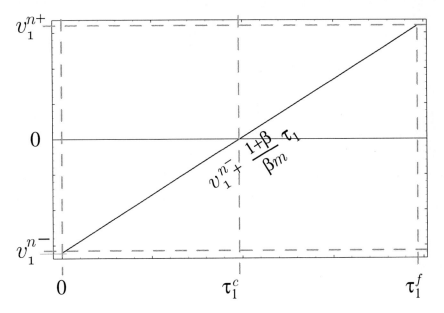

Fig. 20. Impulse diagram for a non-interacting collision at O_1

Then, the post impact velocities can be obtained by substituting the final impulse into the respective velocity expressions, which yields:

$$v_1^{n+} = 2e_1 h \omega_0 \tag{101}$$

$$v_1^{t+} = -\frac{2(1 + e_1) w \omega_0}{1 + \beta} \tag{102}$$

$$v_2^{n+} = \frac{2(1 + e_1) h (\beta - 1) \omega_0}{1 + \beta} \tag{103}$$

$$v_2^{t+} = -\frac{2(1 + e_1) w \omega_0}{1 + \beta} \tag{104}$$

$$\omega^+ = -\frac{[\beta - (1 + 2e_1)] \omega_0}{1 + \beta} \tag{105}$$

ii. Simultaneous collision at both ends:
A simultaneous impact at O_1 and O_2 takes place when the condition in Eq. (98) is violated. During the initial stage of the collision, there is a simultaneous contact at both ends, resulting from an impact at O_1 and contact at O_2. Using Eqs. (69) through (75) one may form the impulse diagrams that are depicted in Fig. (21) by obtaining the following expressions:

$$v_1^n = -2h \omega_0 + \frac{[1 - \alpha + \beta(1 + \alpha)] \tau_1}{\beta m} \tag{106}$$

$$v_1^t = -\frac{w(1-\alpha)\tau_1}{\beta h m} \tag{107}$$

$$v_2^n = \frac{[-1+\alpha+\beta(1+\alpha)]\,\tau_1}{\beta m} \tag{108}$$

$$\Delta v_2^t = -\frac{w(1-\alpha)\tau_1}{\beta h m} \tag{109}$$

$$\omega = \omega_0 - \frac{(1-\alpha)\tau_1}{\beta h m} \tag{110}$$

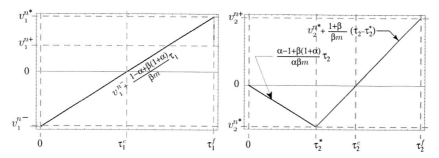

Fig. 21. Impulse diagrams for simultaneous collisions at O_1 and O_2

The maximum compression impulse τ_1^c that can be found by setting $v_1^n = 0$ in Eq. (106) and solving for τ_1, which yields:

$$\tau_1^c = \frac{2hm\beta\omega_0}{1-\alpha+\beta(1+\alpha)} \tag{111}$$

Then, in order to use the energetic definition of the coefficient of restitution we may write:

$$e_1^2 \int_0^{\tau_1^c} v_1^n d\tau_1 + \int_{\tau_1^c}^{\tau_1^f} v_1^n d\tau_1 = 0 \tag{112}$$

where e_1 is the coefficient of restitution at O_1. Solving this equation in terms of τ_1 yields the final impulse for the collision at O_1 as follows:

$$\tau_1^f = 2h\beta m\Gamma_1 \tag{113}$$

where,

$$\Gamma_1 = \frac{(1+e_1)\omega_0}{1-\alpha+\beta(1+\alpha)} \tag{114}$$

Substituting this impulse in the respective velocity expressions one obtains the velocities at the end of the O_1 collision as follows:

$$v_1^{n*} = 2e_1 h\omega_0 \tag{115}$$

$$v_1^{t\,*} = 2(\alpha - 1)w\Gamma_1 \tag{116}$$

$$v_2^{n\,*} = 2\left[\alpha - 1 + \beta\left(1 + \alpha\right)\right]h\Gamma_1 \tag{117}$$

$$v_2^{t\,*} = 2(\alpha - 1)w\Gamma_1 \tag{118}$$

$$w^* = w_0 + 2(\alpha - 1)\Gamma_1 \tag{119}$$

The impact at O_1 comes to an end while the collision at O_2 continues. During the subsequent motion Eqs (91) to (97) describe the relationship among the velocities and the impulse τ_2 at O_2. Now, using the velocities computed at the end of the O_1 impact and the value of the impulse O_2 at that instant, which is given by:

$$\tau_2^* = \alpha\tau_1^f = 2h\beta m\alpha\Gamma_1 \tag{120}$$

one obtains the following:

$$v_1^n = v_1^{n\,*} + \frac{-1+\beta}{\beta m}(\tau_2 - \tau_2^*) \tag{121}$$

$$v_1^t = v_1^{t\,*} + \frac{w}{\beta hm}(\tau_2 - \tau_2^*) \tag{122}$$

$$v_2^n = v_2^{n\,*} + \frac{1+\beta}{\beta m}(\tau_2 - \tau_2^*) \tag{123}$$

$$v_2^t = v_2^{t\,*} + \frac{w}{\beta hm}(\tau_2 - \tau_2^*) \tag{124}$$

$$w = w^* + \frac{1}{\beta hm}(\tau_2 - \tau_2^*) \tag{125}$$

The maximum compression impulse τ_2^c for the collision at O_2 can be found by setting $v_2^n = 0$ in Eq. (123) and solving for τ_2, which yields:

$$\tau_2^c = \frac{2(1-\beta)hm\beta\Gamma_1}{(1+\beta)} \tag{126}$$

Now using the energetic definition of the coefficient of restitution we form the energy equation:

$$e_2^2 \int_0^{\tau_2^c} v_2^n d\tau_2 + \int_{\tau_2^c}^{\tau_2^f} v_2^n d\tau_2 = 0 \tag{127}$$

where e_2 is the coefficient of restitution at O_2. Solving this equation for τ_2^f, the terminal impulse for the O_2 collision results in the following:

$$\tau_2^f = \frac{2(\beta - 1 - \Gamma_2)hm\beta\Gamma_1}{(1+\beta)} \tag{128}$$

where,

$$\Gamma_2 = e_2\sqrt{(1-\beta)\left[1 - \alpha - \beta(1+\alpha)\right]} \tag{129}$$

Substituting the final impulse in the respective velocity expressions, yields the following expressions for the post-impact velocities:

$$v_1^{n+} = 2h\omega_0 \left[1 - \frac{\Gamma_1 \Gamma_2 (1 - \beta + \Gamma_2/e_2)}{(1+\beta)\omega_0}\right] \tag{130}$$

$$v_1^{t+} = \frac{2w(\Gamma_2 - 2\beta)\Gamma_1}{1+\beta} \tag{131}$$

$$v_2^{n+} = 2\Gamma_1 \Gamma_2 \tag{132}$$

$$v_2^{t+} = \frac{2w(\Gamma_2 - 2\beta)\Gamma_1}{1+\beta} \tag{133}$$

$$\omega^+ = \omega_0 + 2(\alpha - 1)\Gamma_1 - \frac{2\left[\alpha - 1 + \beta(\alpha+1) - \Gamma_2\right]\Gamma_1}{1+\beta} \tag{134}$$

The impact sequence ends as long as v_1^n, does not become negative during the O_2 collision. If this velocity becomes negative, then the collision at O_2 triggers a second collision at O_1, leading to another sequence of impacts. The condition of not having another impact at O_1 can be found by setting $v_1^{n+} > 0$ in Eq. (130) and obtaining:

$$\frac{\Gamma_1 \Gamma_2 (1 - \beta + \Gamma_2/e_2)}{(1+\beta)\omega_0} < 1 \tag{135}$$

The next section is dedicated to a method that is developed to solve problems that involve such sequence of collisions.

Finally, we may obtain an upper limit on the impulse correlation ratio by considering inside the square root in Eq. (129). Recalling that $\beta < 1$, one gets,

$$\alpha \le \frac{1-\beta}{1+\beta} \tag{136}$$

when we have unit mass and dimensions $\beta = 2/3$ and thus the limit $\alpha \le 1/5$ which is identical to the value that was calculated in section 3.2.

3.5 Multiple Sequences of Impacts

In the previous section, we considered cases where the collision at O_1 results in one impact at each end. Yet, when the condition in Eq. (135) is violated, additional impacts may emerge.

When the collision at O_2 causes the sign of v_1 to become negative a second collision starts at O_1. The notation has to be modified to accommodate the multiple sequence of impacts. Accordingly, a second index will be added to impulses and landmark velocities to mark the impact sequence.

Figure (22) depicts the impulse diagrams for a case with two impact sequences. The expressions for the two normal velocities shown on the figure can be expressed as follows:

$$v_1^n = \begin{cases} v_1^{n-} + \frac{1-\alpha+\beta(1+\alpha)}{\beta m}\tau_1 & \text{if } 0 \le \tau_1 < \tau_{1,1}^f \\ \frac{-1+\alpha+\beta(1+\alpha)}{\beta m}(\tau_1 - \tau_{1,1}^f) & \text{if } \tau_{1,1}^f \le \tau_1 \le \tau_{1,2}^* \\ v_{1,2}^{n\,*} + \frac{1+\beta}{\beta m}(\tau_1 - \tau_{1,2}^*) & \text{if } \tau_{1,2}^* \le \tau_1 < \tau_{1,2}^f \end{cases} \tag{137}$$

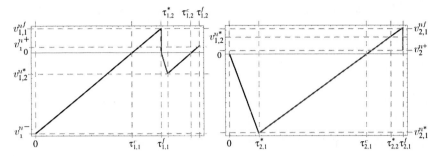

Fig. 22. Normal Velocity - Impulse diagram for a two sequence impact

$$v_2^n = \begin{cases} \dfrac{-1+\alpha+\beta(1+\alpha)}{\beta m}\tau_2 & \text{if } 0 \le \tau_2 < \tau_{2,1}^* \\ v_{2,1}^{n*} + \dfrac{1+\beta}{\beta m}(\tau_2 - \tau_{2,1}^*) & \text{if } \tau_{2,1}^* \le \tau_2 \le \tau_{2,2}^* \\ v_{2,2}^{n*} + \dfrac{1-\alpha+\beta(1+\alpha)}{\beta m}(\tau_2 - \tau_{2,2}^*) & \text{if } \tau_{2,2}^* \le \tau_2 \le \tau_{2,1}^f \end{cases} \qquad (138)$$

Equations (137) and (138) are obtained by classifying the situation during the impact as one of the cases listed in section 3.3. For example, during $0 \le \tau_1 < \tau_{1,1}^f$ (also $0 \le \tau_2 < \tau_{2,1}^*$) the collision at O_1 is driving the collision at O_2, hence we have the first case and use Eqs. (69)-(75) to obtain the respective velocity-impulse relationships. During the interval $\tau_{2,1}^* \le \tau_{2,1}^f$ there is no contact at O_1 and collision at O_2, thus we have the fourth case and use Eqs. (91)-(97). Also, during this interval we observe the sudden drop in v_1^n in the τ_1 diagram, which occurs because of the slowdown caused by the impact at O_2. Next, during $\tau_{1,1}^f \le \tau_1 \le \tau_{1,2}^*$ (also $\tau_{2,1}^* \le \tau_2 \le \tau_{2,1}^f$) the impact at O_2 is driving the impact at O_1, hence the third case and use Eqs. (84)-(90). Finally, during $\tau_{1,2}^* \le \tau_1 < \tau_{1,2}^f$ there is no contact at O_2 and impact at O_1, thus the second case. We use Eqs. (77)-(83) to obtain the velocity expressions. One may observe the sudden change in v_2, which is the result of the impact at O_1. Yet, this drop does not lead to a sign change, and hence the impact sequence ends when the second collision at O_1 is over.

The actual calculation of the impulses $\tau_{1,j}^*$, $\tau_{2,j}^*$, $\tau_{2,j}^f$, $\tau_{2,j}^f$, and the velocities $v_{1,j}^{n*}$, $v_{2,j}^{n*}$, $v_{1,j}^{nf}$, $v_{2,j}^{nf}$, v_1^+, and v_2^+ is carried out by applying the definition of the energetic coefficient of restitution (see section 3.4) to each collision within the respective impulse intervals.

Apparently, one can have a complex sequence of collisions depending on several factors such as geometry, material properties, and initial velocities. As was done for linear chain of balls, this task is automated by writing a Mathematica package.

3.6 Numerical analysis example

Here we consider a numerical example that includes a $100 \times 20 \times 20$ mm block. One end of the block is released from a height of 50 mm while the other end

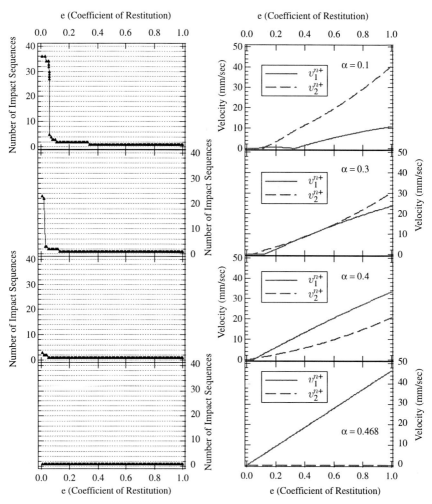

Fig. 23. Effect of coefficient of restitution and Impulse Correlation Ratio on the rebound velocities and the number of impact sequences

is resting on a massive external surface. The density of the block is chosen as $\rho = 7.88 \times 10^{-6}\ kg/m^3$. Figure (23) depicts the post-impact velocities and the number of impact sequences as the coefficient of restitution (assumed identical at both contact point) and the Impulse Correlation Ratio were varied in the intervals of $0 \leq e \leq 1$ and $0 \leq \alpha \leq \frac{1-\beta}{1+\beta} = .468392$ respectively. The post impact velocities shown in Figure (23) exhibit an interesting bimodal pattern. For high values of the Impulse Correlation Ratio, the striking edge bounces with a higher velocity than the resting edge. We describe this as the tendency of the block to "Roll Backward". One the other hand, the block tends to "Roll Forward" for

271

lower values of α. This observation provides a convenient way to check whether the trends predicted by the proposed method are physically valid. One can use a Styrofoam block (for high α) and a metal block (for low α) and physically observe the forward and backward rolling tendencies.

A particularly surprising outcome takes place for collisions where $\alpha = 0.468$ (see Fig. (23)). Here, the post-impact velocity of the resting edge is almost equal to zero even for an elastic impact at this edge ($e_2 = 1$). For this case, the coefficient of τ_2 in velocity v_2 (see Eq. (73)) is almost equal to zero during the initial simultaneous collision phase $(-1 + \alpha + \beta(1 + \alpha) \to 0$ as $\alpha \to \frac{1-\beta}{1+\beta})$. The resulting work done during the compression phase at O_2 is very small (see Fig. (24), which results in almost no rebound at this point.

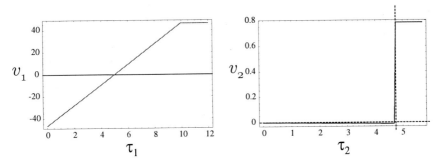

Fig. 24. Velocity-Impulse diagram for elastic collision with $\alpha = 0.468$

Finally, we consider a case that results in four sequences of impact. Figure (25) depicts the velocity impulse diagram for a case with $e = .1$ and $\alpha = .1$. This is a case where the block comes almost to rest as a result of impact. Successive collisions deplete the initial kinetic energy of the block until the edge velocities become equal to zero.

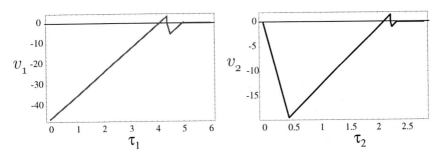

Fig. 25. Velocity-Impulse diagram of a four sequence collision

4 Discussion and conclusion

In this chapter we developed a new method that produces a unique and energetically consistent solution to two classes of collision problems that are difficult to solve using existing methods. First, the N-Ball linear chain problem is considered. The approach is based on the impulse-momentum methods, the energetic coefficient of restitution, and the Impulse Correlation Ratio that is introduced as a new material constant.

Multi-impact problems pose many difficulties and unanswered questions. The linear chain or the Newton's Cradle represents the simplest and the most basic problem of this type. The dynamic problem is simple because one only has to deal with motion of particles, yet it includes the difficulties that are encountered in more complex systems. The solution of the multi-impact problem has been confounded by the lack of sufficient means in the rigid body impact theory to resolve the collisions of stationary bodies that are in contact. This shortcoming manifested itself as the non-uniqueness of solutions obtained using the theory.

Here, we amended the rigid body theory by introducing the Impulse Correlation Ratio. This constant serves as a mechanism to coordinate the force transmission through the chain. Effective use of the energetic coefficient of restitution leads to energetically consistent results. Moreover, the method is the only one we know of that captures the commonly observed grouping of balls through the proposed back propagation process.

A set of experiment were conducted to verify the proposed theoretical methods and procedures. We have shown that the Impulse Correlation Ratio can be measured with relative ease. We have also demonstrated that the predictions of our method produces excellent agreement with the experimentally measured outcomes.

Then, the rocking block problem is considered. This problem was also solved using impulse-momentum methods, the energetic coefficient of restitution, and the Impulse Correlation Ratio. It was shown that the approach also works for this case. The outcomes predicted by the method are energetically and physically consistent.

References

1. Brogliato B., 1996 *Nonsmooth Impact Mechanics : Models, Dynamics and Control*, Springer Verlag, LNCIS 220.
2. Cholet C. 1998 *Chocs de Solids Rigides*, Ph. D. Thesis, University of Paris 6 and LCPC-CNRS, March, 1998)
3. Fremond M. 1995, "Rigid bodies collisions", Physics Letters A, 204, pp 33-41.
4. Goyal, S., Papadopoulos, J. M. 1998, "The Dynamics of Clattering I: Equations of Motion and Examples", *ASME J. of Dynamic Systems, Measurement and Control*, vol. 120, pp. 83-93.
5. Han, I., Gilmore, B. J. 1993, "Multi-Body Impact Motion with Friction-Analysis, Simulation, and Experimental Validation", *ASME J. of Mechanical Design*, vol. 115, pp. 412-422.

273

6. Johnson, W. (1976) "Simple Linear Impact", ImechE. IJMEE, Vol. 4, No.2, pp 167-181.
7. Marghitu D. B., Hurmuzlu Y. 1995, "Three Dimensional Rigid Body Collisions with Multiple Contact Points", *ASME Journal of Applied Mechanics*, 62:725-732.
8. J.J. Moreau, 1994 "Some numerical methods in multibody dynamics : application to granular materials", European J. of Mechanics A/Solids, vol 13, no 4, pp. 93-114.
9. Newby N. D. JR., 1979, "Linear Collisions with Harmonic oscillator Forces: The Inverse Scattering Problem", *American Journal of Phsics*, Vol. 47, No.2, pp. 161-165.
10. Smith, E. A. L., 1955, "Impact and Longitudinal Wave Transmission", *Transactions of the ASME*, pp. 963-973.
11. Stronge, W. J., 1990, "Rigid Body Collisions with Friction", *In Proceedings of Royal Society*, London, A 431, pp. 169-181.
12. Walkiewicz T. A., Newby N. D. JR., 1972, "Linear Collisions", *American Journal of Phsics*, Vol. 40, pp. 133-137.
13. Hurmuzlu Y., Chang, T. 1992, "Rigid Body Collisions of a Special Class of KInematic Chains", *IEEE Transactions in Systems, Man, and Cybernetics*, vol. 22, No. 5, pp.964-971.

Lecture Notes in Physics

For information about Vols. 1–519
please contact your bookseller or Springer-Verlag

Vol. 520: J. A. van Paradijs, J. A. M. Bleeker (Eds.), X-Ray Spectroscopy in Astrophysics. EADN School X. Proceedings, 1997. XV, 530 pages. 1999.

Vol. 521: L. Mathelitsch, W. Plessas (Eds.), Broken Symmetries. Proceedings, 1998. VII, 299 pages. 1999.

Vol. 522: J. W. Clark, T. Lindenau, M. L. Ristig (Eds.), Scientific Applications of Neural Nets. Proceedings, 1998. XIII, 288 pages. 1999.

Vol. 523: B. Wolf, O. Stahl, A. W. Fullerton (Eds.), Variable and Non-spherical Stellar Winds in Luminous Hot Stars. Proceedings, 1998. XX, 424 pages. 1999.

Vol. 524: J. Wess, E. A. Ivanov (Eds.), Supersymmetries and Quantum Symmetries. Proceedings, 1997. XX, 442 pages. 1999.

Vol. 525: A. Ceresole, C. Kounnas, D. Lüst, S. Theisen (Eds.), Quantum Aspects of Gauge Theories, Supersymmetry and Unification. Proceedings, 1998. X, 511 pages. 1999.

Vol. 526: H.-P. Breuer, F. Petruccione (Eds.), Open Systems and Measurement in Relativistic Quantum Theory. Proceedings, 1998. VIII, 240 pages. 1999.

Vol. 527: D. Reguera, J. M. G. Vilar, J. M. Rubí (Eds.), Statistical Mechanics of Biocomplexity. Proceedings, 1998. XI, 318 pages. 1999.

Vol. 528: I. Peschel, X. Wang, M. Kaulke, K. Hallberg (Eds.), Density-Matrix Renormalization. Proceedings, 1998. XVI, 355 pages. 1999.

Vol. 529: S. Biringen, H. Örs, A. Tezel, J.H. Ferziger (Eds.), Industrial and Environmental Applications of Direct and Large-Eddy Simulation. Proceedings, 1998. XVI, 301 pages. 1999.

Vol. 530: H.-J. Röser, K. Meisenheimer (Eds.), The Radio Galaxy Messier 87. Proceedings, 1997. XIII, 342 pages. 1999.

Vol. 531: H. Benisty, J.-M. Gérard, R. Houdré, J. Rarity, C. Weisbuch (Eds.), Confined Photon Systems. Proceedings, 1998. X, 496 pages. 1999.

Vol. 532: S. C. Müller, J. Parisi, W. Zimmermann (Eds.), Transport and Structure. Their Competitive Roles in Biophysics and Chemistry. XII, 400 pages. 1999.

Vol. 533: K. Hutter, Y. Wang, H. Beer (Eds.), Advances in Cold-Region Thermal Engineering and Sciences. Proceedings, 1999. XIV, 608 pages. 1999.

Vol. 534: F. Moreno, F. González (Eds.), Light Scattering from Microstructures. Proceedings, 1998. XII, 300 pages. 2000

Vol. 535: H. Dreyssé (Ed.), Electronic Structure and Physical Properties of Solids: The Uses of the LMTO Method. Proceedings, 1998. XIV, 458 pages. 2000.

Vol. 536: T. Passot, P.-L. Sulem (Eds.), Nonlinear MHD Waves and Turbulence. Proceedings, 1998. X, 385 pages. 1999.

Vol. 537: S. Cotsakis, G. W. Gibbons (Eds.), Mathematical and Quantum Aspects of Relativity and Cosmology. Proceedings, 1998. XII, 251 pages. 1999.

Vol. 538: Ph. Blanchard, D. Giulini, E. Joos, C. Kiefer, I.-O. Stamatescu (Eds.), Decoherence: Theoretical, Experimental, and Conceptual Problems. Proceedings, 1998. XII, 345 pages. 2000.

Vol. 539: A. Borowiec, W. Cegła, B. Jancewicz, W. Karwowski (Eds.), Theoretical Physics. Fin de Siècle. Proceedings, 1998. XX, 319 pages. 2000.

Vol. 540: B. G. Schmidt (Ed.), Einstein's Field Equations and Their Physical Implications. Selected Essays. 1999. XIII, 429 pages. 2000

Vol. 541: J. Kowalski-Glikman (Ed.), Towards Quantum Gravity. Proceedings, 1999. XII, 376 pages. 2000.

Vol. 542: P. L. Christiansen, M. P. Sørensen, A. C. Scott (Eds.), Nonlinear Science at the Dawn of the 21st Century. Proceedings, 1998. XXVI, 458 pages. 2000.

Vol. 543: H. Gausterer, H. Grosse, L. Pittner (Eds.), Geometry and Quantum Physics. Proceedings, 1999. VIII, 408 pages. 2000.

Vol. 544: T. Brandes (Ed.), Low-Dimensional Systems. Interactions and Transport Properties. Proceedings, 1999. VIII, 219 pages. 2000

Vol. 545: J. Klamut, B. W. Veal, B. M. Dabrowski, P. W. Klamut, M. Kazimierski (Eds.), New Developments in High-Temperature Superconductivity. Proceedings, 1998. VIII, 275 pages. 2000.

Vol. 546: G. Grindhammer, B. A. Kniehl, G. Kramer (Eds.), New Trends in HERA Physics 1999. Proceedings, 1999. XIV, 460 pages. 2000.

Vol. 547: D. Reguera, G. Platero, L.L. Bonilla, J.M. Rubí(Eds.), Statistical and Dynamical Aspects of Mesoscopic Systems. Proceedings, 1999. XII, 357 pages. 2000.

Vol. 548: D. Lemke, M. Stickel, K. Wilke (Eds.), ISO Surveys of a Dusty Universe. Proceedings, 1999. XIV, 432 pages. 2000.

Vol. 549: C. Egbers, G. Pfister (Eds.), Physics of Rotating Fluids. Proceedings, 1999. XIV, 439 pages. 2000.

Vol. 550: M. Planat (Ed.), Noise, Oscillators and Algebraic Randomness. Proceedings, 1999. VIII, 417 pages. 2000.

Vol. 551: B. Brogliato (Ed.), Impacts in Mechanical Systems. Analysis and Modelling. Lectures, 1999. IX, 273 pages. 2000.

Vol. 552: Z. Chen, R. E. Ewing, Z.-C. Shi (Eds.), Numerical Treatment of Multiphase Flows in Porous Media. Proceedings, 1999. XXI, 445 pages. 2000.

Monographs
For information about Vols. 1–20
please contact your bookseller or Springer-Verlag

Vol. m 21: G. P. Berman, E. N. Bulgakov, D. D. Holm, Crossover-Time in Quantum Boson and Spin Systems. XI, 268 pages. 1994.

Vol. m 22: M.-O. Hongler, Chaotic and Stochastic Behaviour in Automatic Production Lines. V, 85 pages. 1994.

Vol. m 23: V. S. Viswanath, G. Müller, The Recursion Method. X, 259 pages. 1994.

Vol. m 24: A. Ern, V. Giovangigli, Multicomponent Transport Algorithms. XIV, 427 pages. 1994.

Vol. m 25: A. V. Bogdanov, G. V. Dubrovskiy, M. P. Krutikov, D. V. Kulginov, V. M. Strelchenya, Interaction of Gases with Surfaces. XIV, 132 pages. 1995.

Vol. m 26: M. Dineykhan, G. V. Efimov, G. Ganbold, S. N. Nedelko, Oscillator Representation in Quantum Physics. IX, 279 pages. 1995.

Vol. m 27: J. T. Ottesen, Infinite Dimensional Groups and Algebras in Quantum Physics. IX, 218 pages. 1995.

Vol. m 28: O. Piguet, S. P. Sorella, Algebraic Renormalization. IX, 134 pages. 1995.

Vol. m 29: C. Bendjaballah, Introduction to Photon Communication. VII, 193 pages. 1995.

Vol. m 30: A. J. Greer, W. J. Kossler, Low Magnetic Fields in Anisotropic Superconductors. VII, 161 pages. 1995.

Vol. m 31 (Corr. Second Printing): P. Busch, M. Grabowski, P.J. Lahti, Operational Quantum Physics. XII, 230 pages. 1997.

Vol. m 32: L. de Broglie, Diverses questions de mécanique et de thermodynamique classiques et relativistes. XII, 198 pages. 1995.

Vol. m 33: R. Alkofer, H. Reinhardt, Chiral Quark Dynamics. VIII, 115 pages. 1995.

Vol. m 34: R. Jost, Das Märchen vom Elfenbeinernen Turm. VIII, 286 pages. 1995.

Vol. m 35: E. Elizalde, Ten Physical Applications of Spectral Zeta Functions. XIV, 224 pages. 1995.

Vol. m 36: G. Dunne, Self-Dual Chern-Simons Theories. X, 217 pages. 1995.

Vol. m 37: S. Childress, A.D. Gilbert, Stretch, Twist, Fold: The Fast Dynamo. XI, 406 pages. 1995.

Vol. m 38: J. González, M. A. Martín-Delgado, G. Sierra, A. H. Vozmediano, Quantum Electron Liquids and High-Tc Superconductivity. X, 299 pages. 1995.

Vol. m 39: L. Pittner, Algebraic Foundations of Non-Com-mutative Differential Geometry and Quantum Groups. XII, 469 pages. 1996.

Vol. m 40: H.-J. Borchers, Translation Group and Particle Representations in Quantum Field Theory. VII, 131 pages. 1996.

Vol. m 41: B. K. Chakrabarti, A. Dutta, P. Sen, Quantum Ising Phases and Transitions in Transverse Ising Models. X, 204 pages. 1996.

Vol. m 42: P. Bouwknegt, J. McCarthy, K. Pilch, The W₃ Algebra. Modules, Semi-infinite Cohomology and BV Algebras. XI, 204 pages. 1996.

Vol. m 43: M. Schottenloher, A Mathematical Introduction to Conformal Field Theory. VIII, 142 pages. 1997.

Vol. m 44: A. Bach, Indistinguishable Classical Particles. VIII, 157 pages. 1997.

Vol. m 45: M. Ferrari, V. T. Granik, A. Imam, J. C. Nadeau (Eds.), Advances in Doublet Mechanics. XVI, 214 pages. 1997.

Vol. m 46: M. Camenzind, Les noyaux actifs de galaxies. XVIII, 218 pages. 1997.

Vol. m 47: L. M. Zubov, Nonlinear Theory of Dislocations and Disclinations in Elastic Body. VI, 205 pages. 1997.

Vol. m 48: P. Kopietz, Bosonization of Interacting Fermions in Arbitrary Dimensions. XII, 259 pages. 1997.

Vol. m 49: M. Zak, J. B. Zbilut, R. E. Meyers, From Instability to Intelligence. Complexity and Predictability in Nonlinear Dynamics. XIV, 552 pages. 1997.

Vol. m 50: J. Ambjørn, M. Carfora, A. Marzuoli, The Geometry of Dynamical Triangulations. VI, 197 pages. 1997.

Vol. m 51: G. Landi, An Introduction to Noncommutative Spaces and Their Geometries. XI, 200 pages. 1997.

Vol. m 52: M. Hénon, Generating Families in the Restricted Three-Body Problem. XI, 278 pages. 1997.

Vol. m 53: M. Gad-el-Hak, A. Pollard, J.-P. Bonnet (Eds.), Flow Control. Fundamentals and Practices. XII, 527 pages. 1998.

Vol. m 54: Y. Suzuki, K. Varga, Stochastic Variational Approach to Quantum-Mechanical Few-Body Problems. XIV, 324 pages. 1998.

Vol. m 55: F. Busse, S. C. Müller, Evolution of Spontaneous Structures in Dissipative Continuous Systems. X, 559 pages. 1998.

Vol. m 56: R. Haussmann, Self-consistent Quantum Field Theory and Bosonization for Strongly Correlated Electron Systems. VIII, 173 pages. 1999.

Vol. m 57: G. Cicogna, G. Gaeta, Symmetry and Perturbation Theory in Nonlinear Dynamics. XI, 208 pages. 1999.

Vol. m 58: J. Daillant, A. Gibaud (Eds.), X-Ray and Neutron Reflectivity: Principles and Applications. XVIII, 331 pages. 1999.

Vol. m 59: M. Kriele, Spacetime. Foundations of General Relativity and Differential Geometry. XV, 432 pages. 1999.

Vol. m 60: J. T. Londergan, J. P. Carini, D. P. Murdock, Binding and Scattering in Two-Dimensional Systems. Applications to Quantum Wires, Waveguides and Photonic Crystals. X, 222 pages. 1999.

Vol. m 61: V. Perlick, Ray Optics, Fermat's Principle, and Applications to General Relativity. X, 220 pages. 2000.

Vol. m 62: J. Szabo, Ray Optics, Equivariant Cohomology and Localization of Path Integrals. XII, 315 pages. 2000.

Vol. m 63: R. J. Szabo, Ray Optics, Equivariant Cohomology and Localization of Path Integrals. XII, 315 pages. 2000.

Vol. m 64: I. G. Avramidi, Heat Kernel and Quantum Gravity. X, 143 pages. 2000.